Nutrition and Exercise Interventions on Skeletal Muscle Physiology, Injury and Recovery: From Mechanisms to Therapy

Nutrition and Exercise Interventions on Skeletal Muscle Physiology, Injury and Recovery: From Mechanisms to Therapy

Editors

Sandro Massao Hirabara
Gabriel Nasri Marzuca Nassr
Maria Fernanda Cury Boaventura

Basel • Beijing • Wuhan • Barcelona • Belgrade • Novi Sad • Cluj • Manchester

Editors

Sandro Massao Hirabara
Interdisciplinary
Post-Graduate Program in
Health Sciences, Cruzeiro do
Sul University
São Paulo
Brazil

Gabriel Nasri Marzuca Nassr
Department of Rehabilitation
Sciences, Universidad de La
Frontera
Temuco
Chile

Maria Fernanda Cury
Boaventura
Interdisciplinary
Post-Graduate Program in
Health Sciences, Cruzeiro do
Sul University
São Paulo
Brazil

Editorial Office
MDPI
St. Alban-Anlage 66
4052 Basel, Switzerland

This is a reprint of articles from the Special Issue published online in the open access journal *Nutrients* (ISSN 2072-6643) (available at: https://www.mdpi.com/journal/nutrients/special_issues/A21L47IJ89).

For citation purposes, cite each article independently as indicated on the article page online and as indicated below:

Lastname, A.A.; Lastname, B.B. Article Title. *Journal Name* **Year**, *Volume Number*, Page Range.

ISBN 978-3-7258-0365-1 (Hbk)
ISBN 978-3-7258-0366-8 (PDF)
doi.org/10.3390/books978-3-7258-0366-8

Contents

About the Editors

Sandro Massao Hirabara

Sandro Massao Hirabara has been a full professor of the Interdisciplinary Post-Graduate Program in Health Sciences, Cruzeiro do Sul University, São Paulo, Brazil, since 2007. He graduated with a degree in Biology from the Federal University of Paraná, Curitiba, Brazil (1996–2000), and completed a Ph.D. in Science (Human Physiology) at the University of São Paulo, São Paulo, Brazil (2001–2005), and a Postdoctorate at the University of Sao Paulo (2005–2007) and the University of Geneva, Geneva, Switzerland (2007). He has also served as a collaborative researcher at the University of São Paulo (2008–2017), a visiting scholar scientist at Yale University (2018–2019), and a visiting researcher at the Butantan Institute (2020–2021). Dr. Hirabara has experience in physiology, with an emphasis on endocrine physiology, metabolism, and exercise physiology. He has expertise in the following sub-areas: fatty acids, insulin resistance, skeletal muscle metabolism, cell signaling, mitochondrial function, and physical exercise, with more than 140 research articles published in indexed journals under his name. He is also a scientific adviser of FAPESP, CNPq, and CAPES and an editor and reviewer of several important international journals.

Gabriel Nasri Marzuca Nassr

Gabriel Nasri Marzuca Nassr is the Director of the Skeletal Muscle Mass and Strength Regulation Laboratory. He is currently an associate professor at the Department of Rehabilitation Sciences in the Faculty of Medicine at Universidad de La Frontera, Temuco, Chile. He graduated from Universidad Católica del Maule in Chile with an undergraduate degree and a master's degree in Physiotherapy. Afterward, he graduated from the University of São Paulo, Brazil, with a Ph.D. in Sciences (Human Physiology). His main area of research is skeletal muscle mass and strength regulation investigation strategies to prevent or combat skeletal muscle atrophy due to disuse, aging, or catabolic diseases. He has published more than 60 research articles in international journals. He has also been awarded for his work at national and international congresses. One of the most recent directions of his research is the study of resistance exercise training in very old individuals (80 years and older), as well as in women suffering from breast cancer.

Maria Fernanda Cury Boaventura

Maria Fernanda Cury Boaventura is currently a professor of the Interdisciplinary Post-Graduate Program in Health Sciences at the Cruzeiro do Sul University, Brazil. She graduated in Nutrition at the University of São Paulo, Brazil (2001), and completed her Ph.D. in Health Sciences at the Institute of Biomedical Sciences of the University of São Paulo (2006). During her Ph.D. and Post-doctorate (2007) in Health Sciences at the University of São Paulo, she worked in the research area of lipids and inflammation. Her main research field investigates the role of nutrition and genetics in the molecular mechanisms involved in tissue injury and inflammation induced by physical exercise. She has published more than 70 research articles in international journals, most of them related to tissue injury and inflammation topics.

Preface

Since skeletal muscle is the main tissue of our body, comprising 40–50% of our total body mass, it has a fundamental function in maintaining human health. Maintaining or increasing skeletal muscle mass and its function in response to environmental stimuli (e.g., physical exercise and nutritional interventions) is related to better healthy states, while sarcopenia (decreased skeletal muscle mass and function) is associated with poor-health conditions (e.g., physical inactivity, bad eating habits, obesity, and related diseases). Although great advances in research of skeletal muscle plasticity have been achieved in the last few decades, the precise molecular and cellular mechanisms involved in this process are still under investigation. This Special Issue contributes to our understanding and the recent updates to these mechanisms, bringing together advanced and significant studies performed in the research field and focusing on the modulation of different dietary nutrients, supplements, and physical exercise training protocols on skeletal muscle plasticity, in healthy and pathological conditions, including obesity, diabetes, cardiovascular diseases, chronic obstructive pulmonary disease, sarcopenia, and aging. The editors of this reprint are indebted to the academic/financial support provided by the PRPGP/Cruzeiro do Sul, FAPESP, CNPq, CAPES, and Universidad de La Frontera.

Sandro Massao Hirabara, Gabriel Nasri Marzuca Nassr, and Maria Fernanda Cury Boaventura
Editors

Editorial

Nutrition and Exercise Interventions on Skeletal Muscle Physiology, Injury and Recovery: From Mechanisms to Therapy

Sandro Massao Hirabara [1,*], Gabriel Nasri Marzuca-Nassr [2] and Maria Fernanda Cury-Boaventura [1]

[1] Interdisciplinary Post-Graduate Program in Health Sciences, Cruzeiro do Sul University, São Paulo 01506-000, Brazil; maria.boaventura@cruzeirodosul.edu.br

[2] Department of Rehabilitation Sciences, Universidad de La Frontera, Temuco 4811230, Chile; gabriel.marzuca@ufrontera.cl

* Correspondence: sandro.hirabara@cruzeirodosul.edu.br

Citation: Hirabara, S.M.; Marzuca-Nassr, G.N.; Cury-Boaventura, M.F. Nutrition and Exercise Interventions on Skeletal Muscle Physiology, Injury and Recovery: From Mechanisms to Therapy. *Nutrients* **2024**, *16*, 293. https://doi.org/10.3390/nu16020293

Received: 1 January 2024
Accepted: 11 January 2024
Published: 18 January 2024

Interventional strategies involving nutrition and physical exercise have been widely proposed to positively modulate skeletal muscle function, in both physiological and pathological states, such as obesity, T2DM, inflammatory diseases, cardiovascular diseases, aging, and sarcopenia [1–3]. In this sense, it has been observed that various nutrients or dietary bioactive compounds positively modulate different biological processes, such as metabolism, inflammation, redox balance, mitochondrial function, and gene expression. Several physical exercise protocols have also been proposed to maintain, increase, or recover skeletal muscle function [4–6]. Research groups are focusing potential molecular targets to direct further studies to treat and/or prevent skeletal muscle disorders.

Moreover, the role of nutrients associated with physical exercise in organokines (adipokines and myokines) responsible for metabolic health has been investigated. For instance, in obesity, there is an imbalance in the plasma concentration of adipokines, characterized by increased pro-inflammatory adipokines (IL-6, TNF-alpha, leptin, resistin, ANGPTL2, RBB4, asprosin, chemerin, visfatin, CRP6, SPARC, WISP1, and lipocalin-2), which has been proposed to contribute to the development of insulin resistance and decreased levels anti-inflammatory, anti-atherosclerotic and/or insulin sensitivity-increasing adipokines (vaspin, omentin, adiponectin, ZAG, and SFRP5). Strategies for modulating the levels of these adipokines can benefit cardiometabolic health and prevent obesity-related co-morbidities [7]. Most of these adipokines are expressed in response to elevated adiposity and a pro-inflammatory environment [8].

In the original article from Supriya et al. [9] the authors investigated the effect of spirulina supplementation, a dietary supplement extracted from cyanobacteria with potential antioxidant and anti-inflammatory effects (Calella et al., 2022; Behairy et al., 2023), on the adipokine/cytokine and lipid level profile in obese men submitted to a high-intensity exercise training program for 12 weeks. Interestingly, the association of spirulina supplementation with high-intensity exercise training improved the lipid plasma profile (decreased cholesterol, LDL-cholesterol, and triacylglycerides, and raised HDL-cholesterol) and the adipokine/cytokine level profile (reduced CRP, TNF-alpha, MCP-1, IL-6, and IL-8), suggesting that this interventional combination can be a good strategy to improve several risk markers of metabolic and inflammatory diseases associated with obesity. The same group [10], using a high-intensity exercise training protocol (CrossFit) for 12 weeks, also evaluated the combined effect of astaxanthin supplementation, a dietary supplement derived from *Haematoccus pluvialis* algae with several potential beneficial effects on metabolic and inflammatory diseases, such as obesity, T2DM, cardiovascular diseases, metabolic syndrome, and cancer [11]. This association was able to improve the adipokine level profile concomitantly with ameliorations in body composition and lipid profile in obese male individuals. These results corroborate the findings of Saeidi et al. [12] and suggest that this interventional combination can acts as an effective therapy to decrease the deleterious effects of obesity in metabolism and the adipokine and lipid level profile.

Various myokines have been proposed to regulate muscle function and metabolism, including myogenesis, mitochondrial biogenesis, mitophagy, autophagy, satellite cells activation, anti-inflammatory effect, skeletal muscle fat oxidation, and insulin response [13]. Furthermore, some myokines are hypothesized to be the mediators between myocytes and other cells, improving glucose metabolism regulation, lipid oxidation, and cardioprotective function [14]. In the study by Kysel et al. [15], the authors compared two nutritional interventions in healthy men: (1) a diet with cyclical ketogenic reduction and (2) a diet with a nutritionally balanced decrease. During the nutritional interventions, the participants were submitted to both aerobic and resistance exercise training protocols. The level profile of myokines and cytokines was determined pre- and post-intervention. The administration of a diet with a nutritionally balanced reduction increased performance in the endurance test and raised muscle strength associated with osteonectin and musclin levels. However, the diet with ketogenic reduction did not improve endurance performance induced by exercise training, but it reduced the FGF-21 plasma levels. This last factor is secreted by several tissues, such as the liver, skeletal muscle, adipose tissue, and pancreas, regulating metabolism and playing an important role as a result of nutritional reduction or a ketogenic diet [16].

In the original article from Sierra et al. [17], the authors showed the significant effect of adequate energy, sodium, cholesterol, vitamin C, and fiber dietary consumption on exercise-induced myokine production (myostatin, musclin, irisin, BDNF, apelin, IL-15, and FGF-21) in endurance runners, which is important for cardiometabolic adaptations and adequate tissue repair in response to intense physical exercise. The myostatin pathway was also investigated in the work performed by de Carvalho et al. [18]. In this work, the authors determined whether creatine supplementation (monohydrate) associated with a resistance exercise protocol was able to modulate the myostatin signaling pathway, skeletal muscle morphology and the expression of myosin heavy chain isoforms in the soleus muscle (red, predominantly oxidative fibers) and gastrocnemius muscle (white portion, predominantly glycolytic fibers) of Wistar rats. According to their theory, the authors expected that the combination of creatine supplementation and resistance training would promote greater changes in the white portion of the gastrocnemius muscle compared to the soleus muscle by attenuating the expression of the myostatin. Accordingly, their findings followed this proposition—that is, greater hypertrophy and interstitial remodeling in the white gastrocnemius muscle than in the soleus muscle.

Valero-Breton et al. [19] evaluated the effects of two different cycling training programs (eccentric versus concentric exercise protocols) for 12 weeks in patients with chronic obstructive pulmonary disease who were demonstrated to have muscle redox imbalance and reduced daily activities. In this study, several redox parameters and inflammatory biomarkers, as well as cardiometabolic assessments, were investigated. In accordance with the findings, the authors concluded that patients submitted to the concentric exercise protocol presented with increased antioxidant capacity (at rest and exercise-induced), improved insulin sensitivity, and elevated HDL levels compared to the patients who performed the eccentric exercise protocol.

Sarcopenia is a syndrome characterized by decreased muscle mass and strength, resulting in impaired physical performance. Different strategies have been proposed and analyzed with the aim to reverse the effects of sarcopenia. Cuyul-Vásquez et al. [20], through a systematic review with meta-analysis, verified in older people with sarcopenia the effectiveness of resistance exercise training with or without whey protein supplementation on muscle mass and strength, as well as physical performance. Despite some limitations of the analysis, mainly related to the small effect size and the low/very low quality of evidence from the studies performed so far, the authors concluded that resistance exercise training is more effective in increasing skeletal muscle mass and grip strength when associated with whey protein supplementation than when performed without any supplementation or with placebo supplementation. The authors highlighted the relevance of additional studies to completely comprehend this process.

In the study from Barquilha et al. [21], the authors determined the modulating effects of fish oil supplementation, which has been proposed as having anti-inflammatory effects and increasing skeletal muscle function [22,23], on muscle damage markers and inflammatory cytokines after a unique hypertrophic exercise session in healthy eutrophic men trained using a undulating/strength exercise protocol for six weeks. The authors found that the supplementation reduced plasma indicators of muscle injury (reduced plasma CK and LDH activity), inflammation (lowed C-reactive protein and IL-6), and redox balance (elevated GSH:GSSG ratio). The findings of this study suggest that supplementation with fish oil can be an important nutritional intervention for reducing muscle injury, inflammatory processes, and oxidative stress after an intense strength exercise session, mainly in untrained and beginner young adults with an interest in practicing resistance/strength training protocols. The authors highlighted that further works are required to evaluate the effectiveness of this supplementation in other groups, such as trained and older people, as well as the modulating effect of n-3 PUFAs on muscle regeneration. In this sense, the work of Jannas-Vela et al. (2023) [24] explored recent literature data about the potential effects of this supplementation on muscle regeneration in different inflammatory conditions, in particular, the effects of their lipid mediators, including oxylipins and endocannabinoids. Further works are required to determine the modulating effects of n-3 fatty acids on membrane composition and endocannabinoid and oxylipin generation, in association with the regenerative process after muscle damage in physiological and pathological diseases.

Interestingly, in the work from Franceković and Gliemann [25], an overview was provided on the relevance of dietary nutritional compounds, including n-3 PUFA, probiotics, and vitamin D, for endothelial glycocalyx maintenance and preservation. The authors proposed that these nutritional compounds are of particular importance for chronic inflammatory diseases, where endothelial dysfunction and vascular abnormalities are observed, including obesity, T2DM, dyslipidemias, metabolic syndrome, and cardiovascular diseases.

In summary, various groups are joining efforts to understand the molecular targets and cellular systems implicated in the effects of nutritional interventions or different physical exercise protocols on muscle physiology, recovery, and injury in cellular, animal, and human models. Various studies have been published and great advances have been made related to this topic in recent decades. The present Special Issue has further contributed to the advanced comprehension of the molecular targets and cellular systems implicated in skeletal muscle physiology, injury, and recovery in physiological and pathological conditions.

Author Contributions: S.M.H., G.N.M.-N. and M.F.C.-B. created the first draft and performed the critical revision and editing of this editorial. All authors have read and agreed to the published version of the manuscript.

Funding: The authors received financial support from CAPES, CNPq, FAPESP (2018/09868-7 and 2022/09341-4), and UFRO (FPP22-0016).

Conflicts of Interest: The authors declare no conflicts of interest.

References

1. Yin, Y.H.; Liu, J.Y.W.; Välimäki, M. Effectiveness of non-pharmacological interventions on the management of sarcopenic obesity: A systematic review and meta-analysis. *Exp. Gerontol.* **2020**, *135*, 110937. [CrossRef] [PubMed]
2. Kim, Y.J.; Moon, S.; Yu, J.M.; Chung, H.S. Implication of diet and exercise on the management of age-related sarcopenic obesity in Asians. *Geriatr. Gerontol. Int.* **2022**, *22*, 695–704. [CrossRef]
3. Chang, K.V.; Wu, W.T.; Chen, Y.H.; Chen, L.R.; Hsu, W.H.; Lin, Y.L.; Han, D.S. Enhanced serum levels of tumor necrosis factor-α, interleukin-1β, and -6 in sarcopenia: Alleviation through exercise and nutrition intervention. *Aging* **2023**, *15*, 13471–13485. [CrossRef]
4. Abreu, P.; Mendes, S.V.; Ceccatto, V.M.; Hirabara, S.M. Satellite cell activation induced by aerobic muscle adaptation in response to endurance exercise in humans and rodents. *Life Sci.* **2017**, *170*, 33–40. [CrossRef] [PubMed]
5. Mcleod, J.C.; Currier, B.S.; Lowisz, C.V.; Phillips, S.M. The influence of resistance exercise training prescription variables on skeletal muscle mass, strength, and physical function in healthy adults: An umbrella review. *J. Sport Health Sci.* **2023**, *13*, 47–60. [CrossRef]

6. Zhou, H.H.; Liao, Y.; Zhou, X.; Peng, Z.; Xu, S.; Shi, S.; Liu, L.; Hao, L.; Yang, W. Effects of Timing and Types of Protein Supplementation on Improving Muscle Mass, Strength, and Physical Performance in Adults Undergoing Resistance Training: A Network Meta-Analysis. *Int. J. Sport Nutr. Exerc. Metab.* **2023**, *34*, 54–64. [CrossRef] [PubMed]
7. de Oliveira Dos Santos, A.R.; de Oliveira Zanuso, B.; Miola, V.F.B.; Barbalho, S.M.; Santos Bueno, P.C.; Flato, U.A.P.; Detregiachi, C.R.P.; Buchaim, D.V.; Buchaim, R.L.; Tofano, R.J.; et al. Adipokines, Myokines, and Hepatokines: Crosstalk and Metabolic Repercussions. *Int. J. Mol. Sci.* **2021**, *22*, 2639. [CrossRef]
8. Ren, Y.; Zhao, H.; Yin, C.; Lan, X.; Wu, L.; Du, X.; Griffiths, H.R.; Gao, D. Adipokines, Hepatokines and Myokines: Focus on Their Role and Molecular Mechanisms in Adipose Tissue Inflammation. *Front Endocrinol.* **2022**, *14*, 873699. [CrossRef]
9. Supriya, R.; Delfan, M.; Saeidi, A.; Samaie, S.S.; Al Kiyumi, M.H.; Escobar, K.A.; Laher, I.; Heinrich, K.M.; Weiss, K.; Knechtle, B.; et al. Spirulina Supplementation with High-Intensity Interval Training Decreases Adipokines Levels and Cardiovascular Risk Factors in Men with Obesity. *Nutrients* **2023**, *15*, 4891. [CrossRef]
10. Supriya, R.; Shishvan, S.R.; Kefayati, M.; Abednatanzi, H.; Razi, O.; Bagheri, R.; Escobar, K.A.; Pashaei, Z.; Saeidi, A.; Shahrbanian, S.; et al. Astaxanthin Supplementation Augments the Benefits of CrossFit Workouts on Semaphorin 3C and Other Adipokines in Males with Obesity. *Nutrients* **2023**, *15*, 4803. [CrossRef]
11. Xia, W.; Tang, N.; Kord-Varkaneh, H.; Low, T.Y.; Tan, S.C.; Wu, X.; Zhu, Y. The effects of astaxanthin supplementation on obesity, blood pressure, CRP, glycemic biomarkers, and lipid profile: A meta-analysis of randomized controlled trials. *Pharmacol. Res.* **2020**, *161*, 105113. [CrossRef]
12. Saeidi, A.; Nouri-Habashi, A.; Razi, O.; Ataeinosrat, A.; Rahmani, H.; Mollabashi, S.S.; Bagherzadeh-Rahmani, B.; Aghdam, S.M.; Khalajzadeh, L.; Al Kiyumi, M.H.; et al. Astaxanthin Supplemented with High-Intensity Functional Training Decreases Adipokines Levels and Cardiovascular Risk Factors in Men with Obesity. *Nutrients* **2023**, *15*, 286. [CrossRef] [PubMed]
13. Sabaratnam, R.; Wojtaszewski, J.F.P.; Højlund, K. Factors mediating exercise-induced organ crosstalk. *Acta Physiol.* **2022**, *234*, e13766. [CrossRef]
14. Chow, L.S.; Gerszten, R.E.; Taylor, J.M.; Pedersen, B.K.; van Praag, H.; Trappe, S.; Febbraio, M.A.; Galis, Z.S.; Gao, Y.; Haus, J.M.; et al. Exerkines in health, resilience and disease. *Nat. Rev. Endocrinol.* **2022**, *18*, 273–289. [CrossRef] [PubMed]
15. Kysel, P.; Haluzíková, D.; Pleyerová, I.; Řezníčková, K.; Laňková, I.; Lacinová, Z.; Havrlantová, T.; Mráz, M.; Kasperová, B.J.; Kovářová, V.; et al. Different Effects of Cyclical Ketogenic vs. Nutritionally Balanced Reduction Diet on Serum Concentrations of Myokines in Healthy Young Males Undergoing Combined Resistance/Aerobic Training. *Nutrients* **2023**, *15*, 1720. [CrossRef] [PubMed]
16. Velingkar, A.; Vuree, S.; Prabhakar, P.K.; Kalashikam, R.R.; Banerjee, A.; Kondeti, S. Fibroblast growth factor 21 as a potential master regulator in metabolic disorders. *Am. J. Physiol. Endocrinol. Metab.* **2023**, *324*, E409–E424. [CrossRef] [PubMed]
17. Sierra, A.P.R.; Fontes-Junior, A.A.; Paz, I.A.; de Sousa, C.A.Z.; Manoel, L.A.D.S.; Menezes, D.C.; Rocha, V.A.; Barbeiro, H.V.; Souza, H.P.; Cury-Boaventura, M.F. Chronic Low or High Nutrient Intake and Myokine Levels. *Nutrients* **2022**, *15*, 153. [CrossRef]
18. de Carvalho, M.R.; Duarte, E.F.; Mendonça, M.L.M.; de Morais, C.S.; Ota, G.E.; Gaspar-Junior, J.J.; de Oliveira Filiú, W.F.; Damatto, F.C.; Okoshi, M.P.; Okoshi, K.; et al. Effects of Creatine Supplementation on the Myostatin Pathway and Myosin Heavy Chain Isoforms in Different Skeletal Muscles of Resistance-Trained Rats. *Nutrients* **2023**, *15*, 2224. [CrossRef]
19. Valero-Breton, M.; Valladares-Ide, D.; Álvarez, C.; Peñailillo, R.S.; Peñailillo, L. Changes in Blood Markers of Oxidative Stress, Inflammation and Cardiometabolic Patients with COPD after Eccentric and Concentric Cycling Training. *Nutrients* **2023**, *15*, 908. [CrossRef]
20. Cuyul-Vásquez, I.; Pezo-Navarrete, J.; Vargas-Arriagada, C.; Ortega-Díaz, C.; Sepúlveda-Loyola, W.; Hirabara, S.M.; Marzuca-Nassr, G.N. Effectiveness of Whey Protein Supplementation during Resistance Exercise Training on Skeletal Muscle Mass and Strength in Older People with Sarcopenia: A Systematic Review and Meta-Analysis. *Nutrients* **2023**, *15*, 3424. [CrossRef]
21. Barquilha, G.; Dos Santos, C.M.M.; Caçula, K.G.; Santos, V.C.; Polotow, T.G.; Vasconcellos, C.V.; Gomes-Santos, J.A.F.; Rodrigues, L.E.; Lambertucci, R.H.; Serdan, T.D.A.; et al. Fish Oil Supplementation Improves the Repeated-Bout Effect and Redox Balance in 20-30-Year-Old Men Submitted to Strength Training. *Nutrients* **2023**, *15*, 1708. [CrossRef] [PubMed]
22. Therdyothin, A.; Phiphopthatsanee, N.; Isanejad, M. The Effect of Omega-3 Fatty Acids on Sarcopenia: Mechanism of Action and Potential Efficacy. *Mar. Drugs* **2023**, *21*, 399. [CrossRef] [PubMed]
23. Witard, O.C.; Banic, M.; Rodriguez-Sanchez, N.; van Dijk, M.; Galloway, S.D.R. Long-chain n-3 PUFA ingestion for the stimulation of muscle protein synthesis in healthy older adults. *Proc Nutr Soc.* **2023**, *1*, 1–11. [CrossRef]
24. Jannas-Vela, S.; Espinosa, A.; Candia, A.A.; Flores-Opazo, M.; Peñailillo, L.; Valenzuela, R. The Role of Omega-3 Polyunsaturated Fatty Acids and Their Lipid Mediators on Skeletal Muscle Regeneration: A Narrative Review. *Nutrients* **2023**, *15*, 871. [CrossRef]
25. Francekovic, P.; Gliemann, L. Endothelial Glycocalyx Preservation-Impact of Nutrition and Lifestyle. *Nutrients* **2023**, *15*, 2573. [CrossRef]

 *nutrients*

Article

Spirulina Supplementation with High-Intensity Interval Training Decreases Adipokines Levels and Cardiovascular Risk Factors in Men with Obesity

Rashmi Supriya [1], Maryam Delfan [2,*], Ayoub Saeidi [3], Seyedeh Somayeh Samaie [4], Maisa Hamed Al Kiyumi [5,6], Kurt A. Escobar [7], Ismail Laher [8], Katie M. Heinrich [9,10], Katja Weiss [11], Beat Knechtle [11,12] and Hassane Zouhal [13,14,*]

[1] Center For Health & Exercise Science Research, Department of Sport, Physical Education and Health, Faculty of Social Sciences, Hong Kong Baptist University, Kowloon Tong, Hong Kong SAR 999077, China; rashmisupriya@hkbu.edu.hk
[2] Department of Exercise Physiology, Faculty of Sport Sciences, Alzahra University, Tehran 15847-15414, Iran
[3] Department of Physical Education and Sport Sciences, Faculty of Humanities and Social Sciences, University of Kurdistan, Sanandaj, Kurdistan 66177-15175, Iran; saeidi_as68@yahoo.com
[4] Department of Exercise Physiology, Faculty of Physical Education and Sport Sciences, Alborz Campus, University of Tehran, Tehran 15719-14911, Iran; s.samaie@gmail.com
[5] Department of Family Medicine and Public Health, Sultan Qaboos University, Muscat P.O. Box 35, Oman; drmaysa@squ.edu.om
[6] Department of Family Medicine and Public Health, Sultan Qaboos University Hospital, Muscat P.O. Box 35, Oman
[7] Department of Kinesiology, California State University, Long Beach, CA 90840, USA; kurt.escobar@csulb.edu
[8] Department of Anesthesiology, Pharmacology, and Therapeutics, Faculty of Medicine, University of British Columbia, Vancouver, BC V6T 1Z3, Canada; ismail.laher@ubc.ca
[9] Department of Kinesiology, Kansas State University, Manhattan, KS 66502, USA; kmhphd@ksu.edu
[10] Research Department, the Phoenix, Manhattan, KS 66502, USA
[11] Institute of Primary Care, University of Zurich, 8091 Zurich, Switzerland; katja@weiss.co.com (K.W.); beat.knechtle@hispeed.ch (B.K.)
[12] Medbase St. Gallen Am Vadianplatz, 9000 St. Gallen, Switzerland
[13] M2S (Laboratoire Mouvement, Sport, Santé)—EA 1274, Université de Rennes, 35000 Rennes, France
[14] Institut International des Sciences du Sport (2I2S), 35850 Irodouer, France
* Correspondence: m.delfan@alzahra.ac.ir (M.D.); hassane.zouhal@univ-rennes2.fr (H.Z.)

Citation: Supriya, R.; Delfan, M.; Saeidi, A.; Samaie, S.S.; Al Kiyumi, M.H.; Escobar, K.A.; Laher, I.; Heinrich, K.M.; Weiss, K.; Knechtle, B.; et al. Spirulina Supplementation with High-Intensity Interval Training Decreases Adipokines Levels and Cardiovascular Risk Factors in Men with Obesity. *Nutrients* **2023**, *15*, 4891. https://doi.org/10.3390/nu15234891

Academic Editor: Kenji Doma

Received: 8 September 2023
Revised: 13 November 2023
Accepted: 16 November 2023
Published: 23 November 2023

Abstract: Adiposity, a state characterized by excessive accumulation of body fat, is closely linked to metabolic complications and the secretion of specific adipokines. This study explores the potential of exercise and Spirulina supplementation to mitigate these complications and modulate adipokine release associated with obesity. The primary objective of this investigation was to examine the impact of a 12-week regimen of high-intensity training combined with Spirulina supplementation on adipokine concentrations and lipid profiles in male individuals with obesity (N = 44). The participants were randomly distributed into four groups, each consisting of 11 participants: a control group (CG), a supplement group (SG), a training group (TG), and a training plus supplement group (TSG). The intervention comprised a 12-week treatment involving Spirulina supplementation (6 g capsule daily), a 12-week high-intensity interval training (HIIT) protocol with three sessions per week, or a combined approach. Following the interventions, metabolic parameters, anthropometric measurements, cardiorespiratory indices, and circulating adipokines [CRP, Sema3C, TNF-α, IL-6, MCP1, IL-8] were assessed within 48 h of the before and final training session. Statistical analyses revealed significant differences across all measures among the groups ($p < 0.05$). Notably, post hoc analyses indicated substantial disparities between the CG and the three interventional groups regarding body weight ($p < 0.05$). The combined training and supplementation approach led to noteworthy reductions in low-density lipoprotein (LDL), total cholesterol (TC), and triglyceride (TGL) levels (all $p < 0.0001$), coupled with an elevation in high-density lipoprotein–cholesterol (HDL-C) levels ($p = 0.0001$). Furthermore, adipokine levels significantly declined in the three intervention groups relative to the CG ($p < 0.05$). The findings from this 12-week study demonstrate that Spirulina supplementation in conjunction with high-intensity interval training reduced adipokine levels, improved body weight and BMI, and enhanced lipid profiles. This investigation underscores the potential of Spirulina supplementation and high-intensity interval training as a synergistic strategy to ameliorate obesity-related complications and enhance overall cardiometabolic well-being in obese males.

Keywords: cardiometabolic health; adipocytokines; Spirulina supplementation; metabolic complications

1. Introduction

Obesity exerts detrimental effects on a multitude of physiological systems by impairing the proper functioning of tissues and organs, thereby contributing to the onset of various diseases. It stands as a prominent risk factor for numerous noncommunicable disorders, including type 2 diabetes (T2D), cardiovascular ailments, hypertension, stroke, diverse forms of cancer, and mental health concerns [1]. Key factors in the genesis of obesity and its associated disorders, such as insulin resistance and T2D, encompass alterations in the levels of circulating adipokines and cytokines [2,3]. Adipokines represent bioactive molecules discharged into the bloodstream from adipose tissue, orchestrating metabolic changes in a manner that affects a wide array of tissues and organs [4]. Similarly, cytokines, predominantly released by white blood cells, wield systemic and localized influences [4]. In the context of obesity, inflammatory adipokines and cytokines often display heightened levels and have been implicated in the pathogenesis of various disease processes [1]. Prominent among the elevated obesity-associated circulating adipokines and cytokines are the C-reactive protein (CRP) [5,6], Semaphorin-3C (Sema3C), Tumor Necrosis Factor-alpha (TNF alpha), Interleukin-6 (IL-6) [7], Monocyte Chemoattractant Protein-1 (MCP1) [7], and Interleukin-8 (IL-8) [8].

Oxidative stress plays a pivotal role in mediating the escalation of adipokine-induced inflammation, with reactive oxygen species (ROS) increasing as visceral fat expands during the progression of obesity, thus modifying the expression and secretion of inflammatory adipokines [3,5]. ROS generation can stem from oxidative phosphorylation within the mitochondria, and the extent of production is contingent on mitochondrial function [6]. In physiological circumstances, antioxidant buffering balances ROS production. However, in situations of overnutrition and obesity, ROS production may surpass the capacity of antioxidant buffering, culminating in oxidative stress wherein ROS disrupts cellular and tissue function, including the activation of pathways that augment adipokine and cytokine expression [3,6]. Notably, ROS plays a crucial role in developing obesity-linked cardiometabolic disorders [6].

Consequently, a growing interest has been in mitigating ROS levels in obesity. Several investigations have illustrated that dietary antioxidants, counteracting the pro-oxidative state of cells, might hold therapeutic potential for addressing obesity and its concomitant comorbidities [9]. Exogenous dietary antioxidants synergize with endogenous counterparts to bolster cellular antioxidant capacity [10]. Spirulina, an ancient cyanobacterium and one of the earliest life forms on Earth, boasts antioxidant attributes. Evidence underscores Spirulina's potential to reduce blood lipid levels, body fat, waist circumference, body mass index, and appetite [11–13]. Nonetheless, the impact of Spirulina on oxidative stress and inflammation markers in humans remains a subject of debate [14,15].

Furthermore, the interplay between Spirulina supplementation and exercise, known to alleviate oxidative stress and inflammation markers, remains unclear [16,17]. Although acute exercise initially elevates ROS, regular exercise activates the endogenous antioxidant system, affording protection against oxidative damage [17,18]. A reduction in ROS levels could potentially modulate the altered profiles of adipokines and cytokines witnessed in obesity [17]. Notably, Spirulina supplementation has been shown to enhance acute exercise performance, fat oxidation, and glutathione levels while attenuating the rise in lipid peroxidation prompted by aerobic exercise [19–21]. In a particular study, the combination of High-Intensity Interval Training (HIIT) and Spirulina supplementation positively impacted immunoglobulin levels, cardiorespiratory fitness, and body composition in overweight and obese women, along with an increase in immunoglobulin A (IgA), vital for the immune system [19]. Prior research has also examined Spirulina's influence on nesfatin-1, omentin-1, and lipid profiles among obese and overweight females [22]. The combination of Spirulina and HIIT elevated nesfatin-1 and omentin-1 levels, although Spirulina supplementation alone did not [22]. Another study scrutinized the independent and synergistic effects of Spirulina supplementation (4.5 g per day), with or without engagement in aerobic exercise three days per week and HIIT two days per week, on blood lipids and body mass index (BMI) in 52 sedentary men with excess body weight over six weeks [23]. The findings demonstrated that Spirulina supplementation potentiated the hypolipidemic effects of an intensive physical training regimen in males with excess body weight and dyslipidemia [23]. In past studies, it has been shown that Spirulina alone causes weight loss [13,24], and on the other hand, the combination of Spirulina consumption with exercise training reduces inflammatory indicators in obese people [25].

However, despite these observations pointing toward favorable impacts of Spirulina and HIIT on cardiometabolic health markers among individuals with obesity, alterations in adipokines during such interventions have yet to be investigated.

Consequently, the present study aims to explore the effects of a 12-week regimen combining HIIT and Spirulina supplementation on markers of cardiometabolic health, anthropometric measures, cardiorespiratory fitness, as well as adipokines and cytokines, in comparison to the effects of HIIT and Spirulina interventions in isolation. It is postulated that the combined intervention will yield more pronounced improvements in the measured variables compared to each intervention alone.

2. Methods

2.1. Ethical Approval

The research received approval from the Ethics Committee of the Sport Sciences Research Institute (Ethics code: IR.SSRC.REC.1401.093). All protocols adhered to the most recent iteration of the Declaration of Helsinki [19].

2.2. Participants

Following outreach in public spaces, laboratories, sports clubs, and social networks, 143 adult men volunteered to partake in the current research. Among them, 80 participants were eligible for the study entry criteria. Inclusion criteria: having a BMI exceeding 30 kg/m^2, lacking involvement in regular physical activities over the past six months, having no history of cardiovascular or endocrine disorders, and no smoking and alcohol consumption. Prospective participants with joint ailments, physical disabilities, and those utilizing prescription medications or supplements with potential impacts on muscle and adipose tissue metabolism were excluded from the study.

A subset of 80 individuals was ultimately chosen to participate after thoroughly evaluating the volunteers. The inclusion criteria necessitated that all participants undergo a comprehensive physical examination conducted by a qualified medical professional and clinical exercise physiologist during the initial visit. After the initial evaluation and explanation of different parts of the research, 64 individuals were selected. The sample size was based on the standardized effect size (SES) calculated using the mean and standard deviation values of similar studies reported in the literature [26]. The standardized effect size was placed in the G*Power (3.1.9.4) analysis program [two-sided, $\alpha = 0.05$, power $(1-\beta) = 0.95$, effect size = 1.43]. Accordingly, the minimum sample number per group was determined to be nine. In the present study, we considered 16 subjects for each group. Additionally, participants were required to furnish a signed consent form and complete the Physical Activity Readiness Questionnaire (PAR-Q) [11]. This process ensured adherence to established research standards and ethical guidelines.

2.3. Experimental Design

Participants underwent a familiarization session during which all study procedures were comprehensively elucidated, taking place one week before the initiation of the training regimens. Measurements of height and body mass were conducted for each participant. Subsequently, participants were assigned at random to one of four equitably sized groups (16 participants in each group): the Control Group (CG), the Supplement Group (SG), the Training Group (TG), and the Training + Supplement Group (TSG) Figure 1.

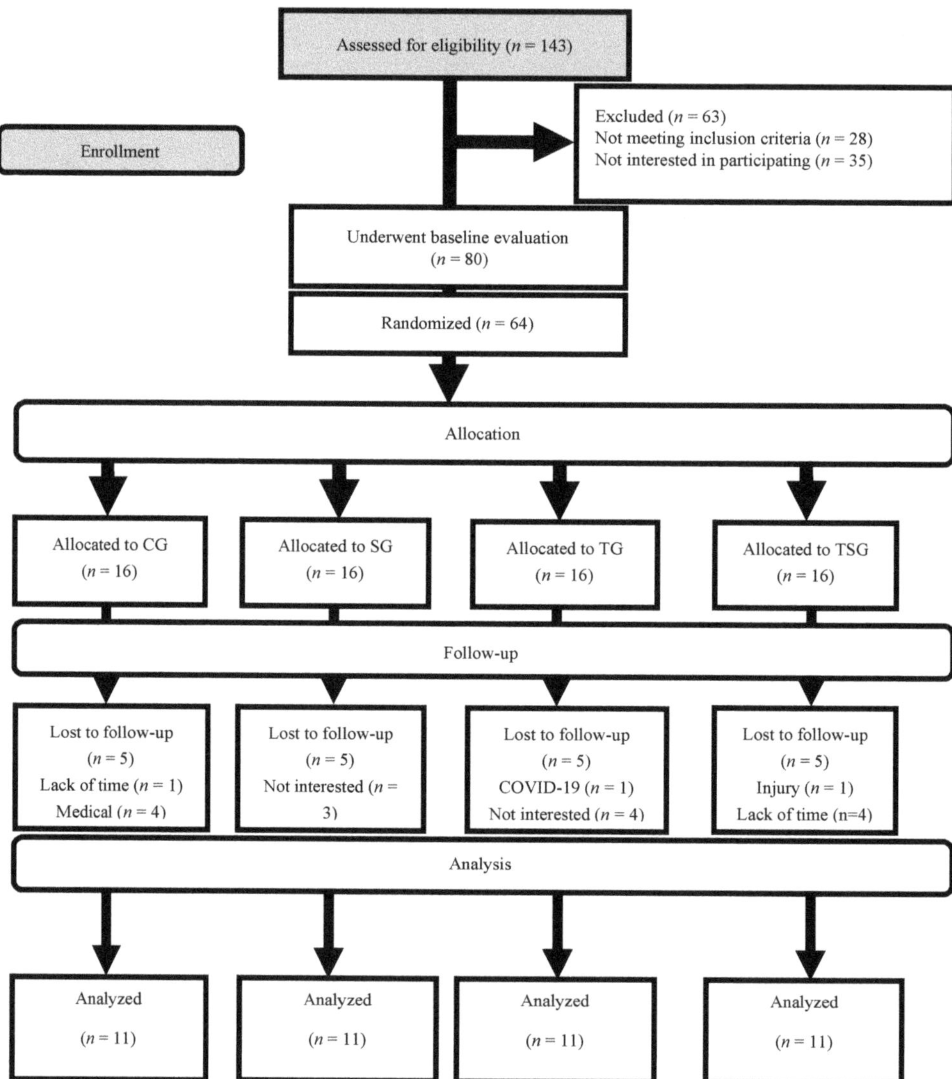

Figure 1. Flow of participant recruitment.

Throughout the study duration, 20 participants withdrew from the study due to medical conditions, work-related challenges, and a lack of sustained interest in the research. As a result, each group had 11 participants remaining for the subsequent analysis. Detailed instructions regarding the execution of training protocols were given to each group during the third session, coinciding with measurements of body mass and VO_{2peak}.

Following the baseline measurements, the two groups engaged in training (TG and TSG) embarked on a 12-week exercise program involving three sessions per week. Participants allocated to all groups were instructed to maintain their lifestyles throughout the study. All measurements were conducted at consistent times of day (with a deviation of approximately 1 h) and under uniform environmental conditions (~20 °C and ~55% humidity).

Baseline assessments were conducted 48 h before the commencement of the training protocols, while post-tests occurred 48 h after the final session for all groups. Those participating in the training protocols were instructed to adhere to the same dietary regimen for 48 h before the baseline assessment and the concluding measurements.

2.4. Anthropometric and Cardio-Respiratory Fitness Assessments

The measurement of participants' body mass and height was carried out using a calibrated scale (Seca, Halmburger, Germany) for weight and a stadiometer (Seca, Halmburger, Germany) for height. These measurements were subsequently utilized to calculate the participants' body mass index (BMI) in kilograms per square meter (kg/m^2). The assessment of VO_{2peak} was conducted employing a modified Bruce protocol within a controlled environment set at a temperature range of 21–23 °C. This protocol has been previously documented in studies involving individuals with overweight and obesity [11,12]. The exercise was performed on a motorized treadmill (H/P/Cosmos, Pulsar med 3p-Sports and Medical, Nussdorf-Traunstein, Germany).

The determination of VO_{2peak} adhered to the physiological criteria outlined by the American College of Sports Medicine (ACSM) guidelines. These criteria included reaching a point of perceived physical exhaustion and maximal effort, as indicated by participants' responses on the Borg scale, or the identification of severe dyspnea, dizziness, and other constraining symptoms by the supervisor, following the ACSM and American Heart Association (AHA) guidelines for cardiopulmonary exercise testing (CPET) [27,28]. A plateau in both VO_2 and respiratory exchange ratio (RER) ≥ 1.10 was used to confirm VO_{2peak} attainment.

Blood pressure readings were obtained using an electronic sphygmomanometer (Kenz BPM AM 300P CE, Nagoya, Japan), and heart rate was continuously monitored throughout the tests using a Polar V800 heart monitor (Finland). Gas analysis was carried out using a gas analyzer system (Metalyzer 3B analyzer, Cortex: biophysics, GmbH, Leipzig, Germany), which underwent calibration before each testing session.

2.5. Training Protocols

VO_{2peak} values were used to prescribe exercise intensity in training sessions, which consisted of treadmill running for 32 min. Prior to each training session, subjects performed 5 min of warm-up activities (stretching movements, walking, and running). In the first week, subjects ran on a treadmill at 65% of VO_{2peak} and in the second week at 75% of VO_{2peak} for 32 min and three sessions a week. In week three, the HIIT sessions began. In the third and fourth weeks, subjects performed intervals of 4 min of running at 75% VO_{2peak} followed by 4 min of inactive recovery for 32 min. In weeks 5, 6, and 7, subjects performed intervals of 4 min at 85% of VO_{2peak} with 4 min active recovery intervals at 15% of VO_{2peak} 32 min. In weeks 8, 9, and 10, subjects performed 4 min intervals at 90% VO_{2peak} with active recovery intervals at 30% VO_{2peak} for 4 min on the treadmill for 32 min. In weeks 11 and 12, subjects completed 4 min intervals at 95% VO_{2peak} with 4 min active rest intervals at 50% VO_{2peak} for 32 min. After completing the training in each session, subjects performed a 5 min cool-down at 50% VO_{2peak} in each session [29]. The control group continued their normal daily activities and were restricted from participating in regular physical activity.

2.6. Supplementation of Spirulina and Placebo

Spirulina samples (Hellenic Spirulina Net: Production unit: Thermopigi, Sidorokastro, Serres, Greece) were encapsulated for administration. Each subject ingested a total of

6 g daily, divided into two doses of 3 g each, one in the morning and the other in the evening, over 12 weeks [30]. Comparable amounts of placebo were also provided to both the Control Group (CG) and Treatment Group (TG). The placebo was formulated using corn starch, tinted with a food-grade green coloring resembling Spirulina powder, and enhanced with the essence of kiwi fruit for flavor. Corn starch, a neutral and inert substance lacking therapeutic properties, was selected for its established safety profile. Widely employed across the culinary and pharmaceutical sectors, it is recognized as a colorless, tasteless, secure, non-toxic, non-irritating, and hypoallergenic powder. Adherence to the supplementation regimen was defined as consumption of =80% of the assigned supplements by each participant.

2.7. Nutrient Intake and Dietary Analysis

Three-day food records (two weekdays and one weekend day) were obtained before and after the study to assess changes in habitual dietary intake over time [31]. Each food item was individually entered into Diet Analysis Plus version 10 (Cengage, Boston, MA, USA), and total energy consumption and the amount of energy derived from proteins, fats, and carbohydrates were determined (Table 1).

Table 1. Mean (±SD) values of nutritional intake in the four study groups.

	CG		SG		TG		TSG	
	Pre	Post	Pre	Post	Pre	Post	Pre	Post
Energy (kcal/day)	2321 ± 47	2342 ± 56	2354 ± 101	2314 ± 100	2349 ± 117	2297 ± 117	2375 ± 157	2301 ± 126
CHO (g/day)	292 ± 30.4	295 ± 31.3	288.4 ± 25.1	278 ± 26.5	298 ± 41.6	270 ± 37.2	297 ± 39.6	269 ± 30.1
Fat (g/day)	91.2 ± 16.0	92 ± 19.8	95.5 ± 17.7	84 ± 16.2	94.4 ± 19.4	84.1 ± 15.2	91 ± 15.87	75.2 ± 18.3
Protein (g/day)	115 ± 17.0	119 ± 19.3	112 ± 15.5	105 ± 16.6	113 ± 13.8	103 ± 11.7	112 ± 11.5	101 ± 12.5

CG: Control group; SG: Supplement group; TG: Training group; TSG: Training and supplement group.

2.8. Blood Markers

All testing was carried out under standard conditions between 8:00 and 10:00 a.m. After 12 h of fasting, venous blood samples were taken from the right arm 48 h before the first exercise session and 48 h after the last session. Blood samples were transferred to EDTA-containing tubes, centrifuged for 10 min at 3000 rpm, and stored at $-80\,^{\circ}$C. Plasma total cholesterol (TC) and triglyceride (TGL) were measured using enzymatic methods (CHOD-PAP); high-density cholesterol (HDL-C) and low-density cholesterol (LDL-C) were determined using a photometric method (Pars Testee's Quantitative Detection kit, Tehran, Iran) with a coefficient and sensitivity of 1.8% and 1 mg/dL and 1.2% and 1 mg/dL respectively. The hs-CRP levels were measured with an ELISA kit (Diagnostic Biochem, London, ON, Canada). Sensitivity: 10 ng/mL. Plasma TNF-α levels were measured with an ELISA kit (Elabscience Biotechnology, Wuhan, China). Catalogue No: E-EL-H0109. Sensitivity: 4.69 pg/mL. Intra-CV = 6.22%, inter-CV = 5.2%. Plasma IL-6 levels were measured with an ELISA kit (Biovendor, Brno, Czech Republic). Catalogue No: RD194015200R. Sensitivity: 0.65 pg/mL. Intra-CV = 4.7%, inter-CV = 4.9%. Plasma Sema3C levels were measured with an ELISA kit (MyBioSource, San Diego, CA, USA). Catalogue No: MBS2883689. Sensitivity: 2.3~40 ng/mL. Plasma MCP-1 levels were measured with an ELISA kit (R&D Systems, USA). Catalogue No: DCP00. Sensitivity: 10 pg/mL. Intra-CV = 7.8%, inter-CV = 6.7%. Plasma IL-8 levels were measured with an ELISA kit (citeab, Bath, UK). Catalogue No: 900-K18. Sensitivity: <7.5%.

2.9. Statistical Analysis

Descriptive statistics (means $\pm$ standard deviation) were used to describe all data. The normality of the data was assessed using the Shapiro–Wilk test. A two-way ANOVA repeated measures test was used to determine the Group $\times$ time interaction. One-way ANOVA and Fisher LSD post hoc tests were used for the evaluation baseline data of four groups. When a significant difference was detected using ANOVA, mean differences were determined using pairwise comparisons. The sample size was calculated to detect a statistical difference between study variables with a 95% confidence interval (CI) equal to or greater than 80% of the power value. Additionally, effect sizes (ES) were reported as partial eta-squared. In accordance with Hopkins et al. (2009) [32], ES were considered trivial (<0.2), small (0.2–0.6), moderate (0.6–1.2), large (1.2–2.0), and very large (2.0–4.0). A p-value of <0.05 was used to indicate statistical significance. All data were evaluated with SPSS software (version 24).

3. Results

3.1. Anthropometry and VO_{2peak}

The four groups' baseline differences were insignificant for body mass ($p = 0.46$) and BMI ($p = 0.46$). Body mass was significantly different from baseline in the TSG ($p = 0.039$); however, it was not in the SG ($p = 0.72$), TG ($p = 0.12$), or CG group ($p = 0.70$) (Table 1). BMI was not significantly different from baseline for any of the four groups ($p > 0.05$) (Table 2). There were no significant interactions between group and time for either weight ($p = 0.28$, $\eta^2 = 0.08$) or BMI ($p = 0.36$, $\eta^2 = 0.07$).

Table 2. Mean ($\pm$SD) values of lipid profile, anthropometric, and VO_{2peak} for the four study groups.

	CG		SG		TG		TSG	
	Pre	Post	Pre	Post	Pre	Post	Pre	Post
Body height (cm)	175.7 ± 4.21	-	171.3 ± 4.17	-	173.3 ± 8.16	-	175.2 ± 6.47	-
Body Mass (kg)	101.22 ± 5.27	102.03 ± 2.48	97.81 ± 4.73	97.05 ± 2.45	99.52 ± 10.21	96.22 ± 2.39	101.48 ± 7.95	96.98 ± 1.93 [a]
BMI (kg/m^2)	32.77 ± 1.18	33.07 ± 1.40	33.31 ± 0.62	33.13 ± 1.99	33.01 ± 0.76	32.16 ± 2.71	33.00 ± 1.00	31.68 ± 2.18
VO_{2peak} (mL·kg^{-1}·min^{-1})	26.58 ± 1.76	25.71 ± 1.73	26.72 ± 1.36	27.92 ± 2.32 [a,b]	26.38 ± 1.30	29.93 ± 2.08 [a,b]	26.46 ± 1.76	30.38 ± 1.97 [a,b,ab]
HDL (mg/dL)	29.76 ± 6.43	31.67 ± 6.76	30.85 ± 4.68	36.40 ± 5.36 [a,b]	31.23 ± 4.32	37.23 ± 7.45 [a,b]	28.19 ± 5.88	42.19 ± 5.48 [a,b,ab]
LDL (mg/dL)	174.0 ± 13.76	173.5 ± 13.49	172.7 ± 13.91	165.9 ± 11.9 [a,b]	174.3 ± 10.63	158.2 ± 8.56 [a,b]	176.4 ± 16.83	151.8 ± 13.87 [a,b,ab]
TC (mg/dL)	264.2 ± 16.23	269.7 ± 12.24	256.9 ± 20.07	245.7 ± 18.3 [a,b]	262.3 ± 13.02	245.0 ± 15.0 [a,b]	258.9 ± 15.77	240.0 ± 11.14 [a,b,ab]
TGL (mg/dL)	260.9 ± 15.51	258.9 ± 12.88	262.8 ± 16.75	258.8 ± 13.5 [a,b]	261.4 ± 20.78	252.0 ± 18.5 [a,b]	265.8 ± 19.17	253.8 ± 16.48 [a,b,ab]

CG: control group; SG: supplement group; TG: training group; TSG: training+ supplement group BMI: body mass index; HDL: high-density lipoprotein; LDL: low-density lipoprotein; TC: total cholesterol; TGL: triglyceride. [a] Indicates significant differences compared to the pre-values ($p < 0.05$). [b] Significant differences compared to the control group ($p < 0.05$). [ab] Significant interaction between time and groups ($p < 0.05$).

There were no significant differences between groups in VO_{2peak} values at baseline ($p = 0.10$). There were significant increases from baseline in the TG ($p = 0.001$) and TSG ($p = 0.0001$) after 12 weeks, but no changes in the CG ($p = 0.29$) or SG ($p = 0.15$). The interaction between time and groups was significant for VO_{2peak} ($p = 0.001$, $\eta^2 = 0.35$). In comparison to the CG, after 12 weeks, VO_{2peak} was significantly higher in the TG ($p = 0.003$) and TSG ($p = 0.001$) but not in the SG ($p = 0.51$). There were no differences in VO_{2peak} between the SG and TG ($p = 0.30$) or between the TSG and TG ($p = 0.99$) or SG ($p = 0.15$) (Table 2).

3.2. Lipid Profiles

There were no differences between groups at baseline for HDL ($p = 0.55$), LDL ($p = 0.94$), TC ($p = 0.72$), and TG ($p = 0.92$). Compared to the baseline, HDL was significantly increased in the SG ($p = 0.0001$), TG ($p = 0.0001$), and TSG ($p = 0.0001$) but not in the CG ($p = 0.07$). The time × group interaction was significant for HDL ($p = 0.0001$, $\eta^2 = 0.65$). The results of the Bonferroni test showed significant differences in the TG ($p = 0.044$) and TSG ($p = 0.0001$) compared to the CG but no difference between the SG ($p = 0.09$) and CG. Also, the changes in HDL in the TSG were significantly higher compared to the SG ($p = 0.0001$) and TG ($p = 0.0001$), but the differences between the SG and TG were not significant ($p = 0.99$) (Table 2).

Post-test values for LDL were significantly lower in the SG ($p = 0.005$), TG ($p = 0.0001$), and TSG ($p = 0.0001$) but not in the CG ($p = 0.82$) compared to baseline. The interaction between time and groups was significant for LDL ($p = 0.0001$, $\eta^2 = 0.63$). Post hoc tests revealed that the reduction in LDL was significantly greater in the TG ($p = 0.0001$) and TSG ($p = 0.0001$) compared to the CG; however, it was not different between the SG and CG ($p = 0.34$). The reductions in LDL in the TG ($p = 0.032$) and TSG ($p = 0.0001$) were significantly greater compared to the SG, while the TSG was not different compared to the TG ($p = 0.066$) (Table 2).

Compared to the baseline, there were decreases in TC in the SG ($p = 0.009$), TG ($p = 0.001$), and TSG ($p = 0.0001$) following the 12-week intervention. There were no significant changes in TC at 12 weeks in the CG ($p = 0.19$). There was a significant interaction between time and groups for TC ($p = 0.0001$, $\eta^2 = 0.36$). Results of the Bonferroni test showed that after 12 weeks, compared to the CG, TC was significantly lower in the SG ($p = 0.039$), TG ($p = 0.002$), and TSG ($p = 0.001$). There were no significant differences in TC at 12 weeks between the SG, TG, and TSG groups ($p = 0.99$) (Table 2).

Compared to baseline, TGL at 12 weeks was significantly lower in the SG ($p = 0.006$), TG ($p = 0.0001$), and TSG ($p = 0.0001$). There was a significant interaction between time and groups for TGL ($p = 0.0001$, $\eta^2 = 0.46$). Results of the Bonferroni test showed that after 12 weeks, TGL was significantly lower in the TG ($p = 0.003$) and TSG ($p = 0.001$) compared to the CG. However, the SG was not significantly lower compared to the CG ($p = 0.99$). The reductions in TGL after 12 weeks were more significant in the TSG compared to the SG ($p = 0.001$) but not compared to TG ($p = 0.99$) (Table 2).

3.3. Adipokines and Cytokines

There were no significant differences between the groups in baseline values for CRP ($p = 0.62$), Sema3C ($p = 0.68$), TNF-α ($p = 0.26$), IL-6 ($p = 0.50$), MCP1 ($p = 0.74$), and IL-8 ($p = 0.78$). Compared to baseline, CRP was lower in the SG ($p = 0.0001$), TG ($p = 0.0001$), and TSG ($p = 0.0001$) but not in the CG ($p = 0.72$) after 12 weeks. There was a significant time × group interaction for CRP ($p = 0.0001$, $\eta^2 = 0.42$). Post hoc test results showed that after 12 weeks, the CRP was lower in the SG ($p = 0.002$), TG ($p = 0.001$), and TSG ($p = 0.0001$) compared to the CG. CRP levels in the TSG were lower compared to the SG ($p = 0.99$) and TG ($p = 0.99$). CRP levels at 12 weeks were not different between the TG and SG ($p = 0.99$) (Figure 2). Sema3C at 12 weeks was significantly lower in the SG ($p = 0.0001$), TG ($p = 0.0001$), and TSG ($p = 0.0001$) compared to the baseline. There was no difference in the CG ($p = 0.37$) at 12 weeks compared to the baseline. Also, the time × groups interaction for Sema3C was significant ($p = 0.0001$, $\eta^2 = 0.40$). The results of the Bonferroni test showed that after 12 weeks, compared to the CG, Sema3C was lower in the SG ($p = 0.001$), TG ($p = 0.003$), and TSG ($p = 0.0001$). However, there were no differences in Sema3C between the SG, TG, and TSG groups ($p = 0.99$) (Figure 3). Following the 12-week intervention, TNF-α was not different in the CG compared to the baseline ($p = 0.17$). However, TNF-α was lower in the SG ($p = 0.0001$), TG (T $= 0.0001$), and TSG ($p = 0.0001$) compared to the baseline. The interaction between time and groups was also significant for TNF-α ($p = 0.0001$, $\eta^2 = 0.52$). The results of the Bonferroni revealed that at 12 weeks, TNF-α was significantly lower in the TG ($p = 0.0001$) and TSG ($p = 0.0001$) compared to the CG. There was no difference between the SG and CG ($p = 0.057$). Compared to the SG, TNF-α was significantly lower in the TG ($p = 0.030$); however, TNF-α levels in the TSG were not different from the SG ($p = 0.057$) and TG ($p = 0.99$) (Figure 4). Compared to the baseline, IL-6 was not different in the CG ($p = 0.46$) at 12 weeks, while concentrations were lower in the SG ($p = 0.008$),

TG ($p = 0.0001$), and TSG ($p = 0.0001$). The time $\times$ groups interaction was also significant for IL-6 ($p = 0.0001$, $\eta^2 = 0.38$). IL-6 was significantly lower in the TG ($p = 0.004$) and TSG ($p = 0.0001$) compared to the CG but not in the SG ($p = 0.096$). The reduction in IL-6 in the TSG was not different compared to the TG ($p = 0.99$) and SG ($p = 0.19$), while there was also no difference between the SG and TG ($p = 0.99$) (Figure 5). Compared to the baseline, MCP-1 at 12 weeks was significantly increased in the CG ($p = 0.46$), while MCP-1 was decreased in the SG ($p = 0.0001$), TG ($p = 0.0001$), and TSG ($p = 0.0001$). Also, a significant interaction between time and groups was observed for MCP-1 ($p = 0.0001$, $\eta^2 = 0.44$). The results of the Bonferroni post hoc test showed that compared to the CG, MCP-1 was lower in the SG ($p = 0.001$), TG ($p = 0.0001$), and TSG ($p = 0.001$). There were no differences in MCP-1 between the SG, TG, and TSG groups ($p = 0.99$) (Figure 6). IL-8 concentrations at 12 weeks were significantly lower in the TG ($p = 0.0001$) and TSG ($p = 0.0001$) but were not different in the SG ($p = 0.06$) or CG ($p = 0.50$). Also, a significant interaction between time and groups was reported for IL-8 ($p = 0.003$, $\eta^2 = 0.30$). The results of the Bonferroni test showed that IL-8 was significantly lower in the TG ($p = 0.005$) and TSG ($p = 0.01$) compared to the CG but not in the SG ($p = 0.46$). There were no significant differences at 12 weeks between the SG, TG, and TSG groups ($p > 0.05$) (Figure 7).

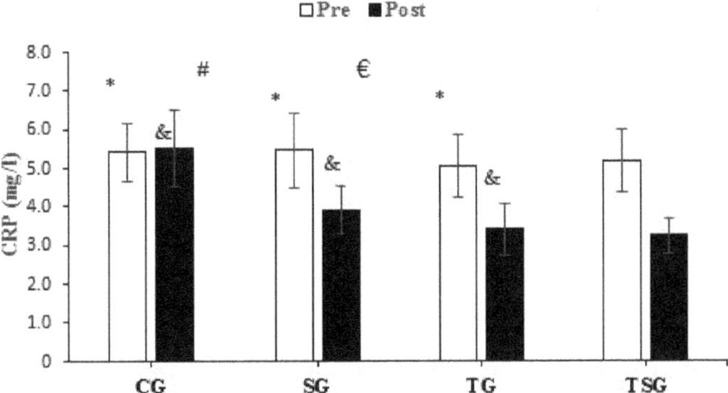

Figure 2. Pre- and post-training values (mean $\pm$ SD) for CRP in the Control (CG), Supplement (SG), Training (TG), and Training + Supplement (TSG) groups. & Significant differences with pretest values ($p < 0.05$). * Significant differences with the control group ($p < 0.05$). # Significant interaction between time and groups ($p < 0.05$). € Significant difference between TG and SG ($p < 0.05$).

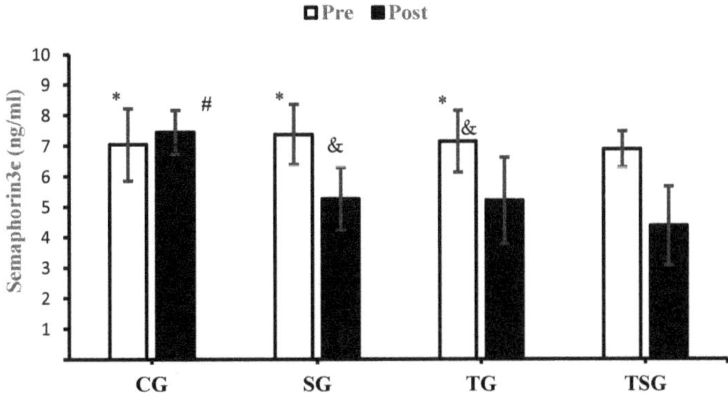

Figure 3. Pre- and post-training values (mean $\pm$ SD) for Semaphorin3c in control (CG), Supplement (SG), Training (TG), and Training + Supplement (TSG) groups. & Significant differences with pretest values ($p < 0.05$). * Significant differences with the control group ($p < 0.05$). # Significant interaction between time and groups ($p < 0.05$).

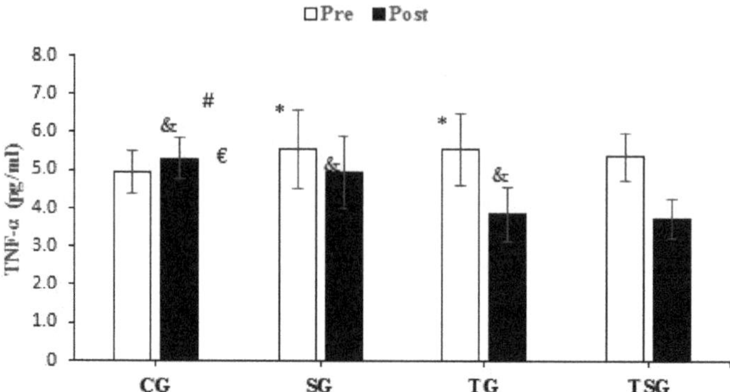

Figure 4. Pre- and post-training values (mean ± SD) for TNF-α in Control (CG), Supplement (SG), training (TG), and training + Supplement (TSG) groups. & Significant differences with pretest values ($p < 0.05$). * Significant differences with the control group ($p < 0.05$). # Significant interaction between time and groups ($p < 0.05$). € Significant difference between TG and SG ($p < 0.05$).

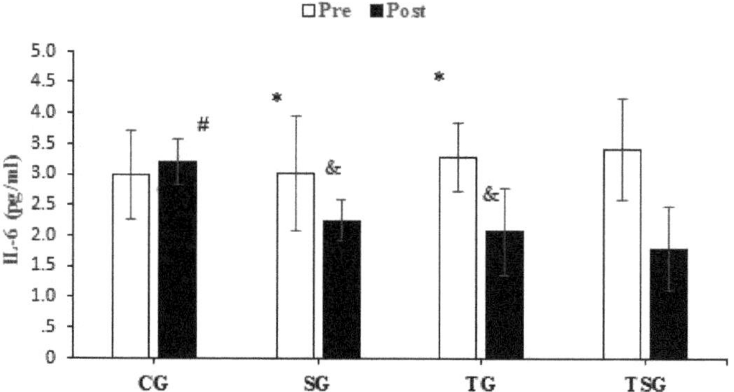

Figure 5. Pre- and post-training values (mean ± SD) for IL-6 in control (CG), Supplement (SG), training (TG), and Training + Supplement (TSG) groups. & Significant differences with pretest values ($p < 0.05$). * Significant differences with the control group ($p < 0.05$). # Significant interaction between time and groups ($p < 0.05$).

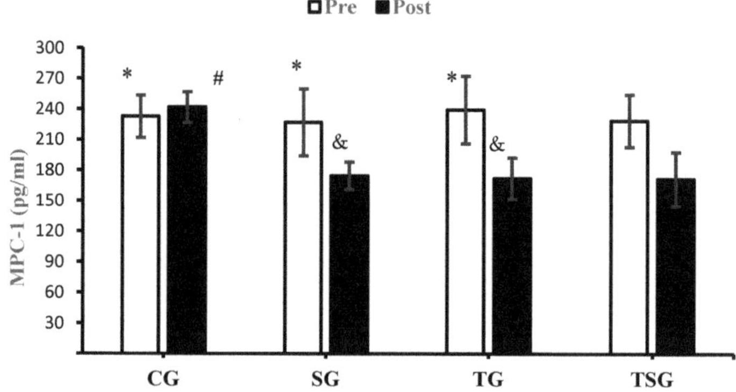

Figure 6. Pre- and post-training values (mean ± SD) for MCP-1 in control (CG), Supplement (SG), training (TG), and Training + Supplement (TSG) groups. & Significant differences with pretest values ($p < 0.05$). * Significant differences with the control group ($p < 0.05$). # Significant interaction between time and groups ($p < 0.05$).

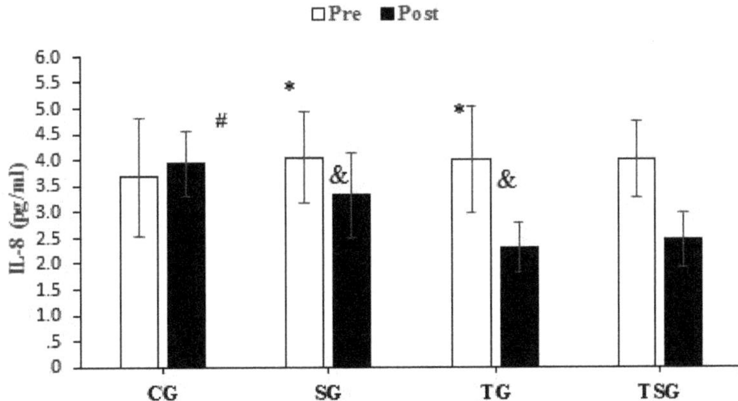

Figure 7. Pre- and post-training values (mean ± SD) for IL-8 in control (CG), Supplement (SG), training (TG), and Training + Supplement (TSG) groups. & Significant differences with pretest values ($p < 0.05$). * Significant differences with the control group ($p < 0.05$). # Significant interaction between time and groups ($p < 0.05$).

4. Discussion

This study demonstrated that 12 weeks of HIIT and Spirulina supplementation, separately and in combination, can improve circulating adipokines levels in obese men. The combination of HIIT and Spirulina supplementation overall led to greater changes in measured outcomes compared to each intervention alone. In this study, it was shown that HIIT and Spirulina supplementation decreased plasma levels of IL-8, IL-6, MCP-1, Semaphorin3c, CRP, LDL, TC, TGL, and TNF-α and increased HDL. We also showed that HIIT and Spirulina supplementation increases VO_{2peak} and decreases BMI.

Metabolic syndrome and other cardiovascular risk factors are strongly correlated with excess body fat [27]. As a result, global healthcare systems annually support new programs to reduce obesity and other cardiovascular hazards [28]. This study provides evidence for the beneficial effects of Spirulina supplementation (6 g/day) and 32 min of HIIT three times per week for 12 weeks on improving anthropometric measurements, cardiometabolic risk factors (TC, LDL, HDL, and TGL), and reducing pro-inflammatory adipokines (CRP, TNF-α, Sema-3C, IL-8, IL-6, and MCP-1) with the combination of HIIT and Spirulina improving cardiometabolic health makers greater than HIIT or Spirulina supplementation individually.

4.1. Favorable Modulations of Metabolic Factors and Cardiorespiratory Parameters

We found that 12 weeks of HIIT and Spirulina supplementation resulted in numerous improved markers of metabolic health. The most significant improvements came when training and supplementation were combined compared to the control group. Cardiorespiratory fitness is strongly related to cardiometabolic health and all-cause mortality [33]. Here, we show that HIIT increased VO_{2peak}; however, the addition of Spirulina did not further increase VO_{2peak} compared to HIIT alone. The current literature on aerobic exercise performance and Spirulina supplementation is equivocal, with some evidence indicating it may possess an ergogenic effect while other evidence suggests it does not [34].

Similarly, blood lipid profiles were improved in all intervention groups; however, the most considerable improvement occurred in TSG vs. TG and SG. Reductions in TC, LDL, and TGL occurred with training and supplementation alone and to a greater degree when combined, while an increase in HDL was observed. While there are data demonstrating lipid-lowering effects of Spirulina supplementation [35] as well as a plethora of literature evidencing the effects of regular exercise, including HIIT on blood lipids in obesity [36], only one study, to our knowledge, has demonstrated a hypolipidemic impact of combined exercise with HIIT and Spirulina supplementation in obese male individuals [23]. Fifty-two inactive males with extra body weight participated in the study, and the researchers looked at the individual and combined effects of Spirulina supplementation (4.5 g/day) with or without physical activity (3 days/week) and HIIT (2 days/week). Similar to our findings, they reported increased HDL levels while TC, TG, and LDL levels decreased [23]. This indicates that when a Spirulina supplement is added with HIIT, it can positively impact cardiometabolic factors, which were affected to a lesser degree when supplementation and HIIT were undertaken alone.

The glycemic effects of Spirulina may stem from the presence of fibers, which lowers glucose absorption from the gut [37]. Additionally, phycocyanin, an antioxidant in Spirulina, has enhanced insulin sensitivity through Akt and AMPK signaling [38]. Exercise is well known to facilitate increased insulin sensitivity and glucose regulation through AMPK signaling, leading to an increased expression of GLUT4 and insulin signaling-related proteins [39–41]. It is interesting to speculate whether the robust effects on glycemia of combining Spirulina supplementation with HIIT may have been mediated by augmented AMPK activity.

The hypocholesterolemic effect of Spirulina may partly be attributed to its g-linolenic acid (GLA) concentration, which is found in Spirulina as it may reduce hepatic lipid accumulation [37,42,43]. Moreover, phycocyanin likely plays a role in improving lipid profiles in that phycocyanin has been shown to reduce intestinal cholesterol absorption and increase lipoprotein lipase (LPL) that is involved in LDL hydrolysis and subsequently reduce LDL levels [44]. While the underlying mechanisms of lipid-improving effects of exercise are still unclear, increased activation of LPL and reverse cholesterol transport are likely involved, which reduce LDL and increase HDL, respectively. Combined, Spirulina supplementation and HIIT appear to have synergistic effects on glycemic function and blood lipids, producing more significant improvements than when completed alone.

4.2. Favorable Modulations of Adipokines

To the authors' knowledge, this study is the first to investigate the effects of HIIT and Spirulina supplementation on adipokines and cytokines in obesity. We found that Spirulina supplementation and HIIT alone and in combination improve markers of inflammation, with the involvement of HIIT (TSG and TG) producing greater improvements in some markers. For example, we found that circulating CRP was reduced with HIIT and Spirulina supplementation with lower TSG concentrations than TG and SG. We also showed that compared to CG, IL-6, IL-8, and TNF-α were lower in TSG and TG but not SG. For Sema3C and MCP1, HIIT and Spirulina reduced circulating concentrations with no significant differences between SG, TG, and TSG.

Chronic inflammation is a central element in the pathogenesis of systemic cardiometabolic dysfunction that occurs in obesity, including insulin resistance [45,46]. Involved in this process is the secretion of adipokines from adipose tissue and cytokines from immune cells. In non-obese conditions, adipose tissue secretes anti-inflammatory adipokines such as adiponectin; however, in obesity, pro-inflammatory macrophages accumulate, and the adipokine secretory profile transitions to pro-inflammatory [46]. Evidence suggests that this shift in the adipokine secretory profile, in part, results from increases in ROS from mitochondrial dysfunction as well as the infiltration of pro-inflammatory macrophages occurring in obesity [4,47,48]. Therefore, treatments aimed at alleviating oxidative stress and subsequent chronic inflammation represent worthwhile investigation. We showed that 12 weeks of HIIT and Spirulina supplementation improved several markers of adipocytokines and were associated with improved cardiometabolic health markers. While this study did not measure ROS or antioxidant capacity, we hypothesize that improvements in oxidative stress are likely involved in the positive alterations in adipo- and cytokines observed following HIIT and Spirulina supplementation. Though acute exercise increases ROS, regular exercise leads to greater resistance against oxidative damage through increased antioxidant capacity [49]. Similarly, Spirulina exerts antioxidant properties and has been shown to increase total antioxidant status in obese males [35]. Both exercise and Spirulina have also been shown to improve inflammatory status, including when performed together [16,22,35]. Four weeks of Spirulina supplementation of 500 mg/d combined with HIIT increased anti-inflammatory nesfatin-1 and omentin-1 to a greater degree than HIIT alone in overweight and obese women [22].

We hypothesized that the combination of HIIT and Spirulina supplementation would promote improvements in adipocytokines to a greater degree than either alone; however, except for CRP, we found that TSG did not produce greater changes in inflammatory markers compared to TG and SG. This is interesting as TSG produced greater improvements in numerous metabolic and anthropometric markers compared to TG and SG. This indicates that either 12 weeks of HIIT or Spirulina alone is sufficient to lead to positive changes in systemic inflammation; however, the combination of HIIT and Spirulina provides additional clinical benefits. Given this, HIIT and Spirulina supplementation appear to promote positive metabolic health outcomes that may not be mediated through inflammatory signaling, such as ATK-AMPK signaling, as mentioned earlier. Reductions in FAT were similar in TSG, TG, and SG, which may explain the similar alterations in adipocytokines as adiposity status influences inflammatory status in obesity. For example, weight loss reduces pro-inflammatory macrophage infiltration in adipose tissue, reducing systemic inflammation [50].

5. Study Limitations

There are various limitations inherent in our investigation. Initially, the processes behind the potential enhancement of adipokine levels by bioactive constituents of Spirolina were not determined. Furthermore, the generalizability of our research is limited due to the exclusion of females in the enrollment of patients. Another limitation of our study is the lack of measurement of blood pressure, heart rate, fat percentage, and MET for the subjects of the present study.

6. Conclusions

This study demonstrated that supplementation of the antioxidant Spirulina and HIIT improves anthropometrics, cardiometabolic health markers, and adipocytokine profiles in obese males. Moreover, we showed that while HIIT and Spirulina alone resulted in similar changes in markers of inflammation, the combination of HIIT and Spirulina led to more significant improvements in cardiometabolic health outcomes. This suggests that while HIIT and Spirulina alone can foster improved inflammation in obesity, the combination of both leads to additional beneficial clinical outcomes that appear to be mediated by mechanisms beyond the modulation of obesity-related inflammation.

Author Contributions: M.D., A.S. and H.Z. designed the study. M.D. and A.S. conducted the study. S.S.S., R.S. and M.H.A.K. analyzed the obtained data. M.D., A.S., K.A.E., K.M.H., M.H.A.K. and H.Z. wrote the first draft of the manuscript. I.L., H.Z., R.S., A.S., K.W. and B.K. read, revised, and approved the final version of the manuscript. All authors have read and agreed to the published version of the manuscript.

Funding: The authors acknowledge the support of Alzahra University.

Institutional Review Board Statement: The research received approval from the Ethics Committee of the Sport Sciences Research Institute (Ethics code: IR.SSRC.REC.1401.093) The date was 18 January 2023. All protocols adhered to the most recent iteration of the Declaration of Helsinki.

Informed Consent Statement: Informed consent was obtained from all subjects involved in the study.

Data Availability Statement: The datasets generated for this study are available upon request from the corresponding authors.

Acknowledgments: The authors thank Jonathan P. Little (British Colombia) for constructive feedback on the manuscript.

Conflicts of Interest: The authors declare no conflict of interest.

References

1. World Health Organization. *WHO European Regional Obesity Report 2022*; World Health Organization Regional Office for Europe: Copenhagen, Denmark, 2022.
2. Syed Ikmal, S.I.Q.; Zaman Huri, H.; Vethakkan, S.R.; Wan Ahmad, W.A. Potential biomarkers of insulin resistance and atherosclerosis in type 2 diabetes mellitus patients with coronary artery disease. *Int. J. Endocrinol.* **2013**, *2013*, 698567. [CrossRef] [PubMed]
3. Nway, N.C.; Sitticharoon, C.; Chatree, S.; Maikaew, P. Correlations between the expression of the insulin sensitizing hormones, adiponectin, visfatin, and omentin, and the appetite regulatory hormone, neuropeptide Y and its receptors in subcutaneous and visceral adipose tissues. *Obes. Res. Clin. Pract.* **2016**, *10*, 256–263. [CrossRef] [PubMed]
4. DeMarco, V.G.; Johnson, M.S.; Whaley-Connell, A.T.; Sowers, J.R. Cytokine abnormalities in the etiology of the cardiometabolic syndrome. *Curr. Hypertens. Rep.* **2010**, *12*, 93–98. [CrossRef]
5. Li, Q.; Wang, Q.; Xu, W.; Ma, Y.; Wang, Q.; Eatman, D.; You, S.; Zou, J.; Champion, J.; Zhao, L. C-reactive protein causes adult-onset obesity through chronic inflammatory mechanism. *Front. Cell Dev. Biol.* **2020**, *8*, 18. [CrossRef] [PubMed]
6. Visser, M.; Bouter, L.M.; McQuillan, G.M.; Wener, M.H.; Harris, T.B. Elevated C-reactive protein levels in overweight and obese adults. *JAMA* **1999**, *282*, 2131–2135. [CrossRef] [PubMed]
7. Panee, J. Monocyte Chemoattractant Protein 1 (MCP-1) in obesity and diabetes. *Cytokine* **2012**, *60*, 1–12. [CrossRef]
8. Straczkowski, M.; Dzienis-Straczkowska, S.; Stêpieň, A.; Kowalska, I.; Szelachowska, M.; Kinalska, I. Plasma interleukin-8 concentrations are increased in obese subjects and related to fat mass and tumor necrosis factor-α system. *J. Clin. Endocrinol. Metab.* **2002**, *87*, 4602–4606. [CrossRef]
9. Tun, S.; Spainhower, C.J.; Cottrill, C.L.; Lakhani, H.V.; Pillai, S.S.; Dilip, A.; Chaudhry, H.; Shapiro, J.I.; Sodhi, K. Therapeutic efficacy of antioxidants in ameliorating obesity phenotype and associated comorbidities. *Front. Pharmacol.* **2020**, *11*, 1234. [CrossRef]
10. Eder, K.; Baffy, N.; Falus, A.; Fulop, A.K. The major inflammatory mediator interleukin-6 and obesity. *Inflamm. Res.* **2009**, *58*, 727–736. [CrossRef]
11. DiNicolantonio, J.J.; Bhat, A.G.; OKeefe, J. Effects of spirulina on weight loss and blood lipids: A review. *Open Heart* **2020**, *7*, e001003. [CrossRef]
12. Calella, P.; Di Dio, M.; Cerullo, G.; Di Onofrio, V.; Galle, F.; Liguori, G. Antioxidant, immunomodulatory, and anti-inflammatory effects of Spirulina in disease conditions: A systematic review. *Int. J. Food Sci. Nutr.* **2022**, *73*, 1047–1056. [CrossRef] [PubMed]
13. Moradi, S.; Ziaei, R.; Foshati, S.; Mohammadi, H.; Nachvak, S.M.; Rouhani, M.H. Effects of Spirulina supplementation on obesity: A systematic review and meta-analysis of randomized clinical trials. *Complement. Ther. Med.* **2019**, *47*, 102211. [CrossRef]
14. Mohiti, S.; Zarezadeh, M.; Naeini, F.; Tutunchi, H.; Ostadrahimi, A.; Ghoreishi, Z.; Ebrahimi Mamaghani, M. Spirulina supplementation and oxidative stress and pro-inflammatory biomarkers: A systematic review and meta-analysis of controlled clinical trials. *Clin. Exp. Pharmacol. Physiol.* **2021**, 1059–1069. [CrossRef] [PubMed]
15. Zeinalian, R.; Farhangi, M.A.; Shariat, A.; Saghafi-Asl, M. The effects of Spirulina Platensis on anthropometric indices, appetite, lipid profile and serum vascular endothelial growth factor (VEGF) in obese individuals: A randomized double blinded placebo controlled trial. *BMC Complement. Altern. Med.* **2017**, *17*, 225. [CrossRef] [PubMed]
16. Beavers, K.M.; Brinkley, T.E.; Nicklas, B.J. Effect of exercise training on chronic inflammation. *Clin. Chim. Acta* **2010**, *411*, 785–793. [CrossRef] [PubMed]
17. Yavari, A.; Javadi, M.; Mirmiran, P.; Bahadoran, Z. Exercise-induced oxidative stress and dietary antioxidants. *Asian J. Sports Med.* **2015**, *6*, e24898. [CrossRef] [PubMed]
18. Araújo, M.M.; Botelho, P.B. Probiotics, prebiotics, and synbiotics in chronic constipation: Outstanding aspects to be considered for the current evidence. *Front. Nutr.* **2022**, *9*, 935830. [CrossRef]
19. Kalafati, M.; Jamurtas, A.Z.; Nikolaidis, M.G.; Paschalis, V.; Theodorou, A.A.; Sakellariou, G.K.; Koutedakis, Y.; Kouretas, D. Ergogenic and antioxidant effects of spirulina supplementation in humans. *Med. Sci. Sports Exerc.* **2010**, *42*, 142–151. [CrossRef]
20. Chaouachi, M.; Vincent, S.; Groussard, C. A Review of the Health-Promoting Properties of Spirulina with a Focus on athletes' Performance and Recovery. *J. Diet. Suppl.* **2023**, 1–32. [CrossRef]
21. Hernández-Lepe, M.A.; López-Díaz, J.A.; Juárez-Oropeza, M.A.; Hernández-Torres, R.P.; Wall-Medrano, A.; Ramos-Jiménez, A. Effect of Arthrospira (Spirulina) maxima supplementation and a systematic physical exercise program on the body composition and cardiorespiratory fitness of overweight or obese subjects: A double-blind, randomized, and crossover controlled trial. *Mar. Drugs* **2018**, *16*, 364. [CrossRef]
22. Golestani, F.; Mogharnasi, M.; Erfani-Far, M.; Abtahi-Eivari, S.H. The effects of spirulina under high-intensity interval training on levels of nesfatin-1, omentin-1, and lipid profiles in overweight and obese females: A randomized, controlled, single-blind trial. *J. Res. Med. Sci.* **2021**, *26*, 10. [CrossRef] [PubMed]
23. Hernández-Lepe, M.A.; Wall-Medrano, A.; López-Díaz, J.A.; Juárez-Oropeza, M.A.; Luqueño-Bocardo, O.I.; Hernández-Torres, R.P.; Ramos-Jiménez, A. Hypolipidemic effect of Arthrospira (Spirulina) maxima supplementation and a systematic physical exercise program in overweight and obese men: A double-blind, randomized, and crossover controlled trial. *Mar. Drugs* **2019**, *17*, 270. [CrossRef] [PubMed]

24. Oriquat, G.A.; Ali, M.A.; Mahmoud, S.A.; Eid, R.M.; Hassan, R.; Kamel, M.A. Improving hepatic mitochondrial biogenesis as a postulated mechanism for the antidiabetic effect of Spirulina platensis in comparison with metformin. *Appl. Physiol. Nutr. Metab.* **2019**, *44*, 357–364. [CrossRef] [PubMed]
25. Nobari, H.; Gandomani, E.E.; Reisi, J.; Vahabidelshad, R.; Suzuki, K.; Volpe, S.L.; Pérez-Gómez, J. Effects of 8 weeks of high-intensity interval training and spirulina supplementation on immunoglobin levels, cardio-respiratory fitness, and body composition of overweight and obese women. *Biology* **2022**, *11*, 196. [CrossRef]
26. Ceylan, H.İ.; Saygın, Ö.; Özel Türkçü, Ü. Assessment of acute aerobic exercise in the morning versus evening on asprosin, spexin, lipocalin-2, and insulin level in overweight/obese versus normal weight adult men. *Chronobiol. Int.* **2020**, *37*, 1252–1268. [CrossRef]
27. Pescatello, L.S. *ACSM's Guidelines for Exercise Testing and Prescription*; Lippincott Williams & Wilkins: Philadelphia, PA, USA, 2014.
28. Myers, J.; Arena, R.; Franklin, B.; Pina, I.; Kraus, W.; McInnis, K.; Balady, G. American Heart Association Committee on Exercise, Cardiac Rehabilitation, and Prevention of the Council on Clinical Cardiology, the Council on Nutrition, Physical Activity, and Metabolism, and the Council on Cardiovascular Nursing. Recommendations for clinical exercise laboratories: A scientific statement from the american heart association. *Circulation* **2009**, *119*, 3144–3161.
29. Soltani, M.; Aghaei Bahmanbeglou, N.; Ahmadizad, S. High-intensity interval training irrespective of its intensity improves markers of blood fluidity in hypertensive patients. *Clin. Exp. Hypertens.* **2020**, *42*, 309–314. [CrossRef]
30. Mazokopakis, E.E.; Papadomanolaki, M.G.; Fousteris, A.A.; Kotsiris, D.A.; Lampadakis, I.M.; Ganotakis, E.S. The hepatoprotective and hypolipidemic effects of Spirulina (*Arthrospira platensis*) supplementation in a Cretan population with non-alcoholic fatty liver disease: A prospective pilot study. *Ann. Gastroenterol.* **2014**, *27*, 387.
31. Thomas, D.T.; Erdman, K.A.; Burke, L.M. Position of the Academy of Nutrition and Dietetics, Dietitians of Canada, and the American College of Sports Medicine: Nutrition and athletic performance. *J. Acad. Nutr. Diet.* **2016**, *116*, 501–528. [CrossRef]
32. Hopkins, W.G.; Marshall, S.W.; Batterham, A.M.; Hanin, J. Progressive statistics for studies in sports medicine and exercise science. *Med. Sci. Sports Exerc.* **2009**, *41*, 3–13. [CrossRef] [PubMed]
33. Booth, F.W.; Roberts, C.K.; Laye, M.J. Lack of exercise is a major cause of chronic diseases. *Compr. Physiol.* **2012**, *2*, 1143. [PubMed]
34. Gurney, T.; Spendiff, O. Algae supplementation for exercise performance: Current perspectives and future directions for spirulina and chlorella. *Front. Nutr.* **2022**, *384*, 865741. [CrossRef]
35. Szulinska, M.; Gibas-Dorna, M.; Miller-Kasprzak, E.; Suliburska, J.; Miczke, A.; Walczak-Gałezewska, M.; Stelmach-Mardas, M.; Walkowiak, J.; Bogdanski, P. Spirulina maxima improves insulin sensitivity, lipid profile, and total antioxidant status in obese patients with well-treated hypertension: A randomized double-blind placebo-controlled study. *Eur. Rev. Med. Pharmacol. Sci.* **2017**, *21*, 2473–2481. [PubMed]
36. Batrakoulis, A.; Jamurtas, A.Z.; Metsios, G.S.; Perivoliotis, K.; Liguori, G.; Feito, Y.; Riebe, D.; Thompson, W.R.; Angelopoulos, T.J.; Krustrup, P. Comparative efficacy of 5 exercise types on cardiometabolic health in overweight and obese adults: A systematic review and network meta-analysis of 81 randomized controlled trials. *Circ. Cardiovasc. Qual. Outcomes* **2022**, *15*, e008243. [CrossRef] [PubMed]
37. Iyer Uma, M.; Sophia, A.; Uliyar, V. Glycemic and lipemic responses of selected spirulina-supplemented rice-based recipes in normal subjects. *Age Years* **1999**, *22*, 17–22.
38. Hao, S.; Li, F.; Li, Q.; Yang, Q.; Zhang, W. Phycocyanin Protects against High Glucose High Fat Diet Induced Diabetes in Mice and Participates in AKT and AMPK Signaling. *Foods* **2022**, *11*, 3183. [CrossRef]
39. Mann, S.; Beedie, C.; Balducci, S.; Zanuso, S.; Allgrove, J.; Bertiato, F.; Jimenez, A. Changes in insulin sensitivity in response to different modalities of exercise: A review of the evidence. *Diabetes/Metab. Res. Rev.* **2014**, *30*, 257–268. [CrossRef]
40. Holten, M.K.; Zacho, M.; Gaster, M.; Juel, C.; Wojtaszewski, J.F.; Dela, F. Strength training increases insulin-mediated glucose uptake, GLUT4 content, and insulin signaling in skeletal muscle in patients with type 2 diabetes. *Diabetes* **2004**, *53*, 294–305. [CrossRef]
41. Kjøbsted, R.; Munk-Hansen, N.; Birk, J.B.; Foretz, M.; Viollet, B.; Björnholm, M.; Zierath, J.R.; Treebak, J.T.; Wojtaszewski, J.F. Enhanced muscle insulin sensitivity after contraction/exercise is mediated by AMPK. *Diabetes* **2017**, *66*, 598–612. [CrossRef]
42. Roughan, P.G. Spirulina: A source of dietary gamma-linolenic acid? *J. Sci. Food Agric.* **1989**, *47*, 85–93. [CrossRef]
43. Dillon, J.; Phuc, A.P.; Dubacq, J. Nutritional value of the alga Spirulina. *Plants Hum. Nutr.* **1995**, *77*, 32–46.
44. Iwata, K.; Inayama, T.; Kato, T. Effects of Spirulina platensis on plasma lipoprotein lipase activity in fructose-induced hyperlipidemic rats. *J. Nutr. Sci. Vitaminol.* **1990**, *36*, 165–171. [CrossRef] [PubMed]
45. Maximus, P.S.; Al Achkar, Z.; Hamid, P.F.; Hasnain, S.S.; Peralta, C.A. Adipocytokines: Are they the theory of everything? *Cytokine* **2020**, *133*, 155144. [CrossRef] [PubMed]
46. Ouchi, N.; Parker, J.L.; Lugus, J.J.; Walsh, K. Adipokines in inflammation and metabolic disease. *Nat. Rev. Immunol.* **2011**, *11*, 85–97. [CrossRef]
47. Otani, H. Oxidative stress as pathogenesis of cardiovascular risk associated with metabolic syndrome. *Antioxid. Redox Signal.* **2011**, *15*, 1911–1926. [CrossRef]
48. Gordon, S. Alternative activation of macrophages. *Nat. Rev. Immunol.* **2003**, *3*, 23–35. [CrossRef]

49. Basak, I.; Patil, K.S.; Alves, G.; Larsen, J.P.; Møller, S.G. microRNAs as neuroregulators, biomarkers and therapeutic agents in neurodegenerative diseases. *Cell. Mol. Life Sci.* **2016**, *73*, 811–827. [CrossRef]
50. Cancello, R.; Henegar, C.; Viguerie, N.; Taleb, S.; Poitou, C.; Rouault, C.; Coupaye, M.; Pelloux, V.; Hugol, D.; Bouillot, J.-L. Reduction of macrophage infiltration and chemoattractant gene expression changes in white adipose tissue of morbidly obese subjects after surgery-induced weight loss. *Diabetes* **2005**, *54*, 2277–2286. [CrossRef]

Article

Astaxanthin Supplementation Augments the Benefits of CrossFit Workouts on Semaphorin 3C and Other Adipokines in Males with Obesity

Rashmi Supriya [1], Sevda Rahbari Shishvan [2], Movahed Kefayati [2], Hossein Abednatanzi [2], Omid Razi [3], Reza Bagheri [4], Kurt A. Escobar [5], Zhaleh Pashaei [6], Ayoub Saeidi [7,*], Shahnaz Shahrbanian [8], Sovan Bagchi [9], Pallav Sengupta [9], Maisa Hamed Al Kiyumi [10,11], Katie M. Heinrich [12,13] and Hassane Zouhal [14,15,*]

[1] Centre for Health and Exercise Science Research, SPEH, Hong Kong Baptist University, Kowloon Tong, Hong Kong SAR 999077, China; rashmisupriya@hkbu.edu.hk

[2] Department of Physical Education and Sport Science, Science and Research Branch, Islamic Azad University, Tehran 15847-15414, Iran; sevdarahbari1370@gmail.com (S.R.S.); movahed.kefayati@yahoo.com (M.K.); h.abednatanzi@yahoo.com (H.A.)

[3] Department of Exercise Physiology, Faculty of Physical Education and Sports Science, Razi University, Kermanshah 94Q5+6G3, Iran; omid.razi.physio@gmail.com

[4] Department of Exercise Physiology, University of Isfahan, Isfahan 81746-73441, Iran; will.fivb@yahoo.com

[5] Department of Kinesiology, California State University, Long Beach, CA 90840, USA; kurt.escobar@csulb.edu

[6] Department of Exercise Physiology, Faculty of Physical Education and Sport Sciences, University of Tabriz, Tabriz 51666-16471, Iran; pashaei.zh@gmail.com

[7] Department of Physical Education and Sport Sciences, Faculty of Humanities and Social Sciences, University of Kurdistan, Sanandaj, Kurdistan 66177-15175, Iran

[8] Department of Sport Science, Faculty of Humanities, Tarbiat Modares University, Tehran 14117-13116, Iran; s.shahrbanian@gmail.com

[9] Department of Biomedical Sciences, College of Medicine, Gulf Medical University, Ajman 4184, United Arab Emirates; dr.sovan@gmu.ac.ae (S.B.); pallav_cu@yahoo.com (P.S.)

[10] Department of Family Medicine and Public Health, Sultan Qaboos University, Muscat P.O. Box 35, Oman; maysa8172@gmail.com

[11] Department of Family Medicine and Public Health, Sultan Qaboos University Hospital, Muscat P.O. Box 35, Oman

[12] Department of Kinesiology, Kansas State University, Manhattan, KS 66506, USA; kmhphd@ksu.edu

[13] Research Department, The Phoenix, Manhattan, KS 66502, USA

[14] M2S (Laboratoire Mouvement, Sport, Santé)—EA 1274, Université de Rennes, 35000 Rennes, France

[15] Institut International des Sciences du Sport (2I2S), 35850 Irodouer, France

* Correspondence: a.saeidi@uok.ac.ir (A.S.); hassane.zouhal@univ-rennes2.fr (H.Z.)

Citation: Supriya, R.; Shishvan, S.R.; Kefayati, M.; Abednatanzi, H.; Razi, O.; Bagheri, R.; Escobar, K.A.; Pashaei, Z.; Saeidi, A.; Shahrbanian, S.; et al. Astaxanthin Supplementation Augments the Benefits of CrossFit Workouts on Semaphorin 3C and Other Adipokines in Males with Obesity. *Nutrients* **2023**, *15*, 4803. https://doi.org/10.3390/nu15224803

Academic Editors: David C. Nieman and Petra Kienesberger

Received: 8 September 2023
Revised: 31 October 2023
Accepted: 6 November 2023
Published: 16 November 2023

Abstract: Regular physical activity and the use of nutritional supplements, including antioxidants, are recognized as efficacious approaches for the prevention and mitigation of obesity-related complications. This study investigated the effects of 12 weeks of CrossFit training combined with astaxanthin (ASX) supplementation on some plasma adipokines in males with obesity. Sixty-eight males with obesity (BMI: 33.6 ± 1.4 kg·m^{-2}) were randomly assigned into four groups: the control group (CG; $n = 11$), ASX supplementation group (SG; $n = 11$), CrossFit group (TG; $n = 11$), and training plus supplement group (TSG; $n = 11$). Participants underwent 12 weeks of supplementation with ASX or placebo (20 mg/day capsule daily), CrossFit training, or a combination of both interventions. Plasma levels of semaphorin 3C (SEMA3C), apelin, chemerin, omentin1, visfatin, resistin, adiponectin, leptin, vaspin, and RBP4 were measured 72 h before the first training session and after the last training session. The plasma levels of all measured adipokines were significantly altered in SG, TG, and TSG groups ($p < 0.05$). The reduction of resistin was significantly higher in TSG than in SG ($p < 0.05$). The plasma levels of omentin1 were significantly higher in both training groups of TG and TSG than SG ($p < 0.05$), although such a meaningful difference was not observed between both training groups ($p > 0.05$). Significant differences were found in the reductions of plasma levels of vaspin, visfatin, apelin, RBP4, chemerin, and SEMA3C between the SG and TSG groups ($p < 0.05$). The study found that a 12-week intervention using ASX supplementation and CrossFit exercises resulted in significant improvements in several adipokines among male individuals with obesity. Notably, the combined approach of supplementation and training had the most pronounced results. The findings presented in this study indicate that the supplementation of ASX and participation in CrossFit exercise have the potential to be effective therapies in mitigating complications associated with obesity and enhancing metabolic health.

Keywords: nutritional supplements; adipokines; semaphorin 3C; CrossFit workouts; adipose tissue

1. Introduction

Obesity is characterized by an excessive accumulation of adipose tissue and is strongly linked to the development and progression of several metabolic disorders [1,2]. Accumulated adipose tissue not only acts as a reservoir for excess energy, but also functions as an endocrine organ that releases molecular proteins known as adipokines [1,3]. Of these adipokines, leptin, resistin, visfatin, apelin, retinol binding protein4 (RBP4), vaspin, and chemerin are associated with obesity, while others such as adiponectin and omentin1 have a negative correlation. These adipokines are involved in various physiological processes such as metabolism and glucose homeostasis, oxidative stress, and the pathophysiology of obesity [4–6]. Leptin exerts its effects on hunger reduction and the restoration of energy balance by acting on central processes, namely by blocking certain leptin-sensitive neurons such as neuropeptide Y and proopiomelanocortin neurons, hence promoting energy homeostasis [7,8].

Engaging in regular physical activity is a potent strategy for enhancing general well-being, preventing and decreasing obesity, and alleviating the adverse health consequences linked to excessive adipose tissue [5,9]. CrossFit is an exercise regimen characterized by the use of diverse functional movements derived from several athletic disciplines, including weightlifting, gymnastics, and powerlifting. These movements are performed in rigorous sessions that emphasize high-intensity training [10]. Previous studies have confirmed the positive effects of CrossFit training on physiological and fitness factors (e.g., body composition, cardiovascular/respiratory fitness, strength, flexibility, power, and balance) [11,12].

At the present time, data are insufficient on the impact of CrossFit training on the adipokines that are the subject of investigation in the current study. However, literature exists on other modes of training for various adipokines [5,13–17]. For example, jogging and step aerobic exercise increased leptin and interleukin-15 (IL-15) while decreasing resistin in overweight women [18]. Jung et al. [15] indicated a significant decrease in blood leptin levels after 12 weeks of engaging in moderate-intensity exercise—namely brisk walking—among both obese men and females. Ouerghi et al. [19] showed that plasma levels of omentin-1 increased after 8 weeks of high-intensity interval training (HIIT) in obese participants, along with reduced obesity, blood lipids, and insulin sensitivity. However, others have found no significant changes in omentin-1 after a training period [20,21].

Antioxidant supplementation can be used to attenuate the negative effects of oxidative stress [22]. Astaxanthin (3,3′-dihydroxy-B, B-carotene4, 4′-dione), which is derived from *Haematoccus pluvialis* algae, has been shown to reduce the effects of oxidative stress on lipid metabolism [23]. Systematic review and meta-analysis studies revealed that Astaxanthin (ASX) supplementation was associated with a decrease in insulin resistance and oxidative stress, an increase in antioxidant capacity and mitochondrial biogenesis in obesity, as well as improvements regarding diabetes, cardiovascular disease, neurodegenerative disorders, chronic inflammatory disease, and some cancers [24,25]. Furthermore, it also improves lipid metabolism [23,26,27]. Moreover, ASX supplementation improves insulin resistance in obese mice by modulating insulin signaling and activating mitochondrial energy metabolism via pathways for AMP-activated protein kinase (AMPK) and peroxisome proliferator-activated receptor γ coactivator1a (Pgc1a) in skeletal muscles [28]. Although there is less available evidence about the precise mechanism by which ASX supplementation acts, it is postulated that the favorable effects of ASX may be attributed to its impact on the secretion of adipokines, akin to other bioactive chemicals such as capsaicin. Although a recent review study found that combined ASX supplementation and exercise did not improve exercise performance [29], it is unknown whether a combination of ASX supplementation and exercise produces beneficial effects on metabolic health in obese individuals. Furthermore, the effects of ASX supplementation on adipokines in the present study are currently unclear [29]. We have previously found that ASX supplementation and CrossFit training improved body composition, metabolic profiles, anthropometric measurements, cardio-respiratory function, and some adipokines (i.e., Cq1/TNF-related protein 9 and 2 [CTRP9 and 2] and growth differentiation factor 8 and 15 [GDF8 and 15]) [30], but the effect of CrossFit training on other adipokines (semaphorin 3C (SEMA3C), apelin, chemerin, omentin1, visfatin, resistin, adiponectin, leptin, vaspin, and RBP4) is unexplored. On the other hand, the current study hypothesized that CrossFit training and ASX supplementation has a positive effect on SEMA3C, apelin, chemerin, omentin1, visfatin, resistin, adiponectin, leptin, vaspin, and RBP4 in males with obesity. Therefore, this study aimed to investigate the effect of 12 weeks of CrossFit training combined with ASX supplementation on SEMA3C, apelin, chemerin, omentin1, visfatin, resistin, adiponectin, leptin, vaspin, and RBP4 in males with obesity.

2. Methods

Participant recruitment has been described previously (see [30]). Study inclusion criteria were body mass index (BMI) > 30 kg·m^{-2}, lack of regular physical exercise in the last six months, absence of cardiovascular, metabolic, or endocrine disorders, and no alcohol intake. The research excluded individuals who had joint disorders, physical limitations, or were using prescription drugs and supplements that might potentially impact muscle and adipose tissue metabolism [31]. The participants were originally presented with a thorough explanation of the research protocols. Subsequently, all participants had a medical examination conducted by a physician and clinical exercise physiologist on their first visit. Additionally, they were required to sign a written consent form and the Physical Activity Readiness Questionnaire (PARQ) [32]. The study was approved by the National Research and Ethics Committee (Ethics code: IR.IAU.DAMGHAN.REC.1401.035) and the Iranian Registry of Clinical Trials (IRCTID: IRCT20151228025732N76). The procedures were conducted in accordance with the most recent version of the Declaration of Helsinki [33].

2.1. Experimental Design

Participants were familiarized with the entire study procedure one week prior to the initiation of the main training protocol. Basic measures including height and body weight were assessed (see [30]). Then, 68 eligible participants (age: 27 ± 8 yrs.; height: 167.8 ± 3.1 cm; body weight: 94.7 ± 2.0 kg, BMI: 33.6 ± 1.4 kg·m^{-2}) were randomly divided into four groups: control group (CG; $n = 17$), ASX supplement group (SG; $n = 17$), CrossFit group (TG; $n = 17$), and training plus supplement group (TSG; $n = 17$). The flow of participant recruitment is outlined in Figure 1. During the study, six individuals per group declined to participate in the remaining protocol procedures due to medical, job, or lack of interest reasons. Each group (collectively, $n = 11$) received instructions on performing the training protocols during the third session. Following baseline measurements, the two training groups (TG and TSG) attended CrossFit training (3 sessions/week) for 12 weeks. The control group participants were provided with instructions to maintain their existing lifestyles throughout the duration of the trial. The measurements for the study were conducted simultaneously, with a time difference of around one hour, under similar climatic circumstances, with a temperature of around 20 °C, and a humidity level of approximately 55%. The pre-and post-test measures were conducted 48 h before initiation and after the end of the last training session, respectively.

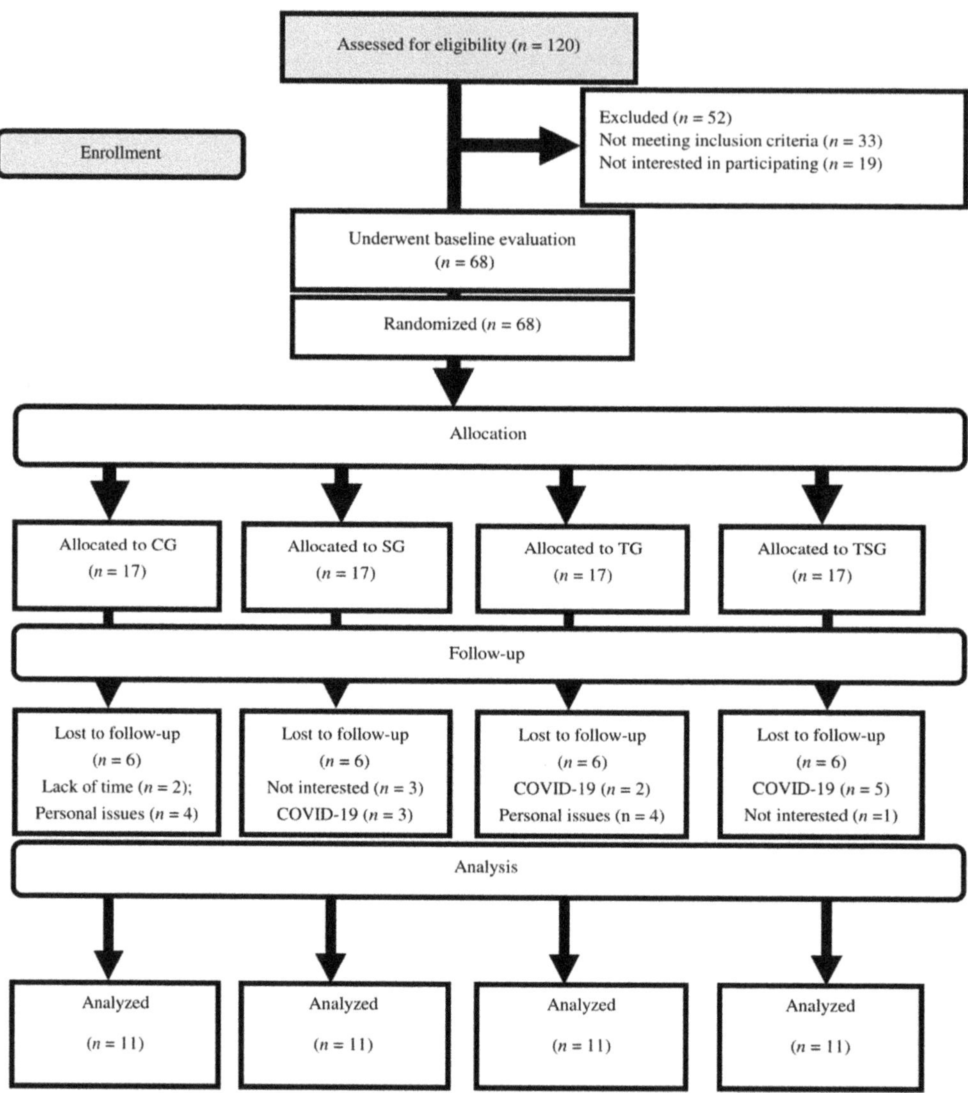

Figure 1. Flow chart of the participant recruitment.

2.2. Training Protocols

In this study, the HIFT program was used, which included CrossFit training in 36 sessions, each session lasting 60 min and performed three times a week. All HIFT sessions were led by a CrossFit Level 1-certified trainer. The first two sessions were designed as an introduction to common movements used in HIFT (air squat, overhead squat, front squat, press, push press, push jerk, deadlift, sumo deadlift high pull, and the medicine ball clean). Starting on the third day, each HIFT session consisted of 10–15 min of stretches and warm-ups; 10–20 min of instruction and practicing methods and movements; and 5–30 min of the workout of the day (WOD), conducted at a vigorous intensity according to each individual's fitness level. Workout modality components included aerobics (e.g., running, jumping rope), body weight (e.g., pull-ups, squats), and weightlifting (e.g., front squats, kettlebell swings) exercises that were continuously varied using the CrossFit training template [34] in single, couplet, or triplet. All weights and movements were prescribed and recorded separately for every HIFT participant [35]. Depending on the structure of the WOD, timings to complete the WOD, rounds, and repetitions performed on the WOD, the weights used and any necessary modifications to the scheduled workout were also noted for each participant. For the HIFT group as a whole, average times for each WOD and the total average WOD time per week were calculated.

2.3. Astaxanthin Supplementation Protocol

The participants in the SG and TSG were randomly allocated to receive a daily dose of 20 mg of ASX capsule (manufactured by Marine Product Tech. Inc., Seongnam, Republic of Korea) or a placebo consisting of a 20 mg dose of a raw corn starch capsule. This administration took place once daily, with breakfast, for a duration of 12 weeks [36].

2.4. Nutrient Intake and Dietary Analysis

To evaluate changes in dietary habits, a set of three-day food records (consisting of two weekdays and one weekend day) was obtained before and after the research. Every meal item was individually inputted into Diet Analysis Plus version 10 (Cengage, Boston, MA, USA) in order to determine the total calorie consumption and the relative distribution of energy derived from fats, proteins, and carbohydrates [31].

2.5. Blood Markers

The procedure of blood testing was performed under standard conditions between 8 and 10 a.m. Samples for fasting blood sugar were drawn from the right arm 12 and 72 h prior to the first exercise session and again at 72 h after the last session. EDTA-containing tubes were used to transfer the blood samples, which were centrifuged for 10 min at 3000 rpm and stored at $-70\,^{\circ}$C. Plasma resistin was measured with an ELISA kit (Biovendor, Czech Republic, Catalogue No: RD191016100. Sensitivity: 0.012 ng/mL. Intra-CV = 5.9%, inter-CV = 7.6%). Plasma leptin was measured with an ELISA kit (Biovendor, Czech Republic, Catalogue No: RD191001100. Sensitivity: 0.2 ng/mL. Intra-CV = 5.9%, inter-CV = 5.6%). Plasma adiponectin was measured with an ELISA kit (Biovendor, Czech Republic, Catalogue No: RD195023100. Sensitivity: 26 ng/mL. Intra-CV = 4.9%, inter-CV = 6.7%). Plasma visfatin was measured with an ELISA kit (Cusabio, China, Catalog No: CSB-E08940h. Sensitivity: 0.156 ng/mL. Intra-CV = < 8%, inter-CV = < 10%). Plasma vaspin was measured with an ELISA kit (Biovendor, Czech Republic, Catalogue No: RD191097200R. Sensitivity: 0.01 ng/mL. Intra-CV = 7.6%, inter-CV = 7.7%). Plasma RBP-4 was measured with an ELISA kit (R&D Systems, USA, and Catalogue No: DRB400. Sensitivity: 0.628 ng/mL. Intra-CV = 7%, inter-CV = 8.6%). Plasma apelin was measured with an ELISA kit (Phoenix Pharmaceuticals, USA, and Catalogue No: EK-057-23. Sensitivity: 0.07 ng/mL. Intra-CV = < 10%, inter-CV = < 15%). Plasma omentin-1 was measured with an ELISA kit (Biovendor, Czech Republic, Catalogue No: RD191100200R. Sensitivity: 0.5 ng/mL. Intra-CV = 3.7%, inter-CV = 4.6%). Plasma chemerin levels were determined using a commercially available ELISA kit (Biovendor, Czech, The intra-assay coefficient of variation of chemerin was 5.1%). The plasma levels of SEMA3C (MBS037239, MBS2883689, MyBioSource, San Diego, CA, USA) were measured by commercially available enzyme-linked immunosorbent assay (ELISA) kits.

2.6. Statistical Analysis

G-power 3.1.9.2 software was used to calculate the sample size, and based on the previous study, it was determined that there was a significant effect of combined training on reducing leptin levels in overweight and obese males [37]. This study utilized the equation for effect size (ES) to determine the impact of combined aerobic and resistance training. In the present study, based on $\alpha = 0.05$, a power $(1-\beta)$ of 0.95, and an effect size (ES) = 1 ((5.4 − 3.6)/1.65), a total sample size of at least 20 participants ($n = 5$ per group) was required to detect significant changes in leptin levels. Nevertheless, given the absence of prior studies investigating the impact of CrossFit on the measured adipokines in the current investigation, along with the potential hindrance of COVID-19 on training and adherence to supplementation, it was deemed necessary to increase the sample size ($n = 17$) to maintain the statistical power of the study. Descriptive statistics (means ± standard deviation) were used to describe all the data. The Shapiro Wilk test and two-way ANOVA were used to assess the normality of the data and determine the group x time interaction, respectively. One-way ANOVA and Fisher LSD post-hoc tests were used for the evaluation of the baseline data of the four groups. In addition, pairwise comparisons were used to determine mean differences when a significant difference between groups was detected by ANOVA. Additionally, effect sizes (ES) were reported as partial eta-squared. In accordance with Hopkins et al. (2009), ES was considered trivial (<0.2), small (0.2–0.6), moderate (0.6–1.2), large (1.2–2.0), and very large (2.0–4.0). Statistical significance was determined using a p-value threshold of less than 0.05. Pearson's linear regressions were performed with a 95% confidence interval (CI). Values ranging from 0 to 0.3 (or 0 to -0.3) are indicative of a weak positive (negative) linear relationship through a shaky linear rule. Values ranging from 0.3 to 0.7 (-0.3 to -0.7) are indicative of a moderate positive (negative) correlation. Values falling within the range of 0.7 to 1.0 (-0.7 and -1.0) are indicative of a strong positive (negative) correlation [38]. The statistical analyses were conducted using SPSS 26, while the generation of figures was carried out using GraphPad Prism (version 8.4.3).

3. Results

3.1. Compliance, Adverse Events, and Nutrient Intakes

Participant compliance was considered when ≥80% of the supplements were consumed. Six participants from each group withdrew due to personal reasons and COVID-19. No adverse events were reported from both training and supplementation procedures. Also, no changes in nutrient intakes were observed throughout the study (Table 1).

Table 1. Mean (±SD) values of BMI, body weight, and nutritional intake throughout the intervention.

	CG		SG		TG		TSG	
	Pre	Post	Pre	Post	Pre	Post	Pre	Post
Energy (kcal/day)	2260 ± 47	2269 ± 56	2278 ± 101	2149 ± 100	2269 ± 117	2141 ± 117	2273 ± 157	2129 ± 126
CHO (g/day)	281 ± 31.4	283 ± 33.3	279.4 ± 27.1	261 ± 27.5	289 ± 48.6	261 ± 39.2	288 ± 38.6	259 ± 29.1
Fat (g/day)	82.2 ± 11.0	81 ± 9.8	86.5 ± 10.7	75 ± 11.2	83.4 ± 12.4	73.1 ± 11.2	80.8 ± 13.87	70.2 ± 11.3
Protein (g/day)	104 ± 12.0	106 ± 11.3	101 ± 13.5	93 ± 12.6	103 ± 14.8	94 ± 11.7	102 ± 14.5	90 ± 13.5
Body Weight (kg)	95.3 ± 1.8	92.1 ± 2.1	94.2 ± 2.6	90.1 ± 2.3 [a]	94.3 ± 0.9	90.1 ± 2.3 [a,b]	95.1 ± 1.9	88.2 ± 2.3 [a,b,ab]
BMI (kg/m^2)	34.1 ± 2.5	33.7 ± 1.4	33.2 ± 1.4	32.4 ± 1.6 [a,b]	33.5 ± 1.7	32.1 ± 1.5 [a,b]	33.8 ± 1.2	31.8 ± 0.6 [a,b,ab]

CG: Control group; SG: Supplement group; TG: Training group; TSG: Training + Supplement group BMI: Body Mass Index. [a] Indicates significant differences compared to the pre-values ($p < 0.05$). [b] Significant differences compared to the control group ($p < 0.05$). [ab] Significant interaction between time and groups ($p < 0.05$).

3.2. Adipokines

Changes in adipokines throughout the intervention are shown in Figure 2. Baseline levels of adiponectin ($p = 0.20$), leptin A ($p = 0.58$), resistin ($p = 0.12$), omentin1, ($p = 0.46$), vaspin ($p = 0.40$), visfatin ($p = 0.24$), apelin ($p = 0.94$), RBP4 ($p = 0.45$), chemerin ($p = 0.89$), and SEMA3C ($p = 0.81$) were not significantly different between groups. Following the 12-week intervention, there were significant group × time interactions for adiponectin ($p = 0.0001$, $\eta^2 = 0.48$, statistical power = 0.999), leptin ($p = 0.0001$, $\eta^2 = 0.49$, statistical power = 0.998), resistin ($p = 0.0001$, $\eta^2 = 0.40$, statistical power = 0.993), omentin-1 ($p = 0.0001$, $\eta^2 = 0.74$, statistical power = 1.00), vaspin ($p = 0.0001$, $\eta^2 = 0.30$, statistical power = 0.936), visfatin ($p = 0.0001$, $\eta^2 = 0.35$, statistical power = 0.937), apelin ($p = 0.0001$, $\eta^2 = 0.43$, statistical power = 0.997), RBP4 ($p = 0.0001$, $\eta^2 = 0.70$, statistical power = 1.00), chemerin ($p = 0.0001$, $\eta^2 = 0.29$, statistical power = 0.856), and SEMA3C ($p = 0.0001$, $\eta^2 = 0.51$, statistical power = 1.00).

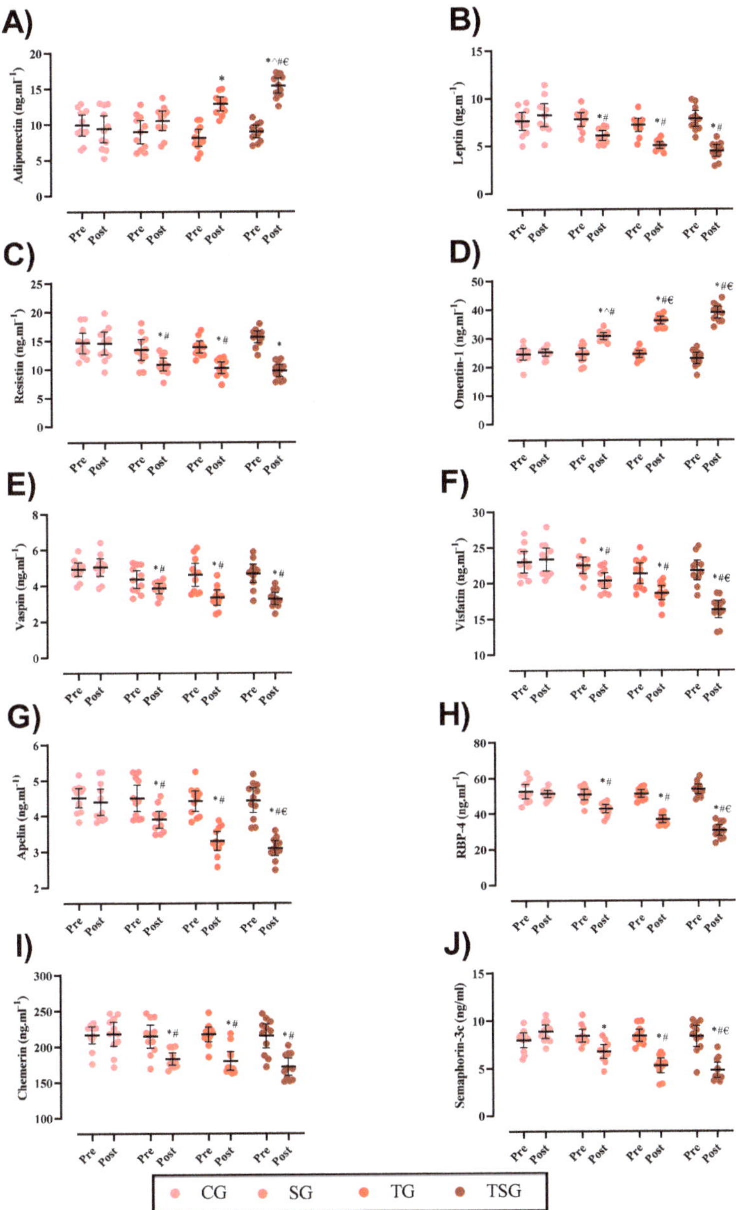

Figure 2. Changes in adipokines throughout the intervention. (**A**) Adiponectin; (**B**) Leptin; (**C**) Resistin; (**D**) Omentin; (**E**) Vaspin; (**F**) Visfatin; (**G**) Apelin; (**H**) Retinol binding protein 4 (RBP-4); (**I**) Chemerin; (**J**) Semaphorin-3c. $n = 11$ per group, error bars represent a 95% confidence interval (CI). * Significantly different from pre-test; # Significantly different than CG; ^ Significantly different than TG; € Significantly different than SG.

In comparison to the baseline, post-intervention values for adiponectin ($p = 0.55$), leptin ($p = 0.22$), resistin ($p = 0.93$), omentin-1 ($p = 0.58$), vaspin ($p = 0.70$), visfatin ($p = 0.69$), Apelin ($p = 0.48$), RBP4 ($p = 0.42$), chemerin ($p = 0.76$) and SEMA3C ($p = 0.10$) were not significantly different in the CG. Changes in adiponectin ($p = 0.80$) and vaspin ($p = 0.09$)

were not significantly different in the SG. Post-test values were significantly different in comparison to the baseline for the rest of the variables in the SG, and also in the TG and TSG for all of the adipokines ($p = 0.001$) in this study.

The increases in plasma adiponectin levels after 12 weeks of intervention in the TG ($p = 0.0001$) and TSG ($p = 0.0001$) were significant, but not in the SG ($p = 0.10$) in comparison to the CG. The differences were non-significant between the TSG and TG ($p = 0.19$), while the differences between the SG and TG ($p = 0.01$) and the SG and TSG ($p = 0.0001$) were statistically significant (Figure 2A). The changes in plasma leptin levels after 12 weeks of intervention in the SG ($p = 0.002$), TG ($p = 0.0001$), and TSG ($p = 0.0001$) were significantly lower in comparison to the CG; the differences between the SG and TSG ($p = 0.02$) were statistically significant (Figure 2B).

Plasma levels of resistin were significantly decreased post-test compared to the baseline in the SG ($p = 0.033$), TG ($p = 0.003$), and TSG ($p = 0.0001$). The differences between the SG and TSG ($p = 0.005$), TSG and CG ($p = 0.0001$), and TG and CG ($p = 0.0001$) were statistically significant (Figure 2C). The plasma levels of omentin-1 were significantly increased among three interventional groups of the SG ($p = 0.001$), TG, and TSG ($p = 0.0001$) in comparison with CG. There were also significant differences in the TG and TSG ($p = 0.001$) compared with SG, as well as between the TG and TSG ($p = 0.007$) (Figure 2D). The plasma levels of vaspin were significantly reduced only in 12-week training groups of the TG ($p = 0.002$) and TSG ($p = 0.001$), while supplementation with (SG) was not different ($p = 0.14$). Vaspin levels were different between the SG and TSG ($p = 0.034$), but not between the TG and TSG ($p = 0.74$), nor between the SG and TG ($p = 0.06$) (Figure 2E). Plasma levels of visfatin were changed in the TG ($p = 0.018$) and TSG ($p = 0.0001$), while there was no difference in the SG ($p = 0.054$). There were no differences between the SG and CG ($p = 0.64$), while the differences between the SG and TSG ($p = 0.011$) and SG and TG ($p = 0.035$) were statistically different (Figure 2F). Plasma levels of apelin were meaningfully reduced in the TG and TSG ($p = 0.0001$), but not in the SG ($p = 0.051$). There were no differences in apelin between the TG and TSG ($p = 0.37$), but there were differences between the SG and TSG ($p = 0.004$) and also between the SG and TG ($p = 0.038$) (Figure 2G). The alterations of plasma RBP4 level were significant in three interventional groups; the SG ($p = 0.007$), TG, and TSG ($p = 0.0001$) following the 12-week interventions. There were, otherwise, significant differences between the SG and TG ($p = 0.009$), SG and TSG ($p = 0.0001$), as well as between the TG and TSG ($p = 0.001$) (Figure 2H). Similarly, 12 weeks of training and/or ASX supplementation altered chemerin in the SG ($p = 0.017$), TG, and TSG ($p = 0.001$). However, there were not any significant differences between the SG and TG ($p = 0.30$), SG and TSG ($p = 0.31$), nor between the TG and TSG ($p = 0.97$) in the changes of chemerin (Figure 2I). The plasma levels of SEMA3C were significantly reduced in three groups of the SG ($p = 0.002$), TG, and TSG ($p = 0.0001$) following 12-week ASX supplementation and CrossFit training. Based on the results of the post-hoc test, there were non-significant differences between the SG and TG ($p = 0.057$) and the TG and TSG ($p = 0.58$), but the changes between SG and TSG ($p = 0.016$) were statistically significant (Figure 2J).

3.3. Weight and BMI

There were no between-group differences in baseline values for weight ($p = 0.46$) and BMI ($p = 0.57$). There were significant group X time interactions for weight ($p = 0.0001$, $\eta^2 = 0.46$, statistical power = 0.999) and BMI ($p = 0.002$, $\eta^2 = 0.30$, statistical power = 0.998) (Table 1).

Body weight reductions after 12 weeks were significant in the SG ($p = 0.008$), TG ($p = 0.0001$), and TSG ($p = 0.0001$) but not in the CG ($p = 0.32$). Furthermore, the post-hoc test for bodyweight shows that after 12 weeks there were significant changes in the CG compared to the TG ($p = 0.004$) and TSG ($p = 0.0001$), and in the TSG compared to the TG ($p = 0.01$) and SG ($p = 0.0001$), while other changes were not significant ($p > 0.05$) (Table 1).

Changes in BMI after 12 weeks were significantly decreased in the SG ($p = 0.019$), TG ($p = 0.0001$), and TSG ($p = 0.0001$) but not in the CG ($p = 0.37$). BMI changes after 12 weeks were significantly decreased in the TG ($p = 0.016$) and the TSG ($p = 0.0001$) compared to the CG. The differences induced by training were significant between the TG and TSG ($p = 0.007$) and between the SG and TSG ($p = 0.007$), while all other differences in BMI between the groups were not significant ($p > 0.05$) (Table 1).

To investigate any potential relationships between training-induced changes in fat mass (Δ FM) and changes in adipokines (Δ marker, independently of groups), initially, a correlation matrix was generated (Figure 3A). Adiponectin (Figure 3B) and omentin1 (Figure 3E) showed moderate negative relationships with Δ FM, while leptin (Figure 3C), vaspin (Figure 3F), visfatin (Figure 3G), apelin (Figure 3H), RBP4 (Figure 3I), chemerin (Figure 3J), and SEMA3C (Figure 3K) showed a moderate positive relationship. Also, resistin (Figure 3D) showed a weak positive relationship. For linear regression of individual Δ (adipokine) as a function of Δ FM, data were examined by the extra sum-of-squares F test to first consider if pooled data could be considered as a single model. Only chemerin and SEMA3C were considered a single group. All data except for resistin showed a significant relationship with changes in FM (a trend was observed; $p = 0.057$).

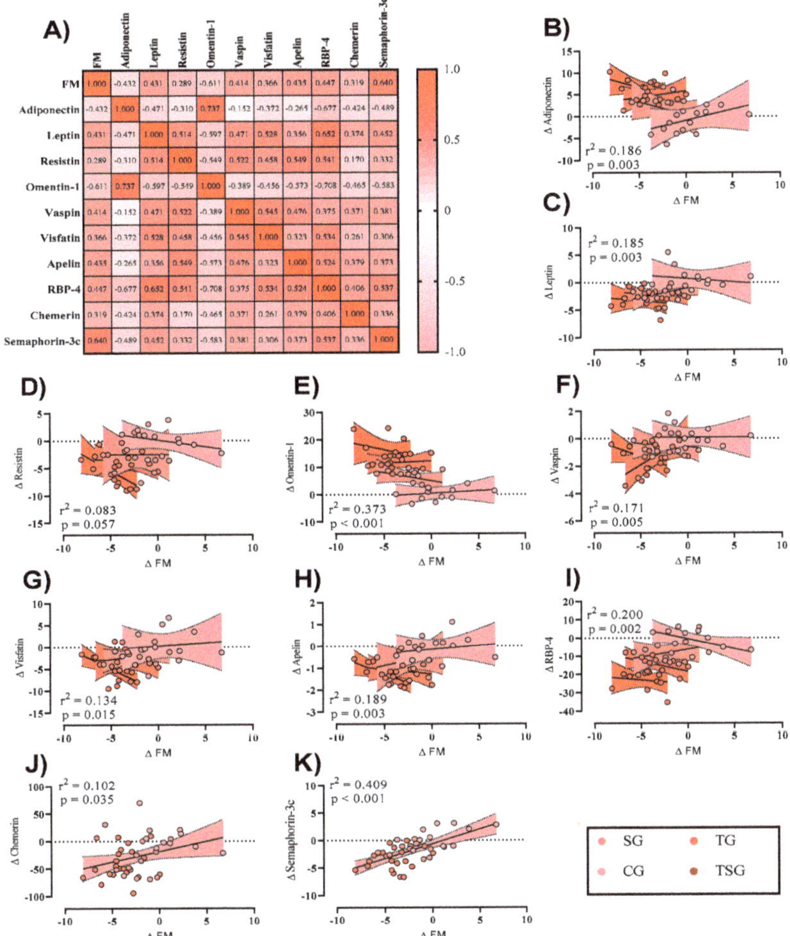

Figure 3. (**A**) Correlation matrix of Δ FM and adipokines, r values as shown. The key indicates the magnitude of r (red = -1 or 1, white = 0). (**B–K**) linear regression (Pearson's) of Δ (adipokine) as a function of Δ FM (kg). Linear regression is indicated by the solid black line, and 95% confidence intervals are indicated by red zones.

4. Discussion

Adipokines play a key role in cardiometabolic health status, and circulating levels are altered in obese states [39]. This study demonstrated that 12 weeks of CrossFit training and ASX supplementation, separately and in combination, can improve circulating adipokines levels in obese men. The combination of CrossFit and ASX supplementation overall led to greater changes in measured outcomes compared to each intervention alone. In our previous study, it was shown that CrossFit training and ASX supplementation decreased the plasma levels of GDF8, GDF15, CTRP2, and CTRP9. We also showed that CrossFit training and ASX supplementation increases high-density lipoprotein (HDL) and VO_{2peak}, and decreases low-density lipoprotein (LDL), total cholesterol (TC), TG, and insulin resistance [30]. This is the first investigation using CrossFit training as a mode of exercise as well as in combination with ASX supplementation on SEMA3C, apelin, chemerin, omentin1, visfatin, resistin, adiponectin, leptin, vaspin, and RBP4 in males with obesity.

Adiponectin has been shown to be inversely associated with insulin resistance and obesity [40]. In the present study, it was shown that CrossFit and ASX supplementation alone caused an increase in adiponectin in obese people, while this increase was greater in the group that took CrossFit training and ASX supplementation together. Some studies confirm the results of our research [41–43], while others do not show any change in adiponectin levels following acute exercise [44,45]. This disagreement may have been related to the laboratory protocols used [40]. Although the mechanism of action of CrossFit training in increasing plasma adiponectin levels is not well understood, the secretion of catecholamines, B-adrenergic receptors activity, and reduction of tumor necrosis factor-alpha (TNFa) may play a role [40]. Also, the increase in HDL and decrease in LDL, cholesterol, TG, and insulin resistance in our research [30] can be one of the reasons for the increase in adiponectin. Previous literature has also shown ASX supplementation increased serum adiponectin levels (~26%) in adults with mild hyperlipidemia [46]. The mechanisms underlying the effect of ASX supplementation on adiponectin are unclear, but one of the possible reasons could be reductions in TNFa through the activity of the peroxisome proliferator-activated receptor gamma (PPARγ) pathway [46].

Plasma leptin levels also decreased in each intervention group after 12 weeks. One piece of research agrees with the results of this study [41,44,47], while others have shown no effect of exercise in altering leptin levels [48–50]. These conflicting results may be due to differences in exercise protocols. The decrease in leptin levels following CrossFit training and ASX supplementation is likely related to the reduction of fat mass [40], which we previously published for this sample [30]. With the advancement of technology along with the improvements in living conditions, chronic diseases such as diabetes have increased as a result of low physical activity levels and improper nutrition. As a result of these processes, it is necessary to find safe interventions without complications [51]. Feng et al. [52] observed that ASX supplementation led to an improvement of insulin sensitivity and glucose tolerance through the suppression of inflammation, which reduced the symptoms of diabetes. Due to the strong antioxidant role of ASX supplementation, its anti-obesity and anti-inflammatory roles have been shown. For example, mechanisms include improving glucose metabolism, lowering blood pressure, improving redox imbalance in lymphocytes, and protecting β cells in the pancreas due to ASX supplementation has proven anti-diabetic and anti-obesity properties [51]. ASX supplementation also accelerates the metabolism of TG and HDL, reduces the incidence of cardiovascular disease, and increases the level of adiponectin, which plays an important role in regulating blood glucose [53]. The researchers observed that ASX reduced the production of nitric oxide (NO), leading to a reduction in insulin resistance through increased serine phosphorylation of insulin receptor substrate 1 (IRS1). In their study, Xia et al. [24] showed that ASX supplementation leads to a decrease in the size of fat cells through the activation of PPARγ. This leads to a decrease in plasma free fatty acid (FFA) levels, which confirms the results of Hussein et al. [54]. Aoi et al. [23] observed in their study that ASX supplementation leads to a reduction in the oxidative damage of carnitine palmitoyl transferase I (CPTI). This factor plays an important role in the oxidation of fatty acids in the mitochondrial membrane of muscle tissue. Contradictory results have been observed regarding the effect of ASX supplementation; some showed a non-significant increase, and others showed no change or decrease [23]. Hossein et al. [54] showed that long-term ASX supplementation (50 mg/kg/day) led to an increase in plasma adiponectin levels. Yoshida et al. [46] also observed these changes. Further research should be performed to determine the most effective dose and duration of treatment to increase blood adiponectin levels under different conditions.

In addition to reducing oxidative stress, ASX supplementation led to an increase in the serum levels of adiponectin in obese rats [55]. The effective mechanism of this process leads to the suppression of liver cancer in obese people [46]. Due to increased hormone-sensitive lipase activity in response to CrossFit training in obese participants, adiponectin levels increased significantly, and this led to body fat regulation. The important role of leptin is in energy balance and appetite control, and as a result, the level of this hormone is low in obese people. The cause of increased leptin in obese people can be attributed to resistance to leptin [56]. The normal level of this hormone is between 1 and 15 ng/mL in normal people, and more than 30 ng/mL in obese people [57]. Chronic high-intensity training has led to a decrease in blood leptin levels in obese participants [58]. The mechanism of this reduction was the reduction of body fat levels. Exercise leads to a decrease in leptin levels and an increase in adiponectin levels [56]. Some of the different results obtained can be due to different training protocols, variables under investigation, and more. However, in general, it has been observed that long-term training has a greater effect on leptin and adiponectin levels [59,60]. Due to its intensity and sufficient duration, CrossFit training leads to a decrease in body weight and a change in body composition through a negative balance created between energy intake and energy consumption [61]. Adiponectin and leptin are among the most well-known cytokines, which are secreted by adipose tissue and play an important role in metabolic and anti-inflammatory processes. In many chronic diseases, low levels of adiponectin and high levels of leptin play an important role in disease progression [40,62].

In the present study, after CrossFit workouts, the level of plasma omentin-1 significantly increased [63]. The effect of aerobic or resistance training on omentin-1 is conflicting, with studies showing increased levels [64,65] or unchanged levels [60]. In our previously published study [30] and other studies [59,60], the mechanism of increase in the level of omentin-1 has been shown, due to the reduction of body weight and improvement of cardiometabolic status. The level of omentin1 decreases in proportion to the increase in obesity [66], and the use of interventions to reduce body weight such as diet and active lifestyle leads to an increase in the level of omentin-1 [67]. This leads to weight loss, and as a result, the level of omentin1 increases [21,68]. Omentin is produced in adipose tissue; it seems that myokines released by muscle cells in response to positive exercise affect omentin1 levels [68].

The possible mechanism of visfatin reduction in different studies may be due to the intensity of training and the amount of changes in body weight and body fat volume. Also, vaspin is an adipokine that improves insulin sensitivity as a result of reducing body fat due to exercise, leading to an increase in its serum levels [69]. SEMA3C is a protein that plays an important role in the development of nervous, cardio-respiratory, kidney systems, and various oncogenesis [70]. This adipokine is secreted from subcutaneous fat tissues, and its level of secretion is related to obesity level, fat cell morphology, and weight changes [70]. Few studies have been performed on the effect of exercise training on the SEMA3C level. Limited research has found a decrease in serum SEMA3C levels following long-term training, and this decrease was significantly associated with improvements in body weight and body fat levels [70]. Also, the increase in HDL and decrease in LDL, TG, and insulin resistance in our research [30] can be one of the reasons for the decreased SEMA3C level. The current investigation shows that a 12-week regimen of CrossFit exercise training, in conjunction with a 20 mg dosage of ASX supplementation, resulted in a substantial decrease in adipokines that are directly associated with obesity.

The observed enhancement was more pronounced in the group that had concurrent Crossfit training and ASX supplementation.

Study Limitations

There are various limitations inherent in our investigation. Initially, the processes behind the potential enhancement of adipokine levels by bioactive constituents of ASX were not determined. Furthermore, the generalizability of our research is limited due to the exclusion of females in the enrollment of patients. Another limitation of our study is the lack of measurement of adipokine levels. Furthermore, it should be noted that the current body of research on the impact of ASX and CrossFit training on adipokines is minimal. Consequently, the precise processes behind this relationship remain undetermined. Therefore, further investigation is needed to elucidate potential pathways.

5. Conclusions

Our research presents novel insights regarding the impact of a combined regimen of CrossFit training and ASX supplementation on adipokines in males with obesity. Our data suggest that non-drug strategies such as ASX supplementation with CrossFit training can reduce SEMA3C, apelin, chemerin, visfatin, RBP4, resistin, vaspin, and leptin, and increase adiponectin and omentin1 in males with obesity. Consequently, individuals with obesity are recommended to include CrossFit exercise in their physical activity regimen and use ASX supplements in their dietary intake.

Author Contributions: A.S., S.S. and H.A. designed and conducted the study. S.R.S., R.B., H.A. and M.K. analyzed the obtained data. O.R., Z.P. and A.S. wrote the first draft of the manuscript. K.A.E., R.S., M.H.A.K., P.S., K.M.H., S.B., P.S., H.Z., R.B. and S.B. read, revised, and approved the final version of the manuscript. All authors have read and agreed to the published version of the manuscript.

Funding: This research did not receive any specific grant from any funding agency in the public, commercial, or not-for-profit sector.

Institutional Review Board Statement: The study was approved by the National Research and Ethics Committee (Ethics code: IR.IAU.DAMGHAN.REC.1401.035) and the Iranian Registry of Clinical Trials (IRCTID: IRCT20151228025732N76). All procedures were performed according to the latest revision of the Declaration of Helsinki.

Informed Consent Statement: All participants completed a physical examination performed by a physician and clinical exercise physiologist on the first visit and provided a written consent form and Physical Activity Readiness Questionnaire (PARQ).

Data Availability Statement: The datasets generated and/or analyzed during the current study are not publicly available but are available from the corresponding author upon reasonable request.

Acknowledgments: The authors would like to thank all the subjects for participating in this study.

Conflicts of Interest: The authors declare no conflict of interest.

References

1. Taylor, E.B. The complex role of adipokines in obesity, inflammation, and autoimmunity. *Clin. Sci.* **2021**, *135*, 731–752. [CrossRef] [PubMed]
2. Tremblay, A.; Clinchamps, M.; Pereira, B.; Courteix, D.; Lesourd, B.; Chapier, R.; Obert, P.; Vinet, A.; Walther, G.; Chaplais, E.; et al. Dietary fibres and the management of obesity and metabolic syndrome: The RESOLVE study. *Nutrients* **2020**, *12*, 2911. [CrossRef] [PubMed]
3. Ohashi, K.; Shibata, R.; Murohara, T.; Ouchi, N. Role of anti-inflammatory adipokines in obesity-related diseases. *Trends Endocrinol. Metab.* **2014**, *25*, 348–355. [CrossRef] [PubMed]
4. de Oliveira Leal, V.; Mafra, D. Adipokines in obesity. *Clin. Chim. Acta* **2013**, *419*, 87–94. [CrossRef]
5. Eskandari, M.; Moghadam, B.H.; Bagheri, R.; Ashtary-Larky, D.; Eskandari, E.; Nordvall, M.; Dutheil, F.; Wong, A. Effects of interval jump rope exercise combined with dark chocolate supplementation on inflammatory adipokine, cytokine concentrations, and body composition in obese adolescent boys. *Nutrients* **2020**, *12*, 3011. [CrossRef]

6. Abdi, A.; Mehrabani, J.; Nordvall, M.; Wong, A.; Fallah, A.; Bagheri, R. Effects of concurrent training on irisin and fibronectin type-III domain containing 5 (FNDC5) expression in visceral adipose tissue in type-2 diabetic rats. *Arch. Physiol. Biochem.* **2022**, *128*, 651–656. [CrossRef]

7. Khan, M.; Joseph, F. Adipose tissue and adipokines: The association with and application of adipokines in obesity. *Scientifica* **2014**, *2014*, 328592. [CrossRef]

8. González Izquierdo, A.; Crujeiras Martínez, A.B.; Casanueva Freijo, F.; Couselo Carreira, M. Leptin, obesity, and leptin resistance: Where are we 25 years later? *Nutrients* **2019**, *11*, 2704. [CrossRef]

9. Ashtary-Larky, D.; Kashkooli, S.; Bagheri, R.; Lamuchi-Deli, N.; Alipour, M.; Mombaini, D.; Baker, J.S.; Ahmadi, A.R.; Wong, A. The effect of exercise training on serum concentrations of chemerin in patients with metabolic diseases: A systematic review and meta-analysis. *Arch. Physiol. Biochem.* **2021**, *129*, 1–10. [CrossRef]

10. Glassman, G. *CrossFit Training: Level 1 Training Guide*; CrossFit, Inc.: Santa Cruz, CA, USA, 2019.

11. Claudino, J.G.; Gabbett, T.J.; Bourgeois, F.; Souza, H.d.S.; Miranda, R.C.; Mezêncio, B.; Soncin, R.; Filho, C.A.C.; Bottaro, M.; Hernandez, A.J.; et al. CrossFit overview: Systematic review and meta-analysis. *Sports Med. Open* **2018**, *4*, 1–14. [CrossRef]

12. Gianzina, E.A.; Kassotaki, O.A. The benefits and risks of the high-intensity CrossFit training. *Sport Sci. Health* **2019**, *15*, 21–33. [CrossRef]

13. Farkhondeh, T.; Llorens, S.; Pourbagher-Shahri, A.M.; Ashrafizadeh, M.; Talebi, M.; Shakibaei, M.; Samarghandian, S. An overview of the role of adipokines in cardiometabolic diseases. *Molecules* **2020**, *25*, 5218. [CrossRef]

14. Hosseini, M.; Bagheri, R.; Nikkar, H.; Baker, J.S.; Jaime, S.J.; Mosayebi, Z.; Zouraghi, M.R.; Wong, A. The effect of interval training on adipokine plasmatic levels in rats with induced myocardial infarction. *Arch. Physiol. Biochem.* **2022**, *128*, 1249–1253. [CrossRef] [PubMed]

15. Jung, S.H.; Park, H.S.; Kim, K.-S.; Choi, W.H.; Ahn, C.W.; Kim, B.T.; Kim, S.M.; Lee, S.Y.; Ahn, S.M.; Kim, Y.K.; et al. Effect of weight loss on some serum cytokines in human obesity: Increase in IL-10 after weight loss. *J. Nutr. Biochem.* **2008**, *19*, 371–375. [CrossRef]

16. Bagheri, R.; Rashidlamir, A.; Ashtary-Larky, D.; Wong, A.; Alipour, M.; Motevalli, M.S.; Chebbi, A.; Laher, I.; Zouhal, H. Does green tea extract enhance the anti-inflammatory effects of exercise on fat loss? *Br. J. Clin. Pharmacol.* **2020**, *86*, 753–762. [CrossRef] [PubMed]

17. Bagheri, R.; Rashidlamir, A.; Ashtary-Larky, D.; Wong, A.; Grubbs, B.; Motevalli, M.S.; Baker, J.S.; Laher, I.; Zouhal, H. Effects of green tea extract supplementation and endurance training on irisin, pro-inflammatory cytokines, and adiponectin concentrations in overweight middle-aged men. *Eur. J. Appl. Physiol.* **2020**, *120*, 915–923. [CrossRef] [PubMed]

18. Düzova, H.; Güllü, E.; Çiçek, G.; Köksal, B.; Kayhan, B.; Güllü, A.; Şahin, İ. The effect of exercise induced weight loss on myokines and adipokines in overweight sedentary females steps aerobics vs. jogging walking exercises. *J. Sports Med. Phys. Fit.* **2016**, *58*, 295–308.

19. Ouerghi, N.; Ben Fradj, M.K.; Bezrati, I.; Feki, M.; Kaabachi, N.; Bouassida, A. Effect of high-intensity interval training on plasma omentin-1 concentration in overweight/obese and normal-weight youth. *Obes. Facts* **2017**, *10*, 323–331. [CrossRef]

20. Urbanová, M.; Dostálová, I.; Trachta, P.; Drápalová, J.; Kaválková, P.; Haluzíková, D.; Matoulek, M.; Lacinová, Z.; Mráz, M.; Kasalický, M.; et al. Serum concentrations and subcutaneous adipose tissue mRNA expression of omentin in morbid obesity and type 2 diabetes mellitus: The effect of very-low-calorie diet, physical activity and laparoscopic sleeve gastrectomy. *Physiol. Res.* **2014**, *63*, 207–218. [CrossRef]

21. Faramarzi, M.; Banitalebi, E.; Nori, S.; Farzin, S.; Taghavian, Z. Effects of rhythmic aerobic exercise plus core stability training on serum omentin, chemerin and vaspin levels and insulin resistance of overweight women. *J. Sports Med. Phys. Fit.* **2015**, *56*, 476–482.

22. Taherkhani, S.; Suzuki, K.; Castell, L. A short overview of changes in inflammatory cytokines and oxidative stress in response to physical activity and antioxidant supplementation. *Antioxidants* **2020**, *9*, 886. [CrossRef]

23. Aoi, W.; Naito, Y.; Takanami, Y.; Ishii, T.; Kawai, Y.; Akagiri, S.; Kato, Y.; Osawa, T.; Yoshikawa, T. Astaxanthin improves muscle lipid metabolism in exercise via inhibitory effect of oxidative CPT I modification. *Biochem. Biophys. Res. Commun.* **2008**, *366*, 892–897. [CrossRef] [PubMed]

24. Xia, W.; Tang, N.; Kord-Varkaneh, H.; Low, T.Y.; Tan, S.C.; Wu, X.; Zhu, Y. The effects of astaxanthin supplementation on obesity, blood pressure, CRP, glycemic biomarkers, and lipid profile: A meta-analysis of randomized controlled trials. *Pharmacol. Res.* **2020**, *161*, 105113. [CrossRef] [PubMed]

25. Fakhri, S.; Abbaszadeh, F.; Dargahi, L.; Jorjani, M. Astaxanthin: A mechanistic review on its biological activities and health benefits. *Pharmacol. Res.* **2018**, *136*, 1–20. [CrossRef]

26. Kim, S.H.; Kim, H. Inhibitory effect of astaxanthin on oxidative stress-induced mitochondrial dysfunction—A mini-review. *Nutrients* **2018**, *10*, 1137. [CrossRef] [PubMed]

27. Nishida, Y.; Nawaz, A.; Kado, T.; Takikawa, A.; Igarashi, Y.; Onogi, Y.; Wada, T.; Sasaoka, T.; Yamamoto, S.; Sasahara, M.; et al. Astaxanthin stimulates mitochondrial biogenesis in insulin resistant muscle via activation of AMPK pathway. *J. Cachexia Sarcopenia Muscle* **2020**, *11*, 241–258. [CrossRef]

28. Murai, T.; Kawasumi, K.; Tominaga, K.; Okada, Y.; Kobayashi, M.; Arai, T. Effects of astaxanthin supplementation in healthy and obese dogs. *Vet. Med. Res. Rep.* **2019**, *10*, 29. [CrossRef]

29. Oharomari, L.K.; Ikemoto, M.J.; Hwang, D.J.; Koizumi, H.; Soya, H. Benefits of Exercise and Astaxanthin Supplementation: Are There Additive or Synergistic Effects? *Antioxidants* **2021**, *10*, 870. [CrossRef]

30. Saeidi, A.; Nouri-Habashi, A.; Razi, O.; Ataeinosrat, A.; Rahmani, H.; Mollabashi, S.S.; Bagherzadeh-Rahmani, B.; Aghdam, S.M.; Khalajzadeh, L.; Al Kiyumi, M.H.; et al. Astaxanthin Supplemented with High-Intensity Functional Training Decreases Adipokines Levels and Cardiovascular Risk Factors in Men with Obesity. *Nutrients* **2023**, *15*, 286. [CrossRef]

31. Moghadam, B.H.; Bagheri, R.; Roozbeh, B.; Ashtary-Larky, D.; Gaeini, A.A.; Dutheil, F.; Wong, A. Impact of saffron (*Crocus sativus* Linn) supplementation and resistance training on markers implicated in depression and happiness levels in untrained young males. *Physiol. Behav.* **2021**, *233*, 113352. [CrossRef]

32. Thomas, S.; Reading, J.; Shephard, R.J. Revision of the physical activity readiness questionnaire (PAR-Q). *Can. J. Sport Sci.* **1992**, *17*, 338–345. [PubMed]

33. Nathanson, V. Revising the declaration of Helsinki. *BMJ* **2013**, *346*, f2837. [CrossRef] [PubMed]

34. Glassman, G. A theoretical template for crossfit's programming. *CrossFit J.* **2003**, *6*, 1–5.

35. Heinrich, K.M.; Patel, P.M.; O'Neal, J.L.; Heinrich, B.S. High-intensity compared to moderate-intensity training for exercise initiation, enjoyment, adherence, and intentions: An intervention study. *BMC Public Health* **2014**, *14*, 1–6. [CrossRef]

36. Choi, H.D.; Kim, J.H.; Chang, M.J.; Kyu-Youn, Y.; Shin, W.G. Effects of astaxanthin on oxidative stress in overweight and obese adults. *Phytother. Res.* **2011**, *25*, 1813–1818. [CrossRef]

37. Annibalini, G.; Lucertini, F.; Agostini, D.; Vallorani, L.; Gioacchini, A.; Barbieri, E.; Guescini, M.; Casadei, L.; Passalia, A.; Del Sal, M.; et al. Concurrent aerobic and resistance training has anti-inflammatory effects and increases both plasma and leukocyte levels of IGF-1 in late middle-aged type 2 diabetic patients. *Oxidative Med. Cell. Longev.* **2017**, *2017*, 3937842. [CrossRef]

38. Ratner, B. The correlation coefficient: Its values range between +1/−1, or do they? *J. Target. Meas. Anal. Mark.* **2009**, *17*, 139–142. [CrossRef]

39. Ashtary-Larky, D.; Lamuchi-Deli, N.; Kashkooli, S.; Mombaini, D.; Alipour, M.; Khodadadi, F.; Bagheri, R.; Dutheil, F.; Wong, A. The effects of exercise training on serum concentrations of chemerin in individuals with overweight and obesity: A systematic review, meta-analysis, and meta-regression of 43 clinical trials. *Arch. Physiol. Biochem.* **2021**, *129*, 1012–1027. [CrossRef]

40. Saeidi, A.; Haghighi, M.M.; Kolahdouzi, S.; Daraei, A.; Ben Abderrahmane, A.; Essop, M.F.; Laher, I.; Hackney, A.C.; Zouhal, H. The effects of physical activity on adipokines in individuals with overweight/obesity across the lifespan: A narrative review. *Obes. Rev.* **2021**, *22*, e13090. [CrossRef]

41. Højbjerre, L.; Rosenzweig, M.; Dela, F.; Bruun, J.M.; Stallknecht, B.; Verkauskiene, R.; Beltrand, J.; Claris, O.; Chevenne, D.; Deghmoun, S.; et al. Acute exercise increases adipose tissue interstitial adiponectin concentration in healthy overweight and lean subjects. *Eur. J. Endocrinol.* **2007**, *157*, 613–623. [CrossRef]

42. Saunders, T.J.; Palombella, A.; McGuire, K.A.; Janiszewski, P.M.; Després, J.-P.; Ross, R. Acute exercise increases adiponectin levels in abdominally obese men. *J. Nutr. Metab.* **2012**, *2012*. [CrossRef] [PubMed]

43. Varady, K.A.; Bhutani, S.; Church, E.C.; Phillips, S.A. Adipokine responses to acute resistance exercise in trained and untrained men. *Med. Sci. Sports Exerc.* **2010**, *42*, 456–462. [CrossRef] [PubMed]

44. Bouassida, A.; Lakhdar, N.; Benaissa, N.; Mejri, S.; Zaouali, M.; Zbidi, A.; Tabka, Z. Adiponectin responses to acute moderate and heavy exercises in overweight middle aged subjects. *J. Sports Med. Phys. Fit.* **2010**, *50*, 330–335.

45. Jamurtas, A.Z.; Theocharis, V.; Koukoulis, G.; Stakias, N.; Fatouros, I.G.; Kouretas, D.; Koutedakis, Y. The effects of acute exercise on serum adiponectin and resistin levels and their relation to insulin sensitivity in overweight males. *Eur. J. Appl. Physiol.* **2006**, *97*, 122–126. [CrossRef]

46. Yoshida, H.; Yanai, H.; Ito, K.; Tomono, Y.; Koikeda, T.; Tsukahara, H.; Tada, N. Administration of natural astaxanthin increases serum HDL-cholesterol and adiponectin in subjects with mild hyperlipidemia. *Atherosclerosis* **2010**, *209*, 520–523. [CrossRef]

47. Legakis, I.N.; Mantzouridis, T.; Saramantis, A.; Lakka-Papadodima, E. Rapid decrease of leptin in middle-aged sedentary individuals after 20 minutes of vigorous exercise with early recovery after the termination of the test. *J. Endocrinol. Investig.* **2004**, *27*, 117–120. [CrossRef]

48. Cruz, I.S.; Rosa, G.; Valle, V.; Mello, D.B.D.; Fortes, M.; Dantas, E.H. Acute effects of concurrent training on serum leptin and cortisol in overweight young adults. *Rev. Bras. Med. Esporte* **2012**, *18*, 81–86. [CrossRef]

49. Racette, S.B.; Coppack, S.W.; Landt, M.; Klein, S. Leptin production during moderate-intensity aerobic exercise. *J. Clin. Endocrinol. Metab.* **1997**, *82*, 2275–2277. [CrossRef]

50. Weltman, A.; Pritzlaff, C.J.; Wideman, L.; Considine, R.V.; Fryburg, D.A.; Gutgesell, M.E.; Hartman, M.L.; Veldhuis, J.D. Intensity of acute exercise does not affect serum leptin concentrations in young men. *Med. Sci. Sports Exerc.* **2000**, *32*, 1556–1561. [CrossRef]

51. Kumar, A.; Dhaliwal, N.; Dhaliwal, J.; Dharavath, R.N.; Chopra, K. Astaxanthin attenuates oxidative stress and inflammatory responses in complete Freund-adjuvant-induced arthritis in rats. *Pharmacol. Rep.* **2020**, *72*, 104–114. [CrossRef]

52. Feng, Y.; Chu, A.; Luo, Q.; Wu, M.; Shi, X.; Chen, Y. The protective effect of astaxanthin on cognitive function via inhibition of oxidative stress and inflammation in the brains of chronic T2DM rats. *Front. Pharmacol.* **2018**, *9*, 748. [CrossRef] [PubMed]

53. Kohandel, Z.; Farkhondeh, T.; Aschner, M.; Samarghandian, S. Nrf2 a molecular therapeutic target for Astaxanthin. *Biomed. Pharmacother.* **2021**, *137*, 111374. [CrossRef] [PubMed]

54. Hussein, G.; Sankawa, U.; Goto, H.; Matsumoto, A.K.; Watanabe, H. Astaxanthin, a carotenoid with potential in human health and nutrition. *J. Nat. Prod.* **2006**, *69*, 443–449. [CrossRef] [PubMed]

55. Boshra, S.A. Astaxanthin Attenuates Adiponectin, Calprotectin, miRNA222 and miRNA378 in Obesity induced by High-Fat Diet in Rats. *Curr. Pharm. Biotechnol.* **2022**, *23*, 609–618. [CrossRef] [PubMed]

56. Vardar, S.A.; Karaca, A.; Güldiken, S.; Palabıyık, O.; Süt, N.; Demir, A.M. High-intensity interval training acutely alters plasma adipokine levels in young overweight/obese women. *Arch. Physiol. Biochem.* **2018**, *124*, 149–155. [CrossRef] [PubMed]

57. Falcão-Pires, I.; Castro-Chaves, P.; Miranda-Silva, D.; Lourenço, A.P.; Leite-Moreira, A.F. Physiological, pathological and potential therapeutic roles of adipokines. *Drug Discov. Today* **2012**, *17*, 880–889. [CrossRef]

58. Racil, G.; Zouhal, H.; Elmontassar, W.; Ben Abderrahmane, A.; De Sousa, M.V.; Chamari, K.; Amri, M.; Coquart, J.B. Plyometric exercise combined with high-intensity interval training improves metabolic abnormalities in young obese females more so than interval training alone. *Appl. Physiol. Nutr. Metab.* **2016**, *41*, 103–109. [CrossRef]

59. Madsen, S.M.; Thorup, A.C.; Bjerre, M.; Jeppesen, P.B. Does 8 weeks of strenuous bicycle exercise improve diabetes-related inflammatory cytokines and free fatty acids in type 2 diabetes patients and individuals at high-risk of metabolic syndrome? *Arch. Physiol. Biochem.* **2015**, *121*, 129–138. [CrossRef]

60. Verheggen, R.J.H.M.; Poelkens, F.; Roerink, S.H.P.P.; Ramakers, R.E.F.S.; Catoire, M.; Hermus, A.R.M.M.; Thijssen, D.H.J.; Hopman, M.T.E. Exercise Improves Insulin Sensitivity in the Absence of Changes in Cytokines. *Med. Sci. Sports Exerc.* **2016**, *48*, 2378–2386. [CrossRef]

61. Feito, Y.; Hoffstetter, W.; Serafini, P.; Mangine, G. Changes in body composition, bone metabolism, strength, and skill-specific performance resulting from 16-weeks of HIFT. *PLoS ONE* **2018**, *13*, e0198324. [CrossRef]

62. Zouhal, H.; Zare-Kookandeh, N.; Haghighi, M.M.; Daraei, A.; de Sousa, M.; Soltani, M.; Ben Abderrahman, A.; Tijani, J.M.; Hackney, A.C.; Laher, I.; et al. Physical activity and adipokine levels in individuals with type 2 diabetes: A literature review and practical applications. *Rev. Endocr. Metab. Disord.* **2021**, *22*, 987–1011. [CrossRef] [PubMed]

63. Nikseresht, M.; Hafezi Ahmadi, M.R.; Hedayati, M. Detraining-induced alterations in adipokines and cardiometabolic risk factors after nonlinear periodized resistance and aerobic interval training in obese men. *Appl. Physiol. Nutr. Metab.* **2016**, *41*, 1018–1025. [CrossRef]

64. AminiLari, Z.; Fararouei, M.; Amanat, S.; Sinaei, E.; Dianatinasab, S.; AminiLari, M.; Daneshi, N.; Dianatinasab, M. The effect of 12 weeks aerobic, resistance, and combined exercises on omentin-1 levels and insulin resistance among type 2 diabetic middle-aged women. *Diabetes Metab. J.* **2017**, *41*, 205. [CrossRef] [PubMed]

65. He, Z.; Tian, Y.; Valenzuela, P.L.; Huang, C.; Zhao, J.; Hong, P.; He, Z.; Yin, S.; Lucia, A. Myokine/adipokine response to "aerobic" exercise: Is it just a matter of exercise load? *Front. Physiol.* **2019**, *10*, 691. [CrossRef] [PubMed]

66. Watanabe, T.; Watanabe-Kominato, K.; Takahashi, Y.; Kojima, M.; Watanabe, R. Adipose tissue-derived omentin-1 function and regulation. *Compr. Physiol.* **2011**, *7*, 765–781.

67. Moreno-Navarrete, J.M.; Catalán, V.; Ortega, F.; Gómez-Ambrosi, J.; Ricart, W.; Frühbeck, G.; Fernández-Real, J.M. Circulating omentin concentration increases after weight loss. *Nutr. Metab.* **2010**, *7*, 27. [CrossRef]

68. Ouerghi, N.; Ben Fradj, M.K.; Duclos, M.; Bouassida, A.; Feki, M.; Weiss, K.; Knechtle, B. Effects of High-Intensity Interval Training on Selected Adipokines and Cardiometabolic Risk Markers in Normal-Weight and Overweight/Obese Young Males: A Pre-Post Test Trial. *Biology* **2022**, *11*, 853. [CrossRef]

69. Kazemi, A.; Rahmati, M.; Dabaghzadeh, R.; Raisi, S.; Aghamolaie, S. The effect of high volume high intensity interval training on serum visfatin and vaspin, insulin resistance, lipid profile and body composition of overweight men. *Daneshvar Med.* **2015**, *23*, 325.

70. Nam, J.S.; Ahn, C.W.; Park, H.J.; Kim, Y.S. Semaphorin 3 C is a novel adipokine representing exercise-induced improvements of metabolism in metabolically healthy obese young males. *Sci. Rep.* **2020**, *10*, 10005. [CrossRef]

nutrients

Article

Effects of Creatine Supplementation on the Myostatin Pathway and Myosin Heavy Chain Isoforms in Different Skeletal Muscles of Resistance-Trained Rats

Marianna Rabelo de Carvalho [1], Ellen Fernandes Duarte [1], Maria Lua Marques Mendonça [1], Camila Souza de Morais [1], Gabriel Elias Ota [1], Jair José Gaspar-Junior [1], Wander Fernando de Oliveira Filiú [2], Felipe Cesar Damatto [3], Marina Politi Okoshi [3], Katashi Okoshi [3], Rodrigo Juliano Oliveira [1], Paula Felippe Martinez [1,4] and Silvio Assis de Oliveira-Junior [1,4,*]

[1] Graduate Program in Health and Development in the Midwestern Region, Federal University of Mato Grosso do Sul (UFMS), Campo Grande 79070-900, MS, Brazil; mariannarabelo5@gmail.com (M.R.d.C.); ellenduarte_12@hotmail.com (E.F.D.); marialuamarques@hotmail.com (M.L.M.M.); nutricionistacamila@outlook.com (C.S.d.M.); gabriel.elias.ota@gmail.com (G.E.O.); gasparjr.ft@gmail.com (J.J.G.-J.); rodrigo.oliveira@ufms.br (R.J.O.); paula.martinez@ufms.br (P.F.M.)
[2] Faculty of Pharmaceutical Sciences, Food and Nutrition, Federal University of Mato Grosso do Sul (UFMS), Campo Grande 79070-900, MS, Brazil; wander.filiu@gmail.com
[3] Internal Medicine Department, Botucatu Medical School, Sao Paulo State University (UNESP), Botucatu 18618-687, SP, Brazil; felipedamatto@hotmail.com (F.C.D.); marina.okoshi@unesp.br (M.P.O.); katashi.okoshi@unesp.br (K.O.)
[4] Graduate Program in Movement Sciences, Federal University of Mato Grosso do Sul (UFMS), Campo Grande 79070-900, MS, Brazil
* Correspondence: silvio.oliveira-jr@ufms.br

Citation: de Carvalho, M.R.; Duarte, E.F.; Mendonça, M.L.M.; de Morais, C.S.; Ota, G.E.; Gaspar-Junior, J.J.; de Oliveira Filiú, W.F.; Damatto, F.C.; Okoshi, M.P.; Okoshi, K.; et al. Effects of Creatine Supplementation on the Myostatin Pathway and Myosin Heavy Chain Isoforms in Different Skeletal Muscles of Resistance-Trained Rats. *Nutrients* 2023, 15, 2224. https://doi.org/10.3390/nu15092224

Academic Editor: Louise Deldicque

Received: 15 February 2023
Revised: 25 April 2023
Accepted: 26 April 2023
Published: 8 May 2023

Abstract: Creatine has been used to maximize resistance training effects on skeletal muscles, including muscle hypertrophy and fiber type changes. This study aimed to evaluate the impact of creatine supplementation on the myostatin pathway and myosin heavy chain (MyHC) isoforms in the slow- and fast-twitch muscles of resistance-trained rats. Twenty-eight male Wistar rats were divided into four groups: a sedentary control (Cc), sedentary creatine supplementation (Cr), resistance training (Tc), and resistance training combined with creatine supplementation (Tcr). Cc and Tc received standard commercial chow; Cr and Tcr received a 2% creatine-supplemented diet. Tc and Tcr performed a resistance training protocol on a ladder for 12 weeks. Morphology, MyHC isoforms, myostatin, follistatin, and ActRIIB protein expressions were analyzed in soleus and white gastrocnemius portion samples. The results were analyzed using two-way ANOVA and Tukey's test. Tc and Tcr exhibited higher performance than their control counterparts. Resistance training increased the ratio between muscle and body weight, the cross-sectional area, as well as the interstitial collagen fraction. Resistance training alone increased MyHC IIx and follistatin while reducing myostatin ($p < 0.001$) and ActRIIB ($p = 0.040$) expressions in the gastrocnemius. Resistance training induced skeletal muscle hypertrophy and interstitial remodeling, which are more evident in the gastrocnemius muscle. The effects were not impacted by creatine supplementation.

Keywords: strength training; myostatin; muscle anabolism; creatine; muscle fibers

1. Introduction

In the diverse world of commercial supplements, creatine monohydrate stands out as being widely used by high-level athletes and physically active individuals [1–3]. Creatine monohydrate consumption can impact skeletal muscle through multiple mechanisms which sustain muscle remodeling [4]. Creatine supplementation is commonly used to maximize the effects of resistance training, including neuromuscular recruitment, increased protein synthesis, and muscle fiber type changes [5–9].

Combinations between exercise training and creatine interventions resulted in greater lean mass and lower muscle protein catabolism [10,11]. Another possible effect of creatine supplementation is the adaptive modulation of myosin heavy chain (MyHC) isoforms in skeletal muscle fibers [12–16]. MyHC isoform profiles determine the muscle fiber phenotype; in general, predominantly MyHC I fibers have greater slow twitch-oxidative characteristics, while fibers containing more MyHC II mostly have a fast glycolytic metabolism [17,18]. Skeletal muscles have a diverse proportion and distribution of highly adaptable fiber types so that phenotypical aspects can be affected by several molecular signaling pathways, which regulate protein synthesis and myogenic activity [19,20].

Myostatin, an extracellular myokine that is primarily expressed in skeletal muscle, plays a crucial role in downregulating muscle mass and modulating fiber-type composition [21,22]. Its expression is inversely associated with the amount of fast glycolytic fibers [23–26]. Myostatin signaling is complex and comprises the activation of several downstream pathways. Mature myostatin binds to the Type IIB activin receptor (ActRIIB) and initiates signaling cascades that upregulate the genes involved in atrophy and downregulate genes involved in myogenesis. Among potential myostatin inhibitors, follistatin is an antagonist of transforming growth factor- β (TGF-β) family members, inhibiting the link of myostatin to ActRIIB. In this context, follistatin overexpression can be associated with skeletal muscle hypertrophy [21,27].

Resistance exercise training has been associated with reduced serum myostatin levels, as well as follistatin modulation [28] and increased muscle mass; these effects were amplified in response to creatine supplementation [29]. In vitro studies have shown that creatine prevents or reverses myostatin-induced muscle atrophy and increases the expression of myostatin-negative regulatory genes [30]. In pigs and chickens, creatine supplementation increased myogenic factor expression and reduced myostatin mRNA levels [31,32]. However, the impact of a combination of resistance training and creatine supplementation on morphology, MyHC isoform expression, and myostatin pathway signaling in skeletal muscles has not been clarified.

The aim of this study was to evaluate the effects of creatine monohydrate supplementation on morphology, MyHC isoform expression, and myostatin pathway signaling in gastrocnemius (white portion) and the soleus muscles of rats submitted to resistance training. Since the gastrocnemius muscle superficial area (white portion) is characterized by the predominance of fast-twitch fibers, while the soleus is a classical slow-twitch muscle, these muscles were used in this study. We hypothesized that a combination of creatine monohydrate supplementation and resistance training could attenuate myostatin expression and modulate downstream targets, promoting more accentuated changes in white gastrocnemius than in the soleus muscle.

2. Materials and Methods

2.1. Animal and Experimental Design

Male Wistar rats (60 days old) were assigned to four groups: a sedentary control (Cc, $n = 7$), sedentary creatine supplementation (Cr, $n = 7$), resistance training (Tc, $n = 7$), and resistance training combined with creatine supplementation (Tcr, $n = 7$). Cc and Tc groups received a daily diet of commercial rodent chow (Nuvilab® CR1, Curitiba, PR, Brazil; 3.6 kcal/g); Cr and Tcr groups received the same standard daily diet but supplemented with creatine monohydrate at a dosage of 2% daily diet weight (3.7 kcal/g) [33,34]. Tc and Tcr groups were submitted to a ladder-climbing resistance training protocol [35]. All animals received water ad libitum and were housed (three rats per cage) under controlled temperature and humidity conditions with 12 h light/dark cycles. The experimental protocol duration was 12 weeks.

All procedures and the experimental protocol were approved by our institution's Animal Ethics Committee (Protocol Number 873/2017) in accordance with the Brazilian College of Animal Experimentation (COBEA) and the US National Institutes of Health "Guide for the Care and Use of Laboratory Animals" (NIH, 2011).

2.2. Climbing Exercise Familiarization

All trained rats were familiarized with the climbing exercise protocol for three days with a 10% animal body weight load. A total of 8 to 12 climbing sets were considered a complete session during familiarization [36,37].

2.3. Initial Maximal Carrying Capacity

All rats were submitted to incremental tests to determine their initial maximum carrying capacity. The climbing test was performed against a 75% animal body weight load. After a 120 s interval, a 30 g load was added to the resistance apparatus for a new climb. The test was completed when the maximal load resulted in exhaustion or the impossibility of additional climbing. The test was also stopped when eight exercise sets were completed. The highest weight carried to the top of the ladder was considered the maximal load [35–39].

2.4. Resistance Training Protocol

Resistance training groups performed resistance training three times per week during the dark cycle for 12 weeks, resulting in a total of 36 sessions. Each session consisted of four to nine climbs. The first four climbs were performed with 50%, 75%, 90%, and 100% of the maximal resistance load achieved during the incremental test. Then, 30 g loads were progressively added to each subsequent climb up to the daily limit of 9 climbs [35]. Resistance loads consisted of lead weights placed in conical plastic flasks, which were attached to the proximal part of the rat tail.

2.5. Final Maximal Carrying Capacity

The maximal carrying capacity test was repeated after 12 weeks to determine the final maximal performance.

2.6. General Characteristics, Tissue Collection, and Serum Biochemical Analysis

Food consumption was measured daily, and body weight (BW) was evaluated weekly. Calorie intake was calculated as follows: daily food consumption x diet energy density. Feed efficiency (the ability to convert calorie intake into BW) was determined by dividing BW gain (g) by the total calorie intake (Kcal) [40]. Ingested creatine was calculated from 2% of the total food intake. At the end of the experiment, the rats were submitted to 6–8 h fasting, were anesthetized with thiopental (80 mg/kg), and euthanized by decapitation. Blood was collected for biochemical analyses, and the serum was separated by centrifugation at $3000\times g$ for 10 min and then stored at $-80\ ^{\circ}\text{C}$ for subsequent assessment. Glucose, cholesterol, triglycerides, albumin, and total protein serum levels were determined by spectrophotometry using enzymatic kits [40,41]. The gastrocnemius and soleus muscles from both pelvic limbs were quickly removed and weighed. After that, muscle samples were immediately frozen in liquid nitrogen and stored at $-80\ ^{\circ}\text{C}$ until analysis. The right tibia was dissected, measured using a pachymeter, and used to normalize muscle mass [41,42].

2.7. Skeletal Muscle Morphology

Total gastrocnemius and soleus muscle weights in both absolute values and respective ratios with body weight and tibia length were used to characterize the macroscopic morphology. Soleus and white (superficial) portions of gastrocnemius muscles were used to obtain cross-sectional histological sections (8-μm-thick) using a cryostat at $-20\ ^{\circ}\text{C}$. The slides were stained with hematoxylin-eosin (HE) to assess the cross-sectional area; at least 150 fibers were measured per animal. Other histological slides were submitted to Picrosirius red (Sirius red F3BA) staining; at least 10 fields per animal were digitalized and used to calculate the interstitial collagen fraction [41,43]. Histological analyzes were performed at $400\times$ magnification using a Leica DM5500B microscope (Wetzlar, Germany) coupled to a digital image projection video camera equipped with Leica Application Suite version 4.0.0 (Heerbrugg, Switzerland). Cross-sectional fiber areas were measured using Image J software (Wayne Rasbandat NIH, Bethesda, MD, USA), and an interstitial collagen fraction was quantified using Image Pro-plus Version 6.0.0.260 (Media Cybernetics, Rockville, MD, USA).

2.8. Western Blotting

Protein levels were analyzed by Western blot according to previous studies [41–44]. Protein was extracted using a RIPA buffer (1 mL/100 mg tissue) containing protease and phosphatase inhibitors. The supernatant protein content was quantified by the Bradford assay [45]. Samples were separated into a polyacrylamide gel and then transferred to a nitrocellulose membrane. After blockade with 5% skimmed milk in TBST for 1 h, the membrane was incubated overnight at 4 °C with specific antibodies: anti-myostatin (GDF-8, sc-6885-R)—dilution 1:200, anti-follistatin (sc-30194)—dilution 1:400, and anti-ActRIIB (sc-376593)—dilution 1:200, Santa Cruz Biotechnology, Inc., Santa Cruz, Dallas, TX, USA). The membrane was then washed with TBS and Tween 20 and incubated with a secondary peroxidase-conjugated antibody (90 min at room temperature). Enhanced Chemio Luminescence (Luminata Crescendo® reagent, Merck Millipore; Darmstadt, Germany) was used to detect bound antibodies [41]. The membrane was then incubated with a ReBlot Plus Strong Antibody Stripping Solution, 10×—Millipore (Burlington, MA, USA), to remove the antibodies attached to the membrane. The blockade process was repeated, and the membrane was incubated overnight at 4 °C with glyceraldehyde-3-phosphate dehydrogenase (GAPDH, sc-32233, Santa Cruz Biotechnology, Inc., Dallas, TX, USA). The procedure continued as described above until a signal was detected. Afterward, respective bands were quantified by densitometry using a Gel-Pro Analyzer 3.1. The results obtained for each protein were normalized to those obtained for GAPDH.

2.9. Myosin Heavy Chain Isoforms

MyHC isoforms were quantified after electrophoresis in a sodium dodecyl sulfate-polyacrylamide gel (SDS-PAGE) [46,47]. Frozen samples were mechanically homogenized in a protein extraction solution containing a 50 mM phosphate potassium lysis buffer 0.5 mL/50 mg tissue. The total protein quantification was performed in supernatant aliquots by the Bradford method. Small volumes of the diluted extracts (10 µL) were loaded onto a 7–10% SDS-PAGE separating gel with a 4% stacking gel, which was run overnight (27 h) at 70 V and stained with Coomassie blue R (Sigma-Aldrich®, St. Louis, MO, USA). MyHC isoforms were quantified by densitometry and identified based on predominant fiber types in studied muscles samples, as reported previously [48–50]. MyHC I and MyHC IIa isoforms were observed in the soleus muscle, and MyHC IIx and MyHC IIb isoforms were found in the gastrocnemius white portion. The relative quantity was expressed as a percentage of the total MyHC expression.

2.10. Statistical Analysis

The results were expressed as the mean ± standard deviations. The data distribution was analyzed using the Kolmogorov–Smirnov test. Student's t-test was used to compare creatine intake between Cr and Tcr groups. Other variables were evaluated using a two-way analysis of variance (Two-Way ANOVA). When significant differences were found ($p < 0.05$) post hoc, Tukey's multiple comparisons test was performed. The level of significance was 5%.

3. Results

3.1. Maximal Carrying Capacity

Before the training protocol, experimental groups exhibited similar performance during the maximal carrying capacity test (Cc 228 ± 36; Cr 225 ± 33; Tc 236 ± 19; Tcr 242 ± 25 g; $p > 0.05$). After training, Tc and Tcr presented a higher load-bearing capacity than their respective controls (Cc 465 ± 38; Cr 434 ± 48; Tc 1284 ± 79; Tcr 1285 ± 127 g; Figure 1).

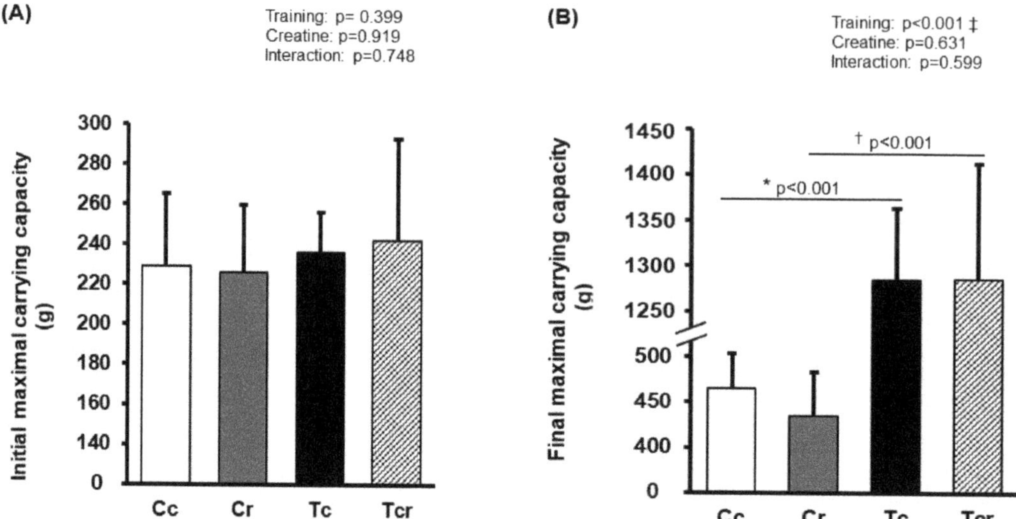

Figure 1. Maximum carrying capacity in climbing test before (**A**) and after (**B**) the training protocol. Cc (n = 7): sedentary control; Cr (n = 7): sedentary creatine supplementation; Tc (n = 7): resistance training; Tcr (n = 7): resistance training combined with creatine supplementation. Two-way ANOVA and Tukey test. ‡ $p < 0.05$, resistance training effect; * $p < 0.05$ vs. Cc; † $p < 0.05$ vs. Cr.

3.2. Nutritional and Serum Biochemical Data

The initial body weight was not different between the groups. At the end of the experimental protocol, exercise training intervention resulted in a lower final body weight; Tc exhibited reduced body weight when compared to the Cc and Tcr groups. The total caloric intake did not differ between groups. Feed efficiency was lower in Tc and Tcr than Cc and Cr, respectively, and was higher in Cr than Cc. Creatine intake did not differ between Cr and Tcr groups (Table 1).

Table 1. Nutritional and serum biochemical data.

	Variables	Cc	Cr	Tc	Tcr	Training	Creatine	Interaction
				Groups			**Factors (p-Value)**	
Nutritional	Initial body weight (g)	240 ± 22	242 ± 25	237 ± 20	251 ± 29	0.740	0.409	0.547
	Final body weight (g)	392 ± 34	407 ± 34	350 ± 71 *	390 ± 72 #	0.044 ‡	0.052	0.374
	Food intake (g)	2132 ± 150	2016 ± 82	2022 ± 160	1983 ± 169	0.162	0.130	0.444
	Creatine intake (g)	-	40.33 ± 1.64	-	39.67 ± 2.32	0.551	-	-
	Calorie intake (Kcal)	7699 ± 544	7421 ± 302	7301 ± 578	7325 ± 444	0.186	0.489	0.414
	Feed efficiency (mg/Kcal)	1.9 ± 0.1	2.2 ± 0.1 *	1.7 ± 0.2 *	1.9 ± 0.1 †	<0.001 ‡	0.007 §	0.437
Biochemical	Glucose (mg/dL)	134 ± 17	131 ± 24	133 ± 24	144 ± 21	0.492	0.629	0.410
	Total cholesterol (mg/dL)	71.3 ± 9.1	77.8 ± 16.6	78.6 ± 9.3	76.5 ± 6.4	0.481	0.600	0.309
	HDL cholesterol (mg/dL)	38.7 ± 6.2	45.1 ± 8.2	42.1 ± 4.3	42.4 ± 4.2	0.880	0.156	0.188
	Non-HDL (mg/dL)	33.9 ± 5.4	32.8 ± 11.4	36.2 ± 8.5	34.1 ± 2.8	0.537	0.602	0.858
	Triglycerides (mg/dL)	95.8 ± 7.7	98.9 ± 6.3	91.0 ± 3.4	94.2 ± 6.4	0.049 ‡	0.190	0.983
	Albumine (mg/dL)	2.30 ± 0.15	2.30 ± 0.12	2.27 ± 0.17	2.44 ± 0.11	0.291	0.119	0.119
	Total protein (mg/dL)	5.16 ± 2.24	5.93 ± 0.41	6.13 ± 0.54	6.46 ± 0.35	0.107	0.231	0.625

HDL: high-density lipoprotein cholesterol; Cc: sedentary control; Cr: creatine supplementation; Tc: resistance training; Tcr: resistance training and creatine supplementation. Mean ± SD. Two-way ANOVA and Tukey test. ‡ $p < 0.05$, resistance training effect; § $p < 0.05$, creatine effect; * $p < 0.05$ vs. Cc; # $p < 0.05$ vs. Tc; † $p < 0.05$ vs. Cr. Creatine intake results analyzed by Student's t-test ($p > 0.05$).

Relative to the serum biochemical data, although groups exhibited similar responses in terms of glycemia, cholesterol, albumin, and protein values, exercise training intervention was independently associated with lower triglyceride serum levels (Sedentary groups, 97.4 ± 1.7; Trained groups, 92.6 ± 1.7 mg/dL; $p < 0.05$, Table 1).

3.3. Morphological Characterization

The total gastrocnemius mass, in absolute values or normalized by the tibia length, did not differ between the groups. The gastrocnemius weight-to-body weight ratio was increased by resistance exercise training; Tc showed a higher gastrocnemius weight-to-body weight ratio than Cc. Soleus mass, in absolute values and in ratios with body weight and tibia length, did not differ between the groups. (Table 2).

Table 2. Macroscopic muscle morphology.

Variables	Groups				Factors (p-Value)		
	Cc	Cr	Tc	Tcr	Training	Creatine	Interaction
Gastrocnemius (mg)	2255 ± 172	2378 ± 106	2433 ± 572	2531 ± 463	0.262	0.451	0.931
Gastrocnemius/BW (mg/g)	5.76 ± 0.24	5.86 ± 0.34	6.92 ± 1.35 *	6.52 ± 1.30	0.020 ‡	0.676	0.498
Gastrocnemius/tibia (mg/mm)	52.9 ± 3.8	56.6 ± 1.6	59.0 ± 13.0	59.7 ± 11.0	0.181	0.512	0.646
Soleus (mg)	141 ± 19	146 ± 20	125 ± 26	153 ± 30	0.633	0.095	0.241
Soleus/BW (mg/g)	0.36 ± 0.03	0.36 ± 0.03	0.35 ± 0.05	0.39 ± 0.05	0.463	0.359	0.307
Soleus/tibia (mg/mm)	3.33 ± 0.47	3.50 ± 0.46	3.05 ± 0.60	3.60 ± 0.62	0.677	0.097	0.362

Cc: sedentary control; Cr: creatine supplementation; Tc: resistance training; Tcr: resistance training and creatine supplementation. Mean ± SD. Two-way ANOVA and Tukey test. ‡ $p < 0.05$, resistance training effect; * $p < 0.05$ vs. Cc.

Representative pictures of hematoxylin-eosin-stained muscle transverse sections are shown in Figure 2. The cross-sectional area of fibers from the gastrocnemius white portion was increased in response to resistance exercise training (Cc 3731 ± 415; Cr 3977 ± 422; Tr 4813 ± 982; Tcr 4273 ± 305 μm^2, Figure 2A). Tc exhibited a higher cross-sectional area than the Cc group. On the other hand, soleus cross-sectional areas did not differ between the groups (Cc 3997 ± 353; Cr 3910 ± 289; Tr 3642 ± 360; Tcr 4247 ± 736 μm^2, Figure 2B).

Picrosirius red-stained muscle transverse sections are shown in Figure 3. Collagen interstitial fraction was increased in response to resistance training in the soleus (Cc 9.97 ± 1.25; Cr 10.55 ± 0.98; Tr 13.85 ± 1.04; Tcr 12.92 ± 2.07%) and gastrocnemius white portion (Cc 4.50 ± 0.91; Cr 6.25 ± 0.86; Tr 9.05 ± 2.62; Tcr 10.18 ± 2.60%) muscles. Tr and Tcr presented a higher collagen interstitial fraction than their respective controls.

3.4. Myosin Heavy Chain Isoforms Distribution

Figures 4 and 5 show MyHC isoform expressions in the gastrocnemius white portion and soleus muscles, respectively. Gastrocnemius showed significant effects from exercise training alone ($p = 0.022$); MyHC IIx expression increased in response to resistance training (Cc 6.25 ± 2.68; Cr 8.88 ± 2.94; Tc 11.47 ± 3.73; Tcr 12.30 ± 6.58%). MyHC IIb expression was lower in the trained groups than in the controls (Cc 93.75 ± 2.68; Cr 91.12 ± 2.94; Tc 88,53 ± 3.73; Tcr 87.70 ± 6.58%).

The soleus muscle's MyHC IIa isoform expression was similar between the groups (Cc 9.43 ± 4.08; Cr 10.19 ± 4.43, Tc 7.87 ⊥ 3.50, Tcr 8.92 ⊥ 3.40%). Likewise, the MyHC I isoform was not affected by dietary intervention or resistance exercise training (Cc 90.57 ± 4.08; Cr 89.81 ± 4.43, Tc 92.13 ± 3.50, Tcr 91.08 ± 3.40%; Figure 5).

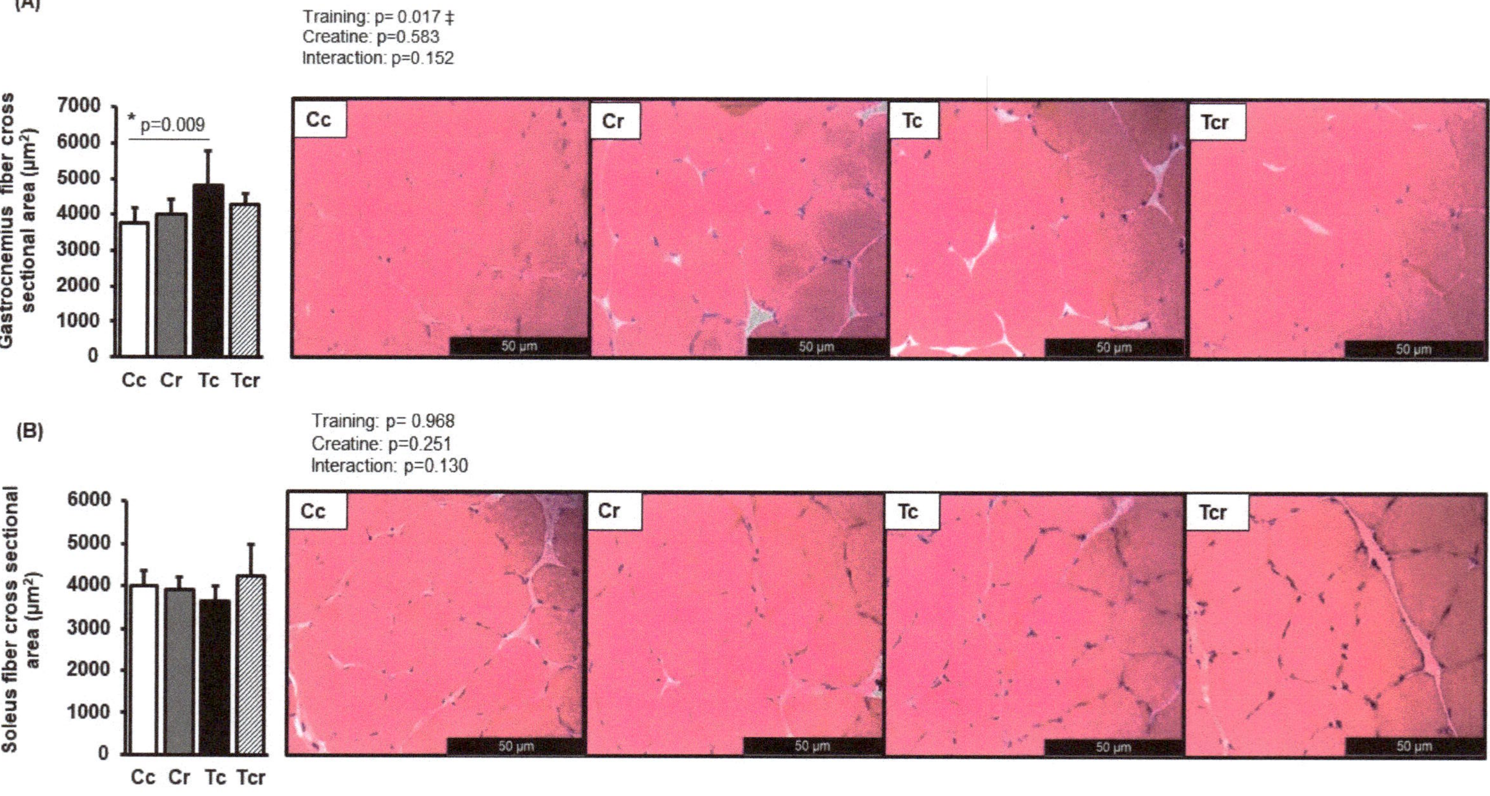

Figure 2. Cross-sectional area and representative transverse histological sections of the gastrocnemius (**A**) and soleus (**B**) muscles (400-fold magnification) stained with hematoxylin-eosin. Cc ($n = 7$): sedentary control; Cr ($n = 7$): sedentary creatine supplementation; Tc ($n = 7$): resistance training; Tcr ($n = 7$): resistance training combined with creatine supplementation. Two-way ANOVA and Tukey test. ‡ $p < 0.05$, resistance training effect; * $p < 0.05$ vs. Cc.

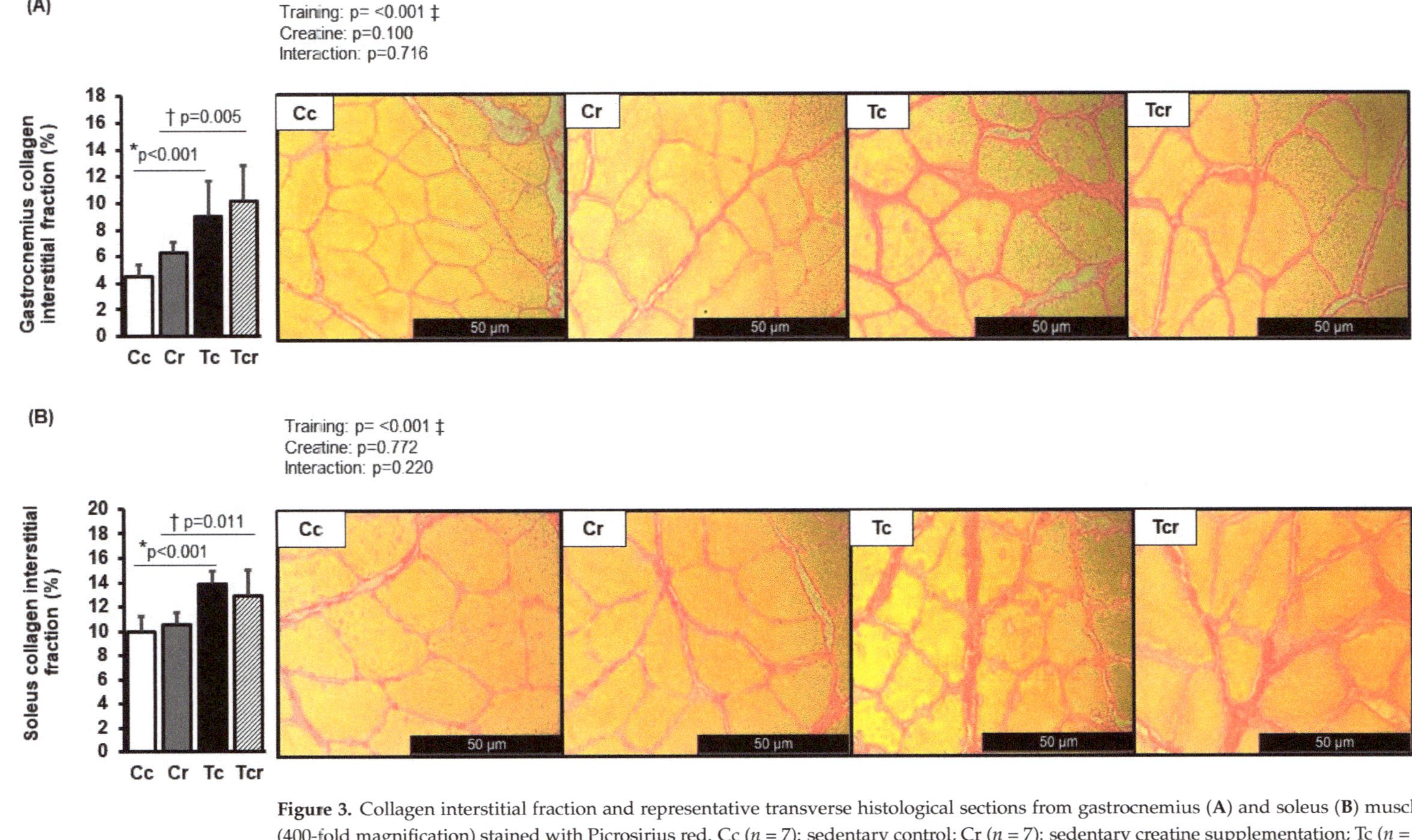

Figure 3. Collagen interstitial fraction and representative transverse histological sections from gastrocnemius (**A**) and soleus (**B**) muscles (400-fold magnification) stained with Picrosirius red. Cc ($n = 7$): sedentary control; Cr ($n = 7$): sedentary creatine supplementation; Tc ($n = 7$): resistance training; Tcr ($n = 7$): resistance training combined with creatine supplementation. Two-way ANOVA and Tukey test. ‡ $p < 0.05$, resistance training effect; * $p < 0.05$ vs. Cc; † $p < 0.05$ vs. Cr.

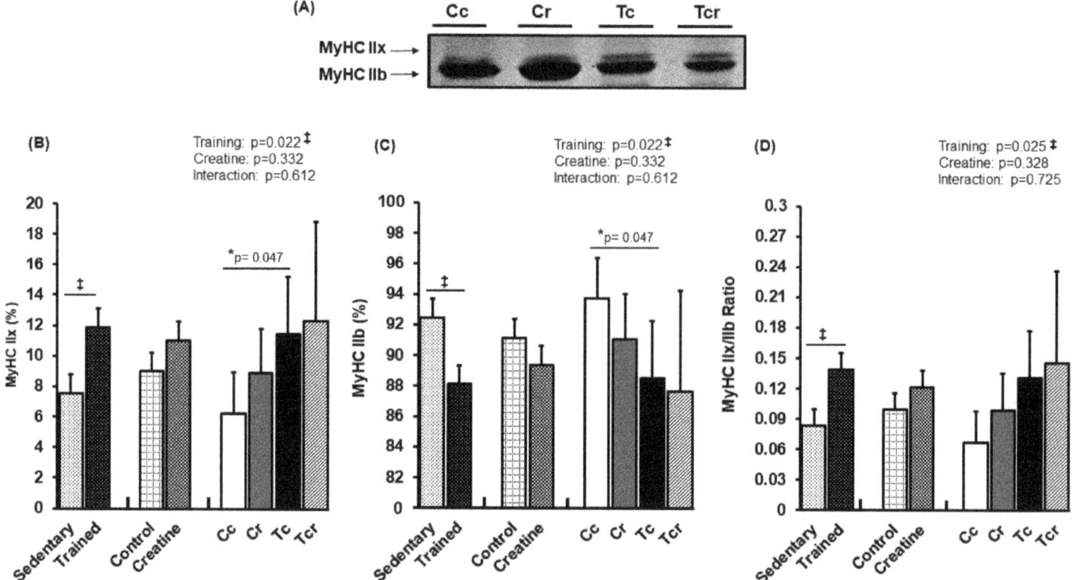

Figure 4. Representative bands of myosin heavy chain isoforms (MyHC IIx and MyHC IIb) (**A**); relative proportions (%) of MyHC IIx (**B**), and MyHC IIb (**C**) isoforms, and MyHC IIx/MyHC IIb ratio (**D**) in the gastrocnemius muscle. Isolated factors: Sedentary; Trained; Control; Creatine. Groups: Cc ($n = 6$): sedentary control; Cr ($n = 6$): sedentary creatine supplementation; Tc ($n = 6$): resistance training; Tcr ($n = 6$): resistance training combined with creatine supplementation. Two-way ANOVA and Tukey test. ‡ $p < 0.05$, resistance training effect; * $p < 0.05$ vs. Cc.

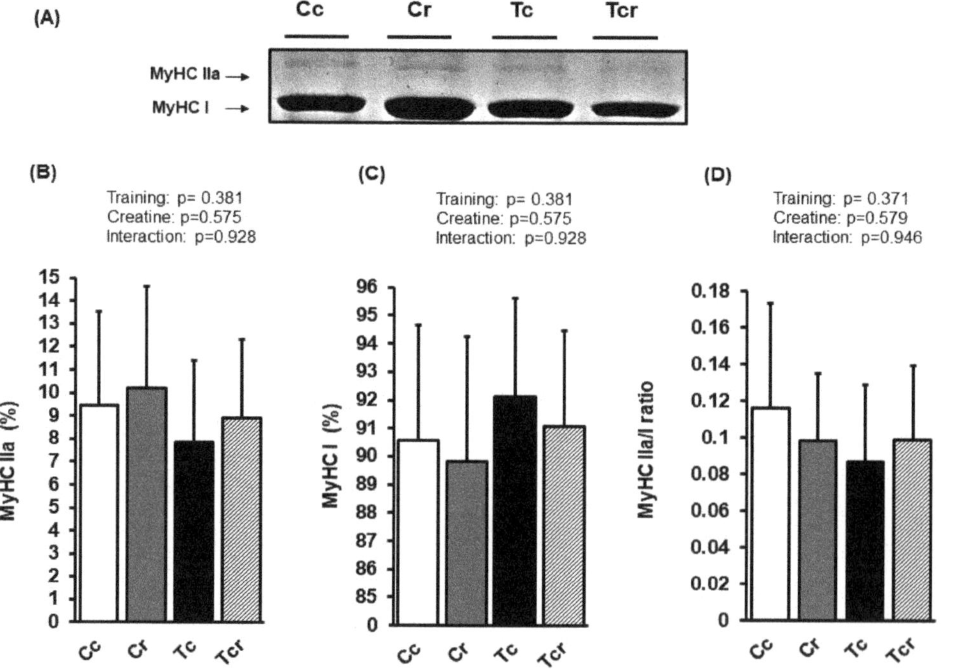

Figure 5. Representative bands of myosin heavy chain isoforms (MyHC IIa and MyHC I) (**A**); relative proportions (%) of MyHC IIa (**B**), and MyHC I (**C**) isoforms, and MyHC IIa/MyHC I ratio (**D**) in the

soleus muscle. Groups: Cc (n = 6): sedentary control; Cr (n = 6): sedentary creatine supplementation; Tc (n = 6): resistance training; Tcr (n = 6): resistance training combined with creatine supplementation. Two-way ANOVA (p > 0.05).

3.5. Protein Expression

Resistance exercise training increased follistatin expression in the gastrocnemius white portion (Figure 6A) and soleus muscles (Figure 7A). Gastrocnemius follistatin expression was higher in Tc than in the Cc and Tcr groups (Cc 0.45 ± 0.16; Cr 0.40 ± 0.17, Tc 0.82 ± 0.33, Tcr 0.52 ± 0.27 arbitrary units; Figure 6A). Both trained groups presented lower gastrocnemius myostatin levels than their respective controls (Cc 1.27 ± 0.32; Cr 1.32 ± 0.17, Tc 0.76 ± 0.22, Tcr 0.88 ± 0.13 arbitrary units; Figure 6B).

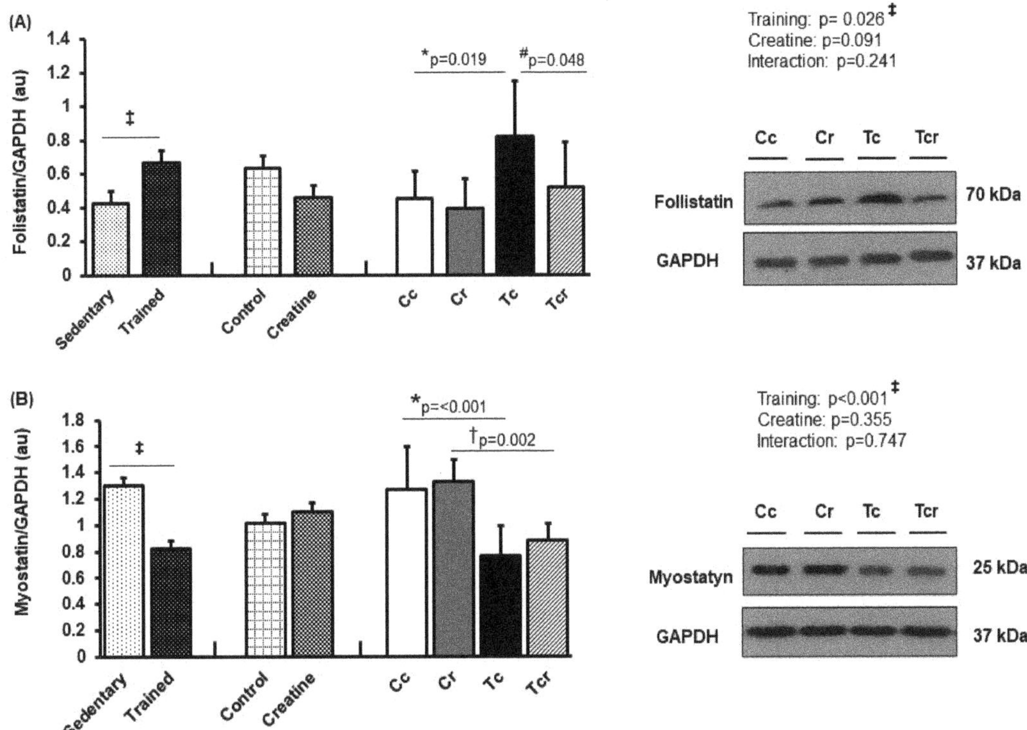

Figure 6. Protein levels and representative Western blots of follistatin (**A**) and myostatin (**B**) expression in the gastrocnemius muscle. Protein levels were normalized to glyceraldehyde-3-phosphate dehydrogenase (GAPDH). Isolated factors: Sedentary; Trained; Control; Creatine. Groups: Cc (n = 7): sedentary control; Cr (n = 7): sedentary creatine supplementation; Tc (n = 7): resistance training; Tcr (n = 7): resistance training combined with creatine supplementation. Two-way ANOVA and Tukey test. ‡ p < 0.05, resistance training effect; * p < 0.05 vs. Cc; † p < 0.05 vs. Cr; # p < 0.05 vs. Tc.

Soleus follistatin was higher in Tcr than Cr (Cc 0.23 ± 0.11; Cr 0.23 ± 0.04, Tc 0.34 ± 0.06, Tcr 0.42 ± 0.12 arbitrary units; Figure 7A), and myostatin expression did not differ between the groups (Cc 1.31 ± 0.27; Cr 0.96 ± 0.21, Tc 1.13 ± 0.38, Tcr 1.07 ± 0.17 arbitrary units; Figure 7B).

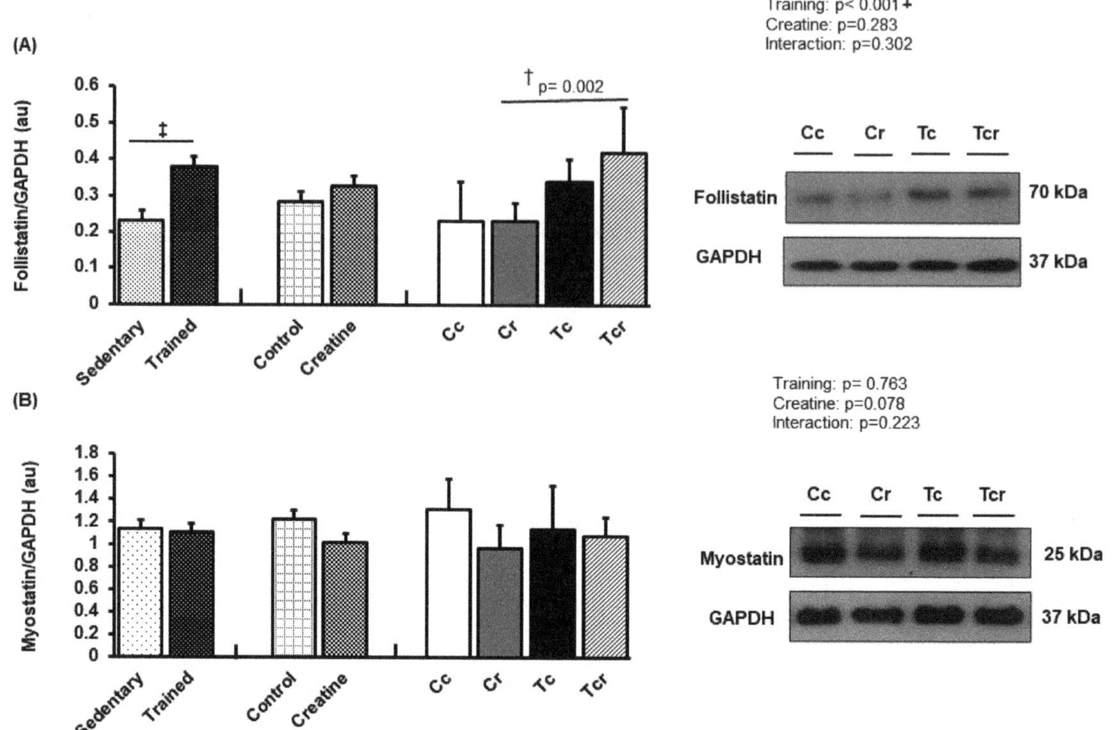

Figure 7. Protein levels and representative Western blots of follistatin (**A**) and myostatin (**B**) expression in the soleus muscle. Protein levels were normalized to glyceraldehyde-3-phosphate dehydrogenase (GAPDH). Isolated factors: Sedentary; Trained; Control; Creatine. Groups: Cc (n = 7): sedentary control; Cr (n = 7): sedentary creatine supplementation; Tc (n = 7): resistance training; Tcr (n = 7): resistance training combined with creatine supplementation. Two-way ANOVA and Tukey test. ‡ $p < 0.05$, resistance training effect; † $p < 0.05$ vs. Cr.

Resistance training alone reduced ActRIIB expression in the gastrocnemius white portion muscle (Cc 0.88 ± 0.17; Cr 0.74 ± 0.20, Tc 0.66 ± 0.17, Tcr 0.64 ± 0.14 arbitrary units; Figure 8A). Soleus ActRIIB expression did not differ between the groups (Cc 0.66 ± 0.22; Cr 0.62 ± 0.13, Tc 0.47 ± 0.21, Tcr 0.50 ± 0.29 arbitrary units; Figure 8B).

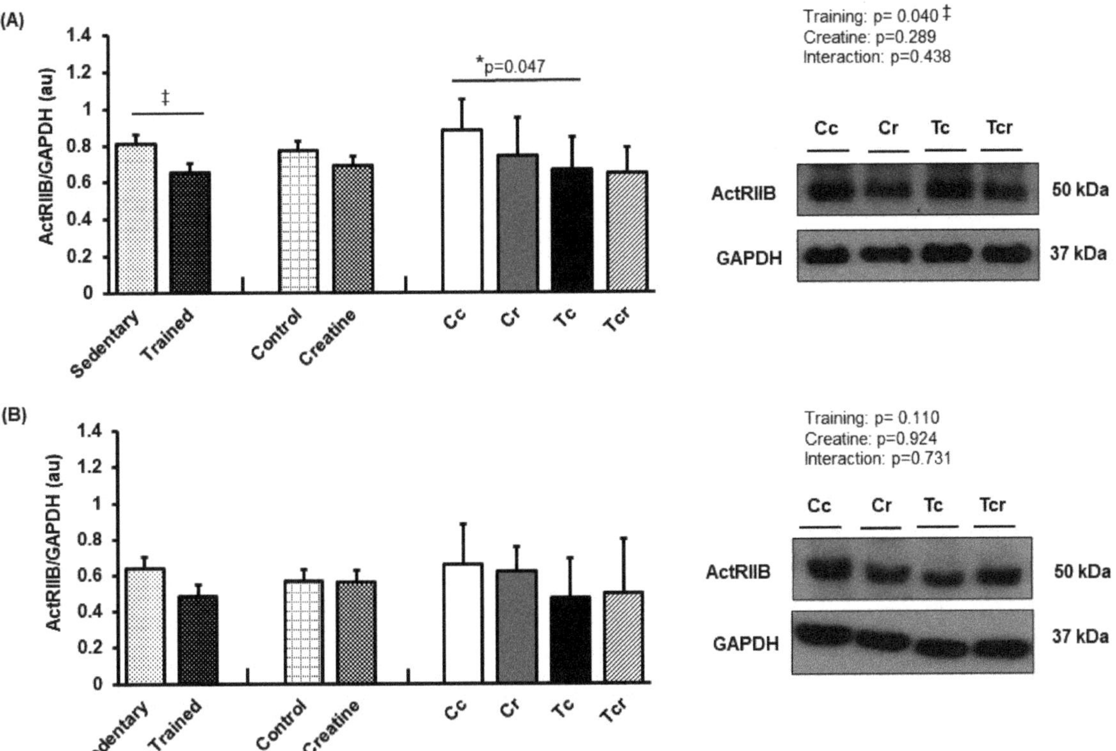

Figure 8. Protein levels and representative Western blots of ActRIIB in (**A**) gastrocnemius and (**B**) the soleus muscle. Protein levels were normalized to glyceraldehyde-3-phosphate dehydrogenase (GAPDH). Isolated factors: Sedentary; Trained; Control; Creatine. Groups: Cc ($n = 7$): sedentary control; Cr ($n = 7$): sedentary creatine supplementation; Tc ($n = 7$): resistance training; Tcr ($n = 7$): resistance training combined with creatine supplementation. Two-way ANOVA and Tukey test. ‡ $p < 0.05$, resistance training effect; * $p < 0.05$ vs. Cc.

4. Discussion

The current study aimed to evaluate the effects of creatine monohydrate supplementation when combined with resistance training on the soleus, a classic slow-twitch muscle, and gastrocnemius white portion, an important fast-twitch muscle type. In contrast to our hypothesis, creatine supplementation, both alone or combined with resistance training, had little impact on morphological and molecular features in both skeletal muscle types. On the other hand, the resistance exercise training intervention resulted in morphological and molecular adaptive effects in both skeletal muscles; these effects were more accentuated in the gastrocnemius than in the soleus muscle, partially confirming the primary hypothesis.

Indeed, resistance exercise training is a non-pharmacological intervention that, when performed continuously and safely, can result in metabolic benefits and promote physiological and structural adaptations that are associated with improved neuromuscular function and motor performance [51–55]. In this study, resistance training was associated with lower triglyceride serum levels and an increased maximal load-carrying capacity, confirming the primary premise as well as findings derived from animal models [56,57]. Moreover, the 12 weeks resistance training protocol increased the mass, cross-sectional area, collagen interstitial fraction, MyHC IIx proportion, and follistatin expression while reducing myostatin and ActRIIB protein levels in the gastrocnemius white portion muscle. Likewise, exercise training resulted in a greater collagen interstitial fraction in the soleus muscle. When combined with creatine supplementation, the ladder resistance training protocol reduced the follistatin level expression in the gastrocnemius muscle (Figure 9).

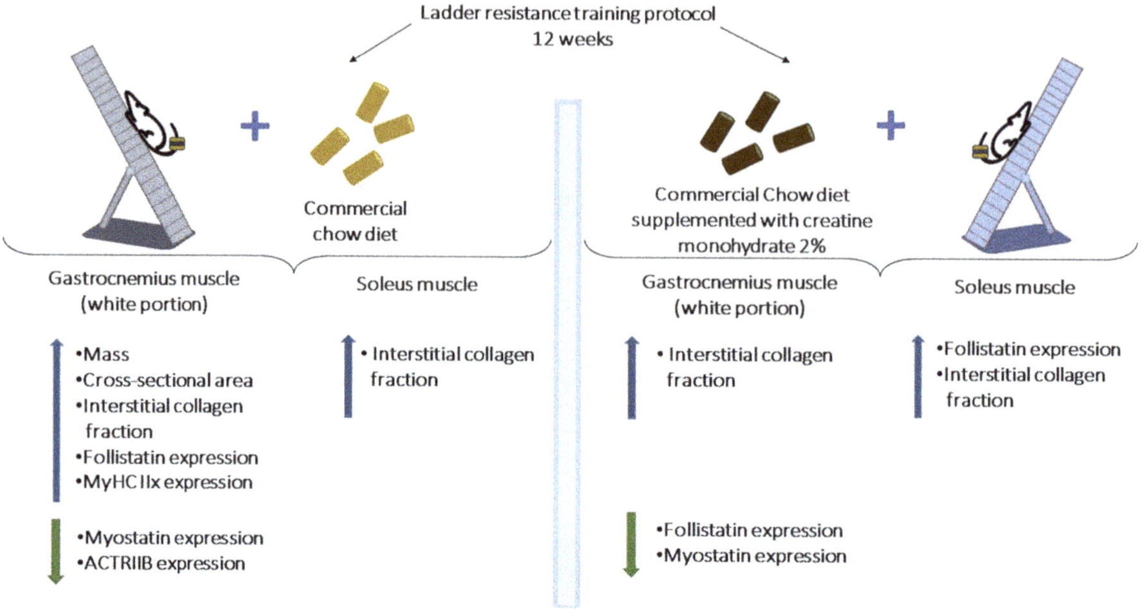

Figure 9. Differential effects of ladder resistance training and creatine monohydrate supplementation on gastrocnemius (white portion) and soleus muscles. Resistance training-induced skeletal muscle remodeling is greater in the fast-twitch than the slow-twitch muscle type.

Skeletal muscle is highly adaptable to increased muscle load, presenting compensatory hypertrophy with improved strength [51,58,59]. In accordance with our results, Ribeiro et al. [60] observed that resistance training on a ladder (3×/week for 12 weeks) induced hypertrophy that was restricted to the gastrocnemius muscle in three-month-old rats; the cross-sectional area in the soleus muscle was unaffected by the intervention. Based on a similar protocol, Deschenes et al. [61] failed to show a significant increase in the cross-sectional area of soleus muscle fibers after resistance training for six weeks. By contrast, rodents that submitted to an alternative training protocol (16 weeks; 5×/week) exhibited a greater soleus cross-sectional area [62]. This is perhaps because soleus remodeling can be associated with greater volume and duration protocols. It is noteworthy that resistance training is predominantly associated with fast-twitch Type II fibers hypertrophy in the skeletal muscle [51,63–65]. Type II muscle fibers respond preferentially to intense training protocols, while Type I muscles are more susceptible to high-volume exercise [66]. Besides these morphological effects, the muscle recruitment pattern in response to exercise training may also be associated with different MyHC stimuli. Accordingly, resistance training preferentially impacted the MyHC IIx proportion in gastrocnemius white portion muscles with no changes in the soleus muscle composition (Figures 4 and 5).

Additionally, resistance training promoted extensive interstitial remodeling in both skeletal muscle types. Recently, Braggion et al. [67] also observed that training performed on a ladder (3×/week for 12 weeks) increased collagen fibers in the soleus skeletal muscle of ovariectomized rats. The upregulation of collagen turnover due to prolonged training could be caused by the increased activity of fibroblasts involved in muscle hypertrophy and regeneration [67,68]. This change may provide mechanical support to muscle fibers, as resistance training promotes a high contractile demand to multiple skeletal muscles, despite different histological, biochemical, and metabolic properties [51].

Considering these molecular mechanisms, resistance training reduced myostatin expression in the gastrocnemius white portion. Myostatin, also known as growth/differentiation factor-8 (GDF-8), is a TGF-β family member [21]. While myostatin overexpression reduces muscle mass, myostatin null animals have an increased muscle mass [69,70]. This effect has been observed with natural myostatin mutations in animals and humans [71–74]. Supporting the findings of our study, myostatin mRNA expression decreased in elderly women after resistance training [75]. Hayashi et al. [76] described how increased myostatin was associated with the reduced expression of the fast MyHC IIx isoform in cattle. Likewise, Ribeiro et al. [60] observed reduced myostatin levels in male rat gastrocnemius in response to 12-week ladder resistance training. Relative to follistatin expression, the results of this study confirm previous evidence [77,78]; resistance exercise training positively modulates follistatin and negatively modulates myostatin and ActRIIB, which are associated with gastrocnemius hypertrophy. In addition to inhibiting myostatin binding to its receptor, follistatin may have anabolic activity, which depends on insulin-like growth factor-1 or insulin [79]. In this context, creatine supplementation could improve the activity of components in the insulin signaling pathway [80]. This possibility may have supported the higher follistatin expression in the soleus skeletal muscle (Figure 7) of trained rats that were creatine supplemented.

In general, exercise resistance training effects are frequently accentuated by creatine supplementation [4]. In previous studies, creatine supplementation promoted an increase in muscle mass during resistance training with a progressive overload increase [12,81]. However, Cooke et al. [82] observed that 12 weeks of creatine supplementation did not modify the lean mass, muscle strength, total myofibrillar protein content, and/or muscle fiber cross-sectional area in men undergoing resistance training. Similarly, creatine supplementation did not promote any direct anabolic effect on the skeletal muscles of rats that were submitted to 5 weeks of intense jumping resistance training [48]. The authors argued for the possible benefits of creatine supplementation in terms of hypertrophic mechanisms activation and additional muscle mass gain in trained rats, which was dependent on a higher workload [48].

Based on the principle of specificity, physiological adaptations due to resistance exercise training are dependent on specific stress characteristics [83]. From this perspective, potential effects depend on a combination of training variables and, generally, are associated with metabolic stress due to glycolysis overload [52,84]. Although creatine monohydrate supplementation could contribute to improvements in glycolysis, it is predominantly beneficial to increase muscle mass through the creatine and phosphocreatine systems [83]. Thus, any anabolic effect from creatine ingestion could be more pronounced when associated with ladder resistance training protocols based on greater workloads resulting from increases in rest periods as well as the number of sets within a session and/or frequency of workouts per week. Therefore, a mismatch between exercise training demand and nutritional supplementation may not be discarded as a possible background for the little effect of creatine consumption in this study. Further investigation is needed to clarify the potential creatine-induced molecular mechanisms modulating the skeletal muscle phenotype.

Moreover, although lactate concentration is commonly used to determine resistance training intensity, this variable was not evaluated to determine functional capacity, which can be considered an additional limitation in this study.

5. Conclusions

Resistance exercise training induces skeletal muscle hypertrophy and interstitial remodeling in rats, which is more evident in the gastrocnemius muscle white portion than in the soleus muscle. This adaptive response can be associated with the myostatin signaling pathway modulation and increased MyHC IIx expression and is not affected by creatine supplementation.

Author Contributions: Conceptualization, M.R.d.C., P.F.M. and S.A.d.O.-J.; methodology, M.R.d.C., P.F.M., K.O., M.P.O. and S.A.d.O.-J.; software, P.F.M. and S.A.d.O.-J.; validation, P.F.M., K.O., M.P.O. and S.A.d.O.-J.; formal analysis, M.R.d.C., P.F.M. and W.F.d.O.F.; investigation, M.R.d.C., E.F.D., M.L.M.M., C.S.d.M., G.E.O., J.J.G.-J. and F.C.D.; resources, K.O., M.P.O. and S.A.d.O.-J.; data curation, M.R.d.C. and E.F.D.; writing—original draft preparation, M.R.d.C. and S.A.d.O.-J.; writing—review and editing, K.O., M.P.O., R.J.O., P.F.M. and S.A.d.O.-J.; visualization, M.R.d.C. and S.A.d.O.-J.; supervision, S.A.d.O.-J.; project administration, P.F.M., R.J.O. and S.A.d.O.-J.; funding acquisition, S.A.d.O.-J. All authors have read and agreed to the published version of the manuscript.

Funding: This research was funded by the Federal University of Mato Grosso do Sul (UFMS), the National Council for Scientific and Technological Development (CNPq; proc. 420495/2016-9), the Foundation to Support the Development of Teaching, Science and Technology of Mato Grosso do Sul (FUNDECT), and the Foundation for Coordinating the Development of Postgraduate Personnel (CAPES; Financial Code 001). The supporters had no role in the study design, data collection, analysis, the decision to publish, or the preparation of the manuscript.

Institutional Review Board Statement: All procedures and the experimental protocol were approved by our institution's Animal Ethics Committee (Protocol Number 873/2017) in accordance with the Brazilian College of Animal Experimentation (COBEA) and the US National Institutes of Health "Guide for the Care and Use of Laboratory Animals" (NIH, 2011).

Informed Consent Statement: Not applicable.

Data Availability Statement: All data that support the findings of this study are available from the corresponding author on reasonable request.

Conflicts of Interest: The authors declare no conflict of interest.

References

1. Butts, J.; Jacobs, B.; Silvis, M. Creatine Use in Sports. *Sport. Health Multidiscip. Approach* **2017**, *10*, 31–34. [CrossRef]
2. Terjung, R.L.; Clarkson, P.; Eichner, E.R.; Greenhaff, P.L.; Hespel, P.J.; Israel, R.G.; Kraemer, W.J.; Meyer, R.A.; Spriet, L.L.; Tarnopolsky, M.A.; et al. Physiological and Health Effects of Oral Creatine Supplementation. *Med. Sci. Sports Exerc.* **2000**, *32*, 706–717. [CrossRef] [PubMed]
3. Wax, B.; Kerksick, C.M.; Jagim, A.R.; Mayo, J.J.; Lyons, B.C.; Kreider, R.B. Creatine for Exercise and Sports Performance, with Recovery Considerations for Healthy Populations. *Nutrients* **2021**, *13*, 1915. [CrossRef] [PubMed]
4. Pinder, M.A.; Myrie, S.B. Creatine Supplementation and Skeletal Muscle Metabolism for Building Muscle Mass-Review of the Potential Mechanisms of Action. *Curr. Protein Pept. Sci.* **2017**, *18*, 1273–1287. [CrossRef]
5. Verdijk, L.B.; Gleeson, B.G.; Jonkers, R.A.M.; Meijer, K.; Savelberg, H.H.C.M.; Dendale, P.; van Loon, L.J. Skeletal Muscle Hypertrophy Following Resistance Training Is Accompanied by a Fiber Type-Specific Increase in Satellite Cell Content in Elderly Men. *J. Gerontol. Ser. A Biol. Sci. Med. Sci.* **2009**, *64*, 332–339. [CrossRef]
6. Mitchell, C.J.; Churchward-Venne, T.A.; Bellamy, L.; Parise, G.; Baker, S.K.; Phillips, S.M. Muscular and Systemic Correlates of Resistance Training-Induced Muscle Hypertrophy. *PLoS ONE* **2013**, *8*, e78636. [CrossRef]
7. Chilibeck, P.D.; Kaviani, M.; Candow, D.G.; Zello, G.A. Effect of creatine supplementation during resistance training on lean tissue mass and muscular strength in older adults: A meta-analysis. *Open Access J. Sport. Med.* **2017**, *8*, 213–226. [CrossRef]
8. Bonilla, D.A.; Kreider, R.B.; Petro, J.L.; Romance, R.; García-Sillero, M.; Benítez-Porres, J.; Vargas-Molina, S. Creatine Enhances the Effects of Cluster-Set Resistance Training on Lower-Limb Body Composition and Strength in Resistance-Trained Men: A Pilot Study. *Nutrients* **2021**, *13*, 2303. [CrossRef]
9. Nouri, H.; Sheikholeslami-Vatani, D.; Moloudi, M.R. Changes in UPR-PERK pathway and muscle hypertrophy following resistance training and creatine supplementation in rats. *J. Physiol. Biochem.* **2021**, *77*, 331–339. [CrossRef]
10. Pinto, C.L.; Botelho, P.B.; Carneiro, J.A.; Mota, J.F. Impact of creatine supplementation in combination with resistance training on lean mass in the elderly. *J. Cachex-Sarcopenia Muscle* **2016**, *7*, 413–421. [CrossRef]
11. Aguiar, A.F.; Januário, R.S.B.; Junior, R.P.; Gerage, A.; Pina, F.L.C.; Nascimento, M.A.D.; Padovani, C.R.; Cyrino, E. Long-term creatine supplementation improves muscular performance during resistance training in older women. *Eur. J. Appl. Physiol.* **2012**, *113*, 987–996. [CrossRef] [PubMed]
12. Willoughby, D.S.; Rosene, J. Effects of oral creatine and resistance training on myosin heavy chain expression. *Med. Sci. Sport. Exerc.* **2001**, *33*, 1674–1681. [CrossRef] [PubMed]
13. Taes, Y.E.; Speeckaert, M.; Bauwens, E.; De Buyzere, M.R.; Libbrecht, J.; Lameire, N.H.; Delanghe, J.R. Effect of Dietary Creatine on Skeletal Muscle Myosin Heavy Chain Isoform Expression in an Animal Model of Uremia. *Nephron Exp. Nephrol.* **2004**, *96*, e103–e110. [CrossRef] [PubMed]

14. Gallo, M.; Gordon, T.; Syrotuik, D.; Shu, Y.; Tyreman, N.; MacLean, I.; Kenwell, Z.; Putman, C.T. Effects of long-term creatine feeding and running on isometric functional measures and myosin heavy chain content of rat skeletal muscles. *Pflug. Arch. Eur. J. Physiol.* **2006**, *452*, 744–755. [CrossRef]

15. Gallo, M.; MacLean, I.; Tyreman, N.; Martins, K.J.B.; Syrotuik, D.; Gordon, T.; Putman, C.T. Adaptive responses to creatine loading and exercise in fast-twitch rat skeletal muscle. *Am. J. Physiol. Integr. Comp. Physiol.* **2008**, *294*, R1319–R1328. [CrossRef]

16. Aguiar, A.F.; Aguiar, D.H.; Felisberto, A.D.; Carani, F.R.; Milanezi, R.C.; Padovani, C.R.; Dal-Pai-Silva, M. Effects of Creatine Supplementation During Resistance Training on Myosin Heavy Chain (MHC) Expression in Rat Skeletal Muscle Fibers. *J. Strength Cond. Res.* **2010**, *24*, 88–96. [CrossRef]

17. Staron, R.S.; Kraemer, W.J.; Hikida, R.S.; Fry, A.C.; Murray, J.D.; Campos, G.E.R. Fiber type composition of four hindlimb muscles of adult Fisher 344 rats. *Histochem.* **1999**, *111*, 117–123. [CrossRef]

18. Bloemberg, D.; Quadrilatero, J. Rapid Determination of Myosin Heavy Chain Expression in Rat, Mouse, and Human Skeletal Muscle Using Multicolor Immunofluorescence Analysis. *PLoS ONE* **2012**, *7*, e35273. [CrossRef]

19. Wigmore, P.M.; Evans, D.J. Molecular and cellular mechanisms involved in the generation of fiber diversity during myogenesis. *Int. Rev. Cytol.* **2002**, *216*, 175–232. [CrossRef]

20. Schiaffino, S.; Reggiani, C. Fiber Types in Mammalian Skeletal Muscles. *Physiol. Rev.* **2011**, *91*, 1447–1531. [CrossRef]

21. Elkina, Y.; Von Haehling, S.; Anker, S.D.; Springer, J. The role of myostatin in muscle wasting: An overview. *J. Cachexia Sarcopenia Muscle* **2011**, *2*, 143–151. [CrossRef] [PubMed]

22. Chen, M.-M.; Zhao, Y.-P.; Zhao, Y.; Deng, S.-L.; Yu, K. Regulation of Myostatin on the Growth and Development of Skeletal Muscle. *Front. Cell Dev. Biol.* **2021**, *9*, 785712. [CrossRef] [PubMed]

23. Artaza, J.N.; Bhasin, S.; Mallidis, C.; Taylor, W.; Ma, K.; Gonzalez-Cadavid, N.F. Endogenous expression and localization of myostatin and its relation to myosin heavy chain distribution in C2C12 skeletal muscle cells. *J. Cell. Physiol.* **2002**, *190*, 170–179. [CrossRef] [PubMed]

24. Hennebry, A.; Berry, C.; Siriett, V.; O'Callaghan, P.; Chau, L.; Watson, T.; Sharma, M.; Kambadur, R. Myostatin regulates fiber-type composition of skeletal muscle by regulating MEF2 and MyoD gene expression. *Am. J. Physiol.-Cell Physiol.* **2009**, *296*, C525–C534. [CrossRef]

25. Wang, M.; Yu, H.; Kim, Y.S.; Bidwell, C.A.; Kuang, S. Myostatin facilitates slow and inhibits fast myosin heavy chain expression during myogenic differentiation. *Biochem. Biophys. Res. Commun.* **2012**, *426*, 83–88. [CrossRef] [PubMed]

26. Xing, X.-X.; Xuan, M.-F.; Jin, L.; Guo, Q.; Luo, Z.-B.; Wang, J.-X.; Luo, Q.-R.; Zhang, G.-L.; Cui, C.-D.; Cui, Z.-Y.; et al. Fiber-type distribution and expression of myosin heavy chain isoforms in newborn heterozygous myostatin-knockout pigs. *Biotechnol. Lett.* **2017**, *39*, 1811–1819. [CrossRef] [PubMed]

27. Han, X.; Møller, L.L.V.; De Groote, E.; Bojsen-Møller, K.N.; Davey, J.; Henríquez-Olguin, C.; Li, Z.; Knudsen, J.R.; Jensen, T.E.; Madsbad, S.; et al. Mechanisms involved in follistatin-induced hypertrophy and increased insulin action in skeletal muscle. *J. Cachex-Sarcopenia Muscle* **2019**, *10*, 1241–1257. [CrossRef] [PubMed]

28. Domin, R.; Dadej, D.; Pytka, M.; Zybek-Kocik, A.; Ruchała, M.; Guzik, P. Effect of Various Exercise Regimens on Selected Exercise-Induced Cytokines in Healthy People. *Int. J. Environ. Res. Public Health* **2021**, *18*, 1261. [CrossRef]

29. Saremi, A.; Gharakhanloo, R.; Sharghi, S.; Gharaati, M.; Larijani, B.; Omidfar, K. Effects of oral creatine and resistance training on serum myostatin and GASP-1. *Mol. Cell. Endocrinol.* **2010**, *317*, 25–30. [CrossRef]

30. Mobley, C.B.; Fox, C.D.; Ferguson, B.S.; Amin, R.H.; Dalbo, V.J.; Baier, S.; Rathmacher, J.A.; Wilson, J.M.; Roberts, M.D. L-leucine, beta-hydroxy-beta-methylbutyric acid (HMB) and creatine monohydrate prevent myostatin-induced Akirin-1/Mighty mRNA down-regulation and myotube atrophy. *J. Int. Soc. Sport. Nutr.* **2014**, *11*, 38. [CrossRef]

31. Young, J.; Bertram, H.; Theil, P.; Petersen, A.-G.; Poulsen, K.; Rasmussen, M.; Malmendal, A.; Nielsen, N.; Vestergaard, M.; Oksbjerg, N. In vitro and in vivo studies of creatine monohydrate supplementation to Duroc and Landrace pigs. *Meat Sci.* **2007**, *76*, 342–351. [CrossRef] [PubMed]

32. Chen, J.; Wang, M.; Kong, Y.; Ma, H.; Zou, S. Comparison of the novel compounds creatine and pyruvateon lipid and protein metabolism in broiler chickens. *Animal* **2011**, *5*, 1082–1089. [CrossRef] [PubMed]

33. Rooney, K.; Bryson, J.; Phuyal, J.; Denyer, G.; Caterson, I.; Thompson, C. Creatine supplementation alters insulin secretion and glucose homeostasis in vivo. *Metab. Clin. Exp.* **2002**, *51*, 518–522. [CrossRef] [PubMed]

34. Ju, J.-S.; Smith, J.L.; Oppelt, P.J.; Fisher, J.S. Creatine feeding increases GLUT4 expression in rat skeletal muscle. *Am. J. Physiol. Endocrinol. Metab.* **2005**, *288*, E347–E352. [CrossRef]

35. Hornberger, T.A., Jr.; Farrar, R.P. Physiological Hypertrophy of the FHL Muscle Following 8 Weeks of Progressive Resistance Exercise in the Rat. *Can. J. Appl. Physiol.* **2004**, *29*, 16–31. [CrossRef]

36. Medeiros, C.S.; Neto, I.V.d.S.; Silva, K.K.S.; Cantuária, A.P.C.; Rezende, T.M.B.; Franco, O.L.; Marqueti, R.d.C.; Freitas-Lima, L.C.; Araujo, R.C.; Yildirim, A.; et al. The Effects of High-Protein Diet and Resistance Training on Glucose Control and Inflammatory Profile of Visceral Adipose Tissue in Rats. *Nutrients* **2021**, *13*, 1969. [CrossRef]

37. Tibana, R.A.; Franco, O.L.; Cunha, G.V.; Sousa, N.M.F.; Neto, I.V.S.; Carvalho, M.M.; Almeida, J.A.; Durigan, J.L.Q.; Marqueti, R.C.; Navalta, J.W.; et al. The Effects of Resistance Training Volume on Skeletal Muscle Proteome. *Int. J. Exerc. Sci.* **2017**, *10*, 1051–1066.

38. De Sousa Neto, I.V.; Tibana, R.A.; da Silva, L.G.O.; de Lira, E.M.; do Prado, G.P.G.; de Almeida, J.A.; Franco, O.L.; Durigan, J.L.Q.; Adesida, A.B.; de Sousa, M.V.; et al. Paternal Resistance Training Modulates Calcaneal Tendon Proteome in the Offspring Exposed to High-Fat Diet. *Front. Cell Dev. Biol.* **2020**, *8*, 380. [CrossRef]

39. De Sousa Neto, I.V.; Durigan, J.L.Q.; Guzzoni, V.; Tibana, R.A.; Prestes, J.; de Araujo, H.S.S.; Marqueti, R.C. Effects of Resistance Training on Matrix Metalloproteinase Activity in Skeletal Muscles and Blood Circulation during Aging. *Front. Physiol.* **2018**, *9*, 190. [CrossRef]
40. Oliveira, S.A., Jr.; Pai-Silva, M.D.; Martinez, P.F.; Lima-Leopoldo, A.P.; Campos, D.H.; Leopoldo, A.S.; Politi, O.M.; Okoshi, M.P.; Okoshi, K.; Roberto, P.C.; et al. Diet-induced obesity causes metabolic, endocrine and cardiac alterations in spontaneously hypertensive rats. *Med. Sci. Monit.* **2010**, *16*, 367–373.
41. Carvalho, M.R.; Mendonça, M.L.M.; Oliveira, J.M.; Romanenghi, R.B.; Morais, C.S.; Ota, G.E.; Lima, A.R.; Oliveira, R.J.; Filiú, W.F.; Okoshi, K.; et al. Influence of high-intensity interval training and intermittent fasting on myocardium apoptosis pathway and cardiac morphology of healthy rats. *Life Sci.* **2020**, *264*, 118697. [CrossRef] [PubMed]
42. Basilio, P.G.; De Oliveira, A.P.C.; De Castro, A.C.F.; De Carvalho, M.R.; Zagatto, A.M.; Martinez, P.F.; Okoshi, M.P.; Okoshi, K.; Ota, G.E.; Dos Reis, F.A.; et al. Intermittent fasting attenuates exercise training-induced cardiac remodeling. *Arq. Bras. Cardiol.* **2020**, *115*, 184–193. [CrossRef] [PubMed]
43. Martinez, P.F.; Bonomo, C.; Guizoni, D.M.; Oliveira-Junior, S.A.; Damatto, R.L.; Cezar, M.D.; Lima, A.R.; Pagan, L.U.; Seiva, F.; Bueno, R.T.; et al. Modulation of MAPK and NF-κB Signaling Pathways by Antioxidant Therapy in Skeletal Muscle of Heart Failure Rats. *Cell. Physiol. Biochem.* **2016**, *39*, 371–384. [CrossRef] [PubMed]
44. Oliveira-Junior, S.A.; Martinez, P.F.; Guizoni, D.M.; Campos, D.H.S.; Fernandes, T.; Oliveira, E.M.; Okoshi, M.P.; Okoshi, K.; Padovani, C.R.; Cicogna, A.C. AT1 Receptor Blockade Attenuates Insulin Resistance and Myocardial Remodeling in Rats with Diet-Induced Obesity. *PLoS ONE* **2014**, *9*, e86447. [CrossRef] [PubMed]
45. Bradford, M.M. A Rapid and sensitive method for the quantitation of microgram quantities of protein utilizing the principle of protein-dye binding. *Anal. Biochem.* **1976**, *72*, 248–254. [CrossRef]
46. Martinez, P.F.; Okoshi, K.; Zornoff, L.A.M.; Carvalho, R.F.; Oliveira, S.A., Jr.; Lima, A.R.R.; Campos, D.H.S.; Damatto, R.L.; Padovani, C.R.; Nogueira, C.R.; et al. Chronic heart failure-induced skeletal muscle atrophy, necrosis, and changes in myogenic regulatiory factors. *Med. Sci. Monit.* **2010**, *16*, 374–383.
47. Lima, A.R.R.; Martinez, P.F.; Damatto, R.L.; Cezar, M.D.M.; Guizoni, D.M.; Bonomo, C.; Oliveira-Junior, S.A.; Silva, M.D.-P.; Zornoff, L.A.M.; Okoshi, K.; et al. Heart Failure-Induced Diaphragm Myopathy. *Cell. Physiol. Biochem.* **2014**, *34*, 333–345. [CrossRef]
48. Aguiar, A.F.; De Souza, R.W.A.; Aguiar, D.H.; Aguiar, R.C.M.; Vechetti, I.J., Jr.; Dal-Pai-Silva, M. Creatine does not promote hypertrophy in skeletal muscle in supplemented compared with nonsupplemented rats subjected to a similar workload. *Nutr. Res.* **2011**, *31*, 652–657. [CrossRef]
49. Damatto, R.; Martinez, P.; Lima, A.; Cezar, M.; Campos, D.; Junior, S.O.; Guizoni, D.; Bonomo, C.; Nakatani, B.; Silva, M.D.P.; et al. Heart failure-induced skeletal myopathy in spontaneously hypertensive rats. *Int. J. Cardiol.* **2013**, *167*, 698–703. [CrossRef]
50. Pereira, J.A.S.A.; De Haan, A.; Wessels, A.; Moorman, A.F.; Sargeant, A.J. The mATPase histochemical profile of rat type IIX fibres: Correlation with myosin heavy chain immunolabelling. *Histochem. J.* **1995**, *27*, 715–722. [CrossRef]
51. Qaisar, R.; Bhaskaran, S.; Van Remmen, H. Muscle fiber type diversification during exercise and regeneration. *Free. Radic. Biol. Med.* **2016**, *98*, 56–67. [CrossRef] [PubMed]
52. Kraemer, W.J.; Ratamess, N.A. Fundamentals of Resistance Training: Progression and Exercise Prescription. *Med. Sci. Sports Exerc.* **2004**, *36*, 674–688. [CrossRef] [PubMed]
53. Schoenfeld, B.J.; Grgic, J.; Ogborn, D.; Krieger, J.W. Strength and Hypertrophy Adaptations Between Low- vs. High-Load Resistance Training: A Systematic Review and Meta-analysis. *J. Strength Cond. Res.* **2017**, *31*, 3508–3523. [CrossRef] [PubMed]
54. Carvalho, L.; Junior, R.M.; Barreira, J.; Schoenfeld, B.J.; Orazem, J.; Barroso, R. Muscle hypertrophy and strength gains after resistance training with different volume-matched loads: A systematic review and meta-analysis. *Appl. Physiol. Nutr. Metab.* **2022**, *47*, 357–368. [CrossRef]
55. Škarabot, J.; Brownstein, C.G.; Casolo, A.; Del Vecchio, A.; Ansdell, P. The knowns and unknowns of neural adaptations to resistance training. *Eur. J. Appl. Physiol.* **2020**, *121*, 675–685. [CrossRef] [PubMed]
56. Stotzer, U.S.; Pisani, G.F.D.; Canevazzi, G.H.R.; Shiguemoto, G.E.; Duarte, A.C.G.D.O.; Perez, S.E.D.A.; Selistre-De-Araújo, H.S. Benefits of resistance training on body composition and glucose clearance are inhibited by long-term low carbohydrate diet in rats. *PLoS ONE* **2018**, *13*, e0207951. [CrossRef]
57. Bae, J.Y. Resistance Exercise Regulates Hepatic Lipolytic Factors as Effective as Aerobic Exercise in Obese Mice. *Int. J. Environ. Res. Public Health* **2020**, *17*, 8307. [CrossRef]
58. Folland, J.P.; Williams, A.G. The adaptations to strength training morphological and neurological contributions to increased strength. *Sport. Med.* **2007**, *37*, 145–168. [CrossRef]
59. Wackerhage, H.; Schoenfeld, B.J.; Hamilton, D.; Lehti, M.; Hulmi, J.J. Stimuli and sensors that initiate skeletal muscle hypertrophy following resistance exercise. *J. Appl. Physiol.* **2019**, *126*, 30–43. [CrossRef]
60. Ribeiro, M.B.T.; Guzzoni, V.; Hord, J.M.; Lopes, G.N.; Marqueti, R.D.C.; de Andrade, R.V.; Selistre-De-Araujo, H.S.; Durigan, J.L.Q. Resistance training regulates gene expression of molecules associated with intramyocellular lipids, glucose signaling and fiber size in old rats. *Sci. Rep.* **2017**, *7*, 8593. [CrossRef]
61. Deschenes, M.R.; Sherman, E.G.; Roby, M.A.; Glass, E.K.; Harris, M.B. Effect of resistance training on neuromuscular junctions of young and aged muscles featuring different recruitment patterns. *J. Neurosci. Res.* **2014**, *93*, 504–513. [CrossRef] [PubMed]

62. Neto, W.K.; Gama, E. Strength training and anabolic steroid do not affect muscle capillarization of middle-aged rats. *Rev. Bras. Med. Esporte* **2017**, *23*, 137–141. [CrossRef]
63. Aagaard, P.; Andersen, J.L.; Dyhre-Poulsen, P.; Leffers, A.-M.; Wagner, A.; Magnusson, S.P.; Halkjaer-Kristensen, J.; Simonsen, E.B. A mechanism for increased contractile strength of human pennate muscle in response to strength training: Changes in muscle architecture. *J. Physiol.* **2001**, *534*, 613–623. [CrossRef] [PubMed]
64. Kim, J.-S.; Park, Y.-M.; Lee, S.-R.; Masad, I.S.; Khamoui, A.V.; Jo, E.; Park, B.-S.; Arjmandi, B.H.; Panton, L.B.; Lee, W.J.; et al. β-hydroxy-β-methylbutyrate did not enhance high intensity resistance training-induced improvements in myofiber dimensions and myogenic capacity in aged female rats. *Mol. Cells* **2012**, *34*, 439–448. [CrossRef]
65. Douglas, J.; Pearson, S.; Ross, A.; McGuigan, M. Chronic Adaptations to Eccentric Training: A Systematic Review. *Sport. Med.* **2016**, *47*, 917–941. [CrossRef]
66. Lourenço, Í.; Neto, W.K.; Amorim, L.D.S.P.; Ortiz, V.M.M.; Geraldo, V.L.; Ferreira, G.H.D.S.; Caperuto, C.; Gama, E.F. Muscle hypertrophy and ladder-based resistance training for rodents: A systematic review and meta-analysis. *Physiol. Rep.* **2020**, *8*, e14502. [CrossRef]
67. Braggion, G.F.; Ornelas, E.D.M.; Cury, J.C.S.; de Sousa, J.P.; Nucci, R.A.B.; Fonseca, F.L.A.; Maifrino, L.B.M. Remodeling of the soleus muscle of ovariectomized old female rats submitted to resistance training and different diet intake. *Acta Histochem.* **2020**, *122*, 151570. [CrossRef]
68. Csapo, R.; Gumpenberger, M.; Wessner, B. Skeletal Muscle Extracellular Matrix—What Do We Know About Its Composition, Regulation, and Physiological Roles? A Narrative Review. *Front. Physiol.* **2020**, *11*, 253. [CrossRef]
69. McPherron, A.C.; Lee, S.-J. Suppression of body fat accumulation in myostatin-deficient mice. *J. Clin. Investig.* **2002**, *109*, 595–601. [CrossRef]
70. Hamrick, M.W.; Pennington, C.; Webb, C.N.; Isales, C.M. Resistance to body fat gain in 'double-muscled' mice fed a high-fat diet. *Int. J. Obes.* **2006**, *30*, 868–870. [CrossRef]
71. McPherron, A.C.; Lee, S.-J. Double muscling in cattle due to mutations in the myostatin gene. *Proc. Natl. Acad. Sci. USA* **1997**, *94*, 12457–12461. [CrossRef] [PubMed]
72. Grobet, L.; Martin, L.J.R.; Poncelet, D.; Pirottin, D.; Brouwers, B.; Riquet, J.; Schoeberlein, A.; Dunner, S.; Ménissier, F.; Massabanda, J.; et al. A deletion in the bovine myostatin gene causes the double–muscled phenotype in cattle. *Nat. Genet.* **1997**, *17*, 71–74. [CrossRef] [PubMed]
73. Mosher, D.S.; Quignon, P.; Bustamante, C.D.; Sutter, N.B.; Mellersh, C.S.; Parker, H.G.; Ostrander, E. A Mutation in the Myostatin Gene Increases Muscle Mass and Enhances Racing Performance in Heterozygote Dogs. *PLoS Genet.* **2007**, *3*, e79. [CrossRef] [PubMed]
74. Schuelke, M.; Wagner, K.R.; Stolz, L.E.; Hübner, C.; Riebel, T.; Kömen, W.; Braun, T.; Tobin, J.F.; Lee, S.-J. Myostatin Mutation Associated with Gross Muscle Hypertrophy in a Child. *N. Engl. J. Med.* **2004**, *350*, 2682–2688. [CrossRef]
75. Raue, U.; Slivka, D.; Jemiolo, B.; Hollon, C.; Trappe, S. Myogenic gene expression at rest and after a bout of resistance exercise in young (18–30 yr) and old (80–89 yr) women. *J. Appl. Physiol.* **2006**, *101*, 53–59. [CrossRef]
76. Hayashi, S.; Miyake, M.; Watanabe, K.; Aso, H.; Hayashi, S.; Ohwada, S.; Yamaguchi, T. Myostatin preferentially down-regulates the expression of fast 2x myosin heavy chain in cattle. *Proc. Jpn. Acad. Ser. B Phys. Biol. Sci.* **2008**, *84*, 354–362. [CrossRef]
77. Santos, A.; Neves, M.; Gualano, B.; Laurentino, G.; Lancha, A.; Ugrinowitsch, C.; Lima, F.; Aoki, M.S. Blood flow restricted resistance training attenuates myostatin gene expression in a patient with inclusion body myositis. *Biol. Sport* **2014**, *31*, 121–124. [CrossRef]
78. Tang, L.; Luo, K.; Liu, C.; Wang, X.; Zhang, D.; Chi, A.; Zhang, J.; Sun, L. Decrease in myostatin by ladder-climbing training is associated with insulin resistance in diet-induced obese rats. *Chin. Med. J.* **2014**, *127*, 2342–2349.
79. Negaresh, R.; Ranjbar, R.; Baker, J.S.; Habibi, A.; Mokhtarzade, M.; Gharibvand, M.M.; Fokin, A. Skeletal muscle hypertrophy, insulin-like growth factor 1, myostatin and follistatin in healthy and sarcopenic elderly men: The effect of whole-body resistance training. *Int. J. Prev. Med.* **2019**, *10*, 29. [CrossRef]
80. Deldicque, L.; Atherton, P.; Patel, R.; Theisen, D.; Nielens, H.; Rennie, M.J.; Francaux, M. Effects of resistance exercise with and without creatine supplementation on gene expression and cell signaling in human skeletal muscle. *J. Appl. Physiol.* **2008**, *104*, 371–378. [CrossRef]
81. Volek, J.S.; Kraemer, W.J.; Bush, J.A.; Boetes, M.; Incledon, T.; Clark, K.L.; Lynch, J.M. Creatine Supplementation Enhances Muscular Performance during High-Intensity Resistance Exercise. *J. Am. Diet. Assoc.* **1997**, *97*, 765–770. [CrossRef] [PubMed]
82. Cooke, M.B.; Brabham, B.; Buford, T.W.; Shelmadine, B.D.; McPheeters, M.; Hudson, G.M.; Stathis, C.; Greenwood, M.; Kreider, R.; Willoughby, D.S. Creatine supplementation post-exercise does not enhance training-induced adaptations in middle to older aged males. *Eur. J. Appl. Physiol.* **2014**, *114*, 1321–1332. [CrossRef] [PubMed]
83. Sun, M.; Jiao, H.; Wang, X.; Li, H.; Zhou, Y.; Zhao, J.; Lin, H. The regulating pathway of creatine on muscular protein metabolism depends on the energy state. *Am. J. Physiol. Physiol.* **2022**, *322*, C1022–C1035. [CrossRef] [PubMed]
84. Mazzetti, S.; Douglass, M.; Yocum, A.; Harber, M. Effect of Explosive versus Slow Contractions and Exercise Intensity on Energy Expenditure. *Med. Sci. Sport. Exerc.* **2007**, *39*, 1291–1301. [CrossRef] [PubMed]

Article

Different Effects of Cyclical Ketogenic vs. Nutritionally Balanced Reduction Diet on Serum Concentrations of Myokines in Healthy Young Males Undergoing Combined Resistance/Aerobic Training

Pavel Kysel [1], Denisa Haluzíková [1], Iveta Pleyerová [2], Kateřina Řezníčková [3], Ivana Laňková [3], Zdeňka Lacinová [2,4], Tereza Havrlantová [2], Miloš Mráz [3], Barbora Judita Kasperová [3], Viktorie Kovářová [3], Lenka Thieme [3], Jaroslava Trnovská [2], Petr Svoboda [2,5], Soňa Štemberková Hubáčková [2], Zdeněk Vilikus [1,*] and Martin Haluzík [3,4,*]

[1] Department of Sports Medicine, First Faculty of Medicine and General University Hospital, 128 08 Prague, Czech Republic
[2] Centre for Experimental Medicine, Institute for Clinical and Experimental Medicine, 140 21 Prague, Czech Republic
[3] Diabetes Centre, Institute for Clinical and Experimental Medicine, 140 21 Prague, Czech Republic
[4] Institute of Medical Biochemistry and Laboratory Diagnostics, First Faculty of Medicine, Charles University and General University Hospital, 121 11 Prague, Czech Republic
[5] Department of Biochemistry and Microbiology, University of Chemistry and Technology Prague, Technicka 5, 166 28 Prague, Czech Republic
* Correspondence: zvili@lf1.cuni.cz (Z.V.); halm@ikem.cz (M.H.)

Citation: Kysel, P.; Haluzíková, D.; Pleyerová, I.; Řezníčková, K.; Laňková, I.; Lacinová, Z.; Havrlantová, T.; Mráz, M.; Kasperová, B.J.; Kovářová, V.; et al. Different Effects of Cyclical Ketogenic vs. Nutritionally Balanced Reduction Diet on Serum Concentrations of Myokines in Healthy Young Males Undergoing Combined Resistance/Aerobic Training. *Nutrients* 2023, 15, 1720. https://doi.org/10.3390/nu15071720

Academic Editors: Sandro Massao Hirabara, Maria Fernanda Cury Boaventura and Gabriel Nasri Marzuca Nassr

Received: 17 February 2023
Revised: 21 March 2023
Accepted: 29 March 2023
Published: 31 March 2023

Abstract: Myokines represent important regulators of muscle metabolism. Our study aimed to explore the effects of a cyclical ketogenic reduction diet (CKD) vs. a nutritionally balanced reduction diet (RD) combined with regular resistance/aerobic training in healthy young males on serum concentrations of myokines and their potential role in changes in physical fitness. Twenty-five subjects undergoing regular resistance/aerobic training were randomized to the CKD ($n = 13$) or RD ($n = 12$) groups. Anthropometric and spiroergometric parameters, muscle strength, biochemical parameters, and serum concentrations of myokines and cytokines were assessed at baseline and after 8 weeks of intervention. Both diets reduced body weight, body fat, and BMI. Muscle strength and endurance performance were improved only by RD. Increased musclin (32.9 pg/mL vs. 74.5 pg/mL, $p = 0.028$) and decreased osteonectin levels (562 pg/mL vs. 511 pg/mL, $p = 0.023$) were observed in RD but not in the CKD group. In contrast, decreased levels of FGF21 (181 pg/mL vs. 86.4 pg/mL, $p = 0.003$) were found in the CKD group only. Other tested myokines and cytokines were not significantly affected by the intervention. Our data suggest that changes in systemic osteonectin and musclin levels could contribute to improved muscle strength and endurance performance and partially explain the differential effects of CKD and RD on physical fitness.

Keywords: body composition; ketogenic diet; strength parameters; endurance; training; myokines; adipokines; cytokines

1. Introduction

A diet with a nutritionally suitable composition is a necessary prerequisite for both professional and amateur sportsmen to enable appropriate energy expenditure and ensure optimal physical performance. For aesthetic and performance reasons, these subjects frequently undergo weight loss in order to reduce body fat amount. However, this may be accompanied by a loss of lean body mass, which can often reach up to 25% of the total weight loss [1,2]. For this purpose, the aim of such approaches is to maximize the reduction of adipose tissue while maintaining muscle mass. To achieve weight loss, it is necessary to increase the energy deficit, which most often means reducing energy intake. There are many types of reduction diets; however, most of them are similar in many points [3]. A caloric reduction of 500–750 calories per day is recommended by many obesity societies [4,5]. To reduce the number of calories, lipids or carbohydrates are usually restricted. Neither macronutrient, fat (low-fat diets) nor carbohydrate (low-carb diets), has been determined to be more important for weight loss as long as a caloric deficit occurs. An extremely low energy intake from carbohydrates (<10%) may result in nutritional ketosis [6]. This type of diet is called the ketogenic diet and is usually accompanied by elevated fat and protein content, with a daily protein intake of 0.8–1.5 g/kg body weight to preserve muscle mass [6,7]. The magnitude and duration of the energy deficit determine the amount of weight loss. Resistance exercise is often used to restrict the loss of lean mass [8].

Skeletal muscle and bone are connected anatomically and physiologically and play a crucial role in human locomotion and metabolism. Historically, the coupling between muscle and bone has been viewed in the light of mechanotransduction, which declares that the mechanical forces applied to muscle are transmitted to the skeleton to initiate bone formation. However, these tissues also communicate through an endocrine route orchestrated by a family of bioactive molecules referred to as myokines (derived from myocytes), osteokines (derived from bone cells), and adipokines (derived from adipose tissue) [9].

Myokines are small proteins, released by skeletal muscle cells in response to muscle contractions, which have important autocrine, paracrine, and/or endocrine effects, including the regulation of energy metabolism. Experimental studies have demonstrated that some of the myokines can be directly involved in the regulation of muscle strength and endurance, thus possibly contributing to the integration of the effects of different diets and various types of exercise on the parameters of physical fitness [10].

One of the myokines with important effects on metabolic regulation and muscle growth and performance is musclin. Musclin is produced by bone and skeletal muscle cells and plays a crucial role in the regulation of bone growth and physical endurance by enhancing the growth of muscle fibers and their regeneration [11–13]. In experimental settings, musclin attenuated inflammation and oxidative stress in doxorubicin (DOX)-induced cardiotoxicity [14].

Osteonectin, also known as secreted protein acidic and rich in cysteine (SPARC), is another important tissue-specific protein, linking the bone mineral and collagen phases and contributing to the initiation of active mineralization in normal skeletal tissue. Osteonectin is an important modulator of the actin cytoskeleton that may be involved in the maintenance of muscular function by binding to actin in regenerating myofibers [15–17]. Several other myokines, such as interleukin 6 (IL6), irisin, insulin growth factor (IGF-1), brain-derived neurotrophic factor (BDNF), myostatin, and fibroblast growth factor 21 (FGF21), exert anabolic/catabolic effects on bone and muscle, and their appropriate balance contributes to the physiological regulation of muscle and bone mass [9]. Especially FGF21 was found to be required for fasting-induced muscle atrophy and weakness [18].

In addition to its effects on body composition, there is growing evidence that a ketogenic diet may be a potential therapeutic approach for Alzheimer's disease or other types of cognitive impairment via inhibition of pro-inflammatory cytokines, such as interleukin-1β (IL-1β) and tumor necrosis factor α (TNF-α) [19–21]. These cytokines are also known to accelerate catabolism, induce contractile dysfunction, disrupt myogenesis, and overall modulate muscle tissue loss (summarized in [22]). Their inhibition can therefore lead to improved muscle function. Moreover, a ketogenic diet may improve insulin sensitivity by reducing carbohydrate intake and promoting weight loss. This can help prevent the development of type 2 diabetes and also, in turn, reduce the risk of sleep apnea and affect a wide variety of other physiological regulations, including the endocannabinoid system [23–26].

Taken together, myokines represent an important interconnection between the effects of diet and exercise and the growth of muscle fibers and their regeneration, modulation of angiogenesis, and also the regulation of glucose and lipid metabolism [27]. We have previously demonstrated that a nutritionally balanced reduction diet (RD), but not cyclical ketogenic diet (CKD), in healthy young males undergoing regular resistance/aerobic training improved muscle strength and endurance performance, while both reduced body weight to a similar degree [28].

Here, we explored possible mechanisms behind the different effects of RD vs. CKD on physical performance. The main aim of this study was to test the hypothesis that exercise-induced myokines are produced differently according to the type of diet protocol selected and thus differentially affect physical performance. To this end, we measured serum concentrations of selected myokines and cytokines in healthy young males on RD or CKD undergoing regular resistance/aerobic training at baseline and after eight weeks of training combined with their respective diets.

2. Materials and Methods

2.1. Study Subjects

In our earlier paper [28], the specific study subjects' characteristics and the training regimen were laid forth in detail. In summary, 25 healthy young male volunteers with a wide range of fitness and at least a year of resistance training experience were recruited through physical education colleges and through a website with people interested in nutrition and exercise. Ages between 18 and 30 years and at least a year of resistance and aerobic training experience were requirements for inclusion. Participants who expressed interest in taking part were assessed to make sure they met the requirements for enrollment in this study.

Exclusion criteria included the existence of cardiovascular disorders, diabetes mellitus, arterial hypertension, or any other illnesses that required pharmacological treatment. These ailments could impact athletes' ability to perform in sports or put them at risk for subsequent injuries. Additionally, participants were required to stop using any performance-enhancing supplements (such as creatine, β-hydroxy-β-methyl butyrate, caffeine, protein powder, weight gainer, thermogenic, etc.) at least one week before baseline testing and to refrain from doing so for the duration of this study.

This study was authorized by the Human Ethics Review Board, First Faculty of Medicine and General University Hospital, Prague, Czech Republic (ethical approval code 764/18 S-IV) and performed in agreement with the principles of the WMA Declaration of Helsinki—Ethical Principles for Medical Research Involving Human Subjects [29]. All subjects were required to sign informed consent forms before randomization.

For eight weeks, participants were randomized by an electronic randomization system to follow either CKD or RD (both requiring a 500 kcal/day reduction in total calorie intake) while engaging in three resistance exercise sessions and three cardio exercises per week (30 min run, heart rate around 130–140 beats/min). From a balanced hypocaloric diet with a reduction of energy intake of 500 to 1000 kcal from the regular caloric intake, a total caloric intake reduction of 500 kcal/day was calculated. Such diets are suggested by the U.S. Food and Drug Administration (FDA) as the "standard therapy" for clinical trials [30].

2.2. Baseline and Postinterventional Testing

Medical history, anthropometric measurements, power performance tests, cycling spiroergometry, blood draws while fasting to gather laboratory data, and data collection during baseline and post-intervention testing after 8 weeks were all part of the data collection process. By applying the height measurement, the scale determined the body mass index (BMI). A simple stadiometer was used to measure height accurately (Seca 222, Seca Co., Birmingham, UK).

2.2.1. Biochemical and Anthropometric Examination

At the beginning of this study, the BMIs of all subjects were computed. InBody Body Composition Analyzers were used to measure body composition (InBody230, InBody Co., Ltd., Seoul, Republic of Korea). Body weight and other measurements of body composition, such as lean body mass, body fat mass, BMI, water content, and body fat percentage, were taken, with measurements of the subject's weight being taken to the nearest 0.5 kg.

Blood samples were taken before this study began and at its conclusion, eight weeks following the diet, for biochemical measurements. After centrifuging the serum, samples were kept in aliquots at 80 °C until additional analysis. The longest period of storage was 8 months.

Biochemical parameters were estimated by spectrophotometric methods using the ARCHITECT c Systems device (Abbott Park, IL, USA) in the Department of Biochemistry of the Institute for Clinical and Experimental Medicine in Prague. Serum levels of myokines (brain-derived neurotrophic factor (BDNF), fatty acid binding protein 3 (FABP3), fractalkine, follistatin-like protein 1 (FSTL1), IL6, musclin, oncostatin M, osteonectin) were analyzed by the multiplex assay MILLIPLEX® Human Myokine Magnetic Bead Panel (HMYOMAG-56K; Merck KGaA, Darmstadt, Germany) and cytokines/chemokines (Interferon gamma (IFNγ), IL8, IL10, IL23, TNFα were measured by the multiplex assay MILLIPLEX® Human Cytokine/Chemokine Magnetic Bead Panel (HCYTOMAG-60K; Merck KGaA, Darmstadt, Germany), both with using MAGPIX system (Luminex corporate, Austin, TX, USA). Serum levels of FGF21, FGF19, and C-reactive protein (CRP) were estimated by the highly sensitive ELISA kits (EH188RB, Invitrogen for FGF21; RD191107200R, BioVendor for FGF19; and BMS288INST, Invitrogen for CRP) using an Epoch microplate spectrophotometer (Agilent, Santa Clara, CA, USA).

2.2.2. Strength and Aerobic Performance Testing

The training protocol was delineated in our primary paper [28]. In order to evaluate maximal power performance, the bench press, lat pull-down, and leg press were used as the three exercises in a strength performance evaluation. Aerobic performance testing was conducted by bicycle spiroergometry along with an analyzer of respiratory gases (Quark CPET, Cosmed, Concord, CA, USA). The participants were advised to work out until they became involuntarily exhausted and to keep their pedal cadence between 70 and 90 rpm. According to Gordon et al., we applied a modified exercise step regimen of 0.33 W·min^{-1}. The subject's inability to keep up a 40-rpm pedal cadence led the test to an end.

2.3. Diet Protocol

For eight weeks, subjects were randomly assigned, using an automated randomization mechanism, to either the CKD or the RD group. Prior to the start of the trial, subjects were required to attend an obligatory diet consultation with a nutritionist, who gave them thorough instructions on accurately recording their dietary food intake. The DietSystem software was used to enter and evaluate all food record data (DietSystem App, DietSystem App, s.r.o., Brno, Czech Republic). Detailed directions on what items would work for both types of diets were supplied to each participant. Additionally, according to randomization, participants undergo either an 8-week low-carb diet plan or a meal plan for a reduction diet. Each participant's total daily calorie intake was lowered by 500 kcal based on their lifestyle (individually determined based on somatotype, physical activity, type of work, etc.).

Every week, a nutritionist checked on the participants' overall diet compliance. Moreover, blood-hydroxybutyrate measurements at the completion of the research and twice-daily urinary ketone measures were used to assess adherence to CKD.

2.3.1. Cyclical Ketogenic Reduction Diet

The CKD protocol included a five-day low-carbohydrate phase to induce and maintain ketosis, followed by a two-day carbohydrate phase (weekends) (nutrient ratio: carbs up to 30 g; proteins 1.6 g/kg; fats: computation of energy intake to substitute carbohydrates): (Intake of proteins 15%; carbs 8–10 g/1 kg of non–fat tissue 70%; and fat 15%).

2.3.2. Nutritionally Balanced Reduction Diet

The RD diet protocol was established on the concepts of a healthy diet (nutrient ratio of carbohydrates 55%, fat 30%, and proteins 15% of total energy intake).

2.4. Training Protocol

The training protocol was delineated in our primary publication [28]. It consisted of a predefined combination of resistance training to support strength skills and aerobic training to increase endurance skills. A sport tester for aerobic performance and required check-in procedures at a gym were used to verify training compliance (TomTom Runner Cardio, TomTom, Amsterdam, The Netherlands).

2.5. Post-Intervention Testing

The methods used to gather data for the post-intervention testing were the same as those used for the baseline testing. The same researcher who conducted power measurements and performance testing for each subject at baseline did so to ensure reliability. Additionally, participants took their tests at the same time each day. After data analysis, test results were given to participants and compared to their baseline values.

2.6. Statistical Analysis

SigmaPlot 13.0 software (SPSS Inc., Chicago, IL, USA) was used for statistical analysis. Normally distributed data are listed as mean $\pm$ standard deviation (SD); non-parametric data are listed as median (interquartile range). The normality of all the data was estimated by the Shapiro–Wilk test. Intragroup differences were calculated by paired samples *t*-test or Wilcoxon signed rank test (non-parametric test). Two-way repeated measures ANOVA with Sidak's post hoc test was utilized for testing intergroup differences. In all statistical tests, p values < 0.05 were deemed significant.

3. Results

3.1. The Influence of CKD vs. RD on Anthropometric, Biochemical and Hormonal Characteristics

Anthropometric parameters of subjects on CKD or RD at baseline (V1) and after 8 weeks of diet (V2) are delineated in Table 1. No significant differences between CKD and RD groups in anthropometric or body composition parameters were detected at baseline. Both CKD and RD lowered body weight, body fat mass, and body mass index (BMI), with similar effects of both diets. Lean body mass and body water content were significantly lowered by CKD, while they were not influenced by RD (Table 1).

Table 1. Anthropometric parameters of study subjects with CKD or RD at baseline (V1) and after 8 weeks of diet (V2).

Parameters	CKD		RD	
	V1	**V2**	**V1**	**V2**
Age (year)	23 $\pm$ 5		24 $\pm$ 4	
Height (cm)	181 $\pm$ 6		186 $\pm$ 10	
BMI (kg/m^2)	26.1 $\pm$ 3.7	24.6 $\pm$ 3.3 *	26.9 $\pm$ 4.3	25.5 $\pm$ 4.2 *
WEIGHT (kg)	85.6 $\pm$ 13.4	81.0 $\pm$ 12.0 *	93.0 $\pm$ 17.5	88.5 $\pm$ 17.4 *
MUSCLES (kg)	41.8 $\pm$ 4.5	40.0 $\pm$ 4.6 *	43.5 $\pm$ 5.3	43.1 $\pm$ 5.3
FAT (kg)	12.9 $\pm$ 6.9	11.0 $\pm$ 5.8 *	17.6 $\pm$ 9.8	13.6 $\pm$ 9.0 *
% FAT	14.5 $\pm$ 5.5	13.0 $\pm$ 5.1 *	17.9 $\pm$ 6.9	14.2 $\pm$ 6.9 *
WATER (kg)	53.2 $\pm$ 5.6	51.0 $\pm$ 5.6 *	55.1 $\pm$ 6.4	54.8 $\pm$ 6.5

Data are shown as mean $\pm$ SD; statistical significance of intragroup differences is from paired samples *t*-test or Wilcoxon signed rank test (V1—baseline testing vs. V2—testing after 8 weeks of diet). * $p < 0.05$ V1 vs. V2. BMI: Body mass index.

Biochemical and hormonal parameters of study subjects with CKD or RD at V1 and V2 are shown in Table 2. No differences between CKD and RD groups in biochemical and hormonal parameters were found at V1. CKD significantly increased total, LDL, and

HDL cholesterol as well as β-OH-butyrate concentrations. CKD also decreases blood glucose levels. In the RD group, total cholesterol and uric acid concentrations decreased significantly after the intervention, while other parameters did not change. Total and LDL cholesterol levels at the end of the intervention (V2) were significantly lower in the RD vs. CKD group.

Table 2. Biochemical and hormonal parameters of study subjects with CKD or RD at baseline (V1) and after 8 weeks of diet (V2).

Parameters	CKD		RD	
	V1	V2	V1	V1
Total cholesterol (mmol/L)	3.92 ± 0.52	4.92 ± 0.74 *	4.47 ± 0.67	4.03 ± 0.80 *$^\Delta$
Triglycerides (mmol/L)	0.89 (0.65–1.11)	0.8 (0.66–0.93)	0.92 (0.71–1.03)	0.81 (0.66–1.01)
LDL cholesterol (mmol/L)	2.37 ± 0.41	3.14 ± 0.64 *	2.73 ± 0.52	2.43 ± 0.63 $^\Delta$
HDL cholesterol (mmol/L)	1.15 ± 0.24	1.39 ± 0.28 *	1.30 ± 0.39	1.20 ± 0.30
Fasting glucose (mmol/L)	5.28 ± 0.34	4.96 ± 0.47 *	5.26 ± 0.36	5.22 ± 0.36
Insulin (μIU/mL)	5.37 (3.99–8.29)	5.86 (3.46–7.42)	6.35 (3.13–13.34)	8.44 (4.76–10.98)
Uric acid (mmol/L)	357 (312.5–430.5)	350 (324.5–421.5)	397 ± 63	368 ± 53 *
CK (ukat/L)	4.40 ± 2.81	2.81 ± 1.21	3.80 ± 2.03	3.03 ± 2.03
LDH (ukat/L)	2.68 ± 0.60	2.47 ± 0.42	2.74 ± 0.44	2.55 ± 0.33
β-OH-butyrate (mmol/L)	0.1 (0.1–0.2)	0.2 (0.1–0.6) *	0.1 (0.1–0.3)	0.1 (0.1–0.2)

Normally distributed data are shown as mean ± SD; non-parametric data are expressed as median (interquartile range). Statistical significance of intragroup differences is from paired samples *t*-test or Wilcoxon signed rank test (V1—baseline testing vs. V2—testing after 8 weeks of diet). * $p < 0.05$ V1 vs. V2; for intergroup differences, significance is from a Two-way repeated measures ANOVA with Sidak's post hoc test. $^\Delta$ $p < 0.05$ V2 CKD vs. V2 RD. LDL cholesterol: low-density lipoprotein cholesterol; HDL cholesterol: high-density lipoprotein cholesterol; CK: Creatine kinase; LDH: Lactate dehydrogenase; β-OH-butyrate—β-hydroxy-butyrate.

3.2. The Influence of CKD vs. RD on Serum Concentrations of Myokines

Serum concentrations of selected myokines, adipokines, and cytokines in subjects on CKD or RD at baseline (V1) and after 8 weeks of diet (V2) are shown in Table 3. No differences in serum myokine concentrations between the CKD and RD groups were found at V1. At V2, serum osteonectin levels were significantly lower in the RD group as compared to the CKD group.

Table 3. Changes in serum myokines, adipokines, and cytokines of study subjects on CKD or RD at baseline (V1) and after 8 weeks of diet (V2).

Parameters	CKD		RD	
	V1	V2	V1	V2
Oncostatin M (pg/mL)	8.26 (5.16–10.6)	8.75 (5.75–11.2)	10.5 (8.63–12.5)	10.9 (9.35–17.9)
Musclin (pg/mL)	48.6 (26.8–80)	55.8 (36.5–83.2)	32.9 (12.2–85.8)	74.5 (34.7–95.4) *
Osteonectin (pg/mL)	630 (489–701)	596 (529–803)	562 (490–665)	511 (484–568) *$^\Delta$
BDNF (ng/mL)	11.9 (10.9–13.3)	12.9 (10.5–13.5)	11.6 (10.1–13.5)	11.9 (11.2–13)
FABP3 (ng/mL)	1.01 (0.87–1.34)	1.06 (0.87–1.55)	1.27 (0.79–1.98)	1.17 (0.88–1.97)
FSTL1 (ng/mL)	2.75 (1.1–4.89)	2.95 (2.07–4.84)	3.64 (1.58–7.09)	3.79 (1.63–7.29)
FGF19 (pg/mL)	194 (134–327)	207 (119–292)	165 (120–210)	133 (120–222)
CRP (mg/L)	1.02 (0.3–2.5)	0.85 (0.19–2.34)	0.69 (0.28–1.6)	0.71 (0.16–1.29)
FGF21 (pg/mL)	181 (112–709)	86.4 (45.1–571) *	272 (176–1138)	193 (144–1142)
Fractalkin (pg/mL)	241 (213–315)	208 (183–316)	211 (191–247)	222 (202–300)
IFNγ (pg/mL)	17.6 ± 8.2	16.5 ± 8.0	16.3 ± 5.1	17.5 ± 5.1
IL10 (pg/mL)	11.5 ± 6.1	10.5 ± 6.5	12.1 ± 8.4	11.6 ± 7.2
IL23 (pg/mL)	265 ± 134	246 ± 126	272 ± 107	284 ± 122
IL6 (pg/mL)	3.46 (1.16–5.82)	2.72 (1.02–4.13)	2.13 (1.01–4.24)	2.68 (1.44–5.52)

Table 3. *Cont.*

Parameters	CKD		RD	
	V1	**V2**	**V1**	**V2**
IL8 (pg/mL)	9.8 (8–20.6)	11.4 (7.5–18.4)	9.7 (7.4–11.1)	11.2 (8.3–22.3)
TNFα (pg/mL)	8.85 (6.93–12.19)	9.14 (7.32–12.74)	9.07 (7.46–10.41)	11.38 (6.94–16.56)

Normally distributed data are shown as mean ± SD; non-parametric data are expressed as median (interquartile range). Statistical significance of intragroup differences is from paired samples *t*-test or Wilcoxon signed rank test (V1—baseline testing vs. V2—testing after 8 weeks of diet). * $p < 0.05$ V1 vs. V2; for intergroup differences, significance is from a Two-way repeated measures ANOVA with Sidak's post hoc test. $^\Delta$ $p < 0.05$ V2 CKD vs. V2 RD. BDNF: Brain-derived neurotrophic factor; FABP3: Fatty acid binding protein 3; FSTL1: Follistatin-related protein 1; FGF19: Fibroblast growth factor 19; CRP: c-reactive protein; FGF21: Fibroblast growth factor 21; IFNγ: Interferon gamma; IL10/23/6/8: Interleukin 10/23/6/8; TNFα: tumor necrosis factor α.

In the RD group, serum osteonectin concentrations significantly decreased after the intervention, while no change in the concentrations of this myokine was found in the CKD group. Increased levels of musclin were found at V2 in subjects from the RD group compared to V1, while no such changes were detected in the CKD group. In contrast, the FGF21 level significantly decreased at V2 in the CKD group while it remained unchanged in the RD group. No significant changes in the rest of the parameters were detected (Table 3).

4. Discussion

The interactions between diet and physical activity play a major role in the long-term regulation of body weight and physical fitness. Endocrine factors, including not only classical hormones but also adipokines, cytokines, and myokines, were shown to be very active players in this process [10,31,32]. The aim of our study was to comprehensively analyze the changes in body composition, physical fitness, and serum concentrations of selected myokines and cytokines in healthy young males undergoing combined resistance/aerobic training in combination with two dietary protocols: CKD or RD.

This study follows up on our earlier findings [28], where eight weeks of regular exercise training combined with either CKD or RD decreased body weight and body fat to a similar degree, while muscle strength and endurance performance were improved only in the RD group. These changes were seen on the background of slightly reduced lean body mass and body water content. It was significantly reduced in CKD subjects, while RD had no effect on it. The large decline in body water in the CKD group was most likely caused by low carbohydrate intake-related glycogen depletion.

Myokines, muscle-secreted molecules, play a crucial role in the regulation of muscle metabolism and function. Out of the myokine/adipokine/cytokine panel analyzed, the serum concentrations of three myokines were affected by the intervention. Musclin is a peptide produced by bone and skeletal muscle cells that plays an important role in the regulation of bone growth and physical endurance by enhancing the growth of muscle fibers [11–13]. Given that musclin has been demonstrated to contribute to muscle growth and regeneration [33], it is tempting to speculate that its increased concentration in the RD group could have contributed to the increased lean body mass and improved exercise performance detected in the RD group. In support of this hypothesis, musclin levels were unchanged in the CKD group, where no changes in muscle strength or endurance performance were detected. Recently, it has been reported that runners with sufficient carbohydrate intake have higher levels of musclin, which may contribute to improved adaptations to exercise, such as improved glucose homeostasis and browning of adipose tissue [34]. These findings are again in line with the increased musclin levels and enhanced endurance performance observed in our RD subjects.

Osteonectin is an acidic extracellular matrix glycoprotein that plays an important role in bone mineralization and collagen binding [35]. In our study, serum osteonectin levels decreased in the RD group while they remained unaffected in the CKD group. The interconnection between osteonectin, exercise, and metabolic and energy homeostasis regulation is still not clear. In experimental studies, osteonectin deficiency was accompanied by an accelerated aging phenotype and reduced physical activity [36]. These changes were attributed to reduced extracellular matrix mass and decreased collagen maturity. In other studies, osteonectin has also been described as a metabolic enhancer of skeletal muscle growth, with its mRNA expression being induced by exercise in both mice and humans [37]. The reason for decreased osteonectin levels after 8 weeks of exercise in our study is thus unclear. One possible explanation is that most of the studies published so far were performed in rodents, which may not be representative of the regulation of osteonectin levels in humans. Furthermore, in the human study by Aoi and colleagues [37], osteonectin levels were increased only when measured shortly after exercise and returned to baseline values six hours after exercise. Therefore, as our study measured osteonectin levels in the fasting state, we would not be able to detect any post-exercise changes in its levels.

FGF21 is a novel metabolic regulator produced primarily by the liver, but to a certain degree also by other tissues such as white and brown adipose tissue, skeletal muscle, and the pancreas [38]. FGF21 has potent antidiabetic and lipid-lowering effects in animal models of obesity and type 2 diabetes mellitus. This hormone contributes to body weight regulation and is strongly involved in the response to nutritional deprivation and the ketogenic state in mice [39]. A systematic review of clinical studies exploring the effect of exercise on FGF21 levels concluded that acute exercise tended to increase circulatory levels of FGF21, while chronic exercise with a duration over 4 weeks had rather the opposite effect [40]. It was also described that higher levels of daily physical activity could decrease circulating FGF21 levels [41]. Moreover, it has been reported that carbohydrate intake correlates with levels of FGF21 [34]. In our study, FGF21 levels decreased in the CKD group but remained unchanged in the RD group, suggesting that the effects of the ketogenic diet could have interacted with exercise.

In addition to the important role of myokines in the regulation of muscle regeneration and physical performance, these molecules may also represent a promising therapeutic target for metabolic diseases [42]. A sedentary lifestyle without a low level of physical activity is then directly proportional to the development of many chronic diseases accompanied by systemic inflammation [43,44]. Increased glucose levels in obese patients with type 2 diabetes can disrupt the regulation of vascular tension and, among other things, humoral and cellular inflammatory processes [45]. Several studies have confirmed the role of ketogenic diet consumption on body weight reduction, body composition change, and regulation of cytokine and adiponectin levels [46–50]. The lack of significant changes in pro-inflammatory cytokines and adipokines in the CKD group in our study as compared to its decrease in patients with obesity and type 2 diabetes after the ketogenic diet may be due to the fact that these positive effects are seen in the context of metabolic impairments and subclinical inflammation but not in metabolically healthy lean subjects. Taken together, the ketogenic diet and nutritionally balanced reduction diet are two different approaches to weight loss and overall health improvement, with some overlap in their benefits but also differences in the advantages and disadvantages of each approach (see Table 4 for the summary).

Table 4. Comparison of advantages and disadvantages of the CKD and RD.

CKD		RD	
Advantages	**Disadvantages**	**Advantages**	**Disadvantages**
↓ fasting glucose	↔ VO$_{2max}$	sustainable weight loss	slow initial weight loss
↓ weight	↔ TTE	↓ adipose tissue	feel hunger and cravings
↓ adipose tissue	↓ strength	↑ strength	
feel satiety	↓ LBM	↑ endurance	
↑ cognitive function	↑ keto flu	all nutrients	
	↑ LDL cholesterol	variation of food	
	↓ fiber	↑ adherence	

Table 4. *Cont.*

CKD		RD	
Advantages	**Disadvantages**	**Advantages**	**Disadvantages**
	nutrient deficiencies		
	↓ adherence		
	↑ depression		

TTE: time to exhaustion; LBM: lean body mass; LDL cholesterol: low density cholesterol; VO_2max: peak oxygen uptake. ↓ decrease, ↑ increase, ↔ unchanged.

For proper interpretation of the results of our study within the context of other published data, it is important to consider its strengths and limitations. The randomized design and the good compliance of the subjects to the dietary and treatment regimens are strong points of our trial. On the other hand, the limitations include a relatively short duration, a low number of subjects, and the inclusion of only male participants. Another limitation may be the use of bioelectrical impedance analysis instead of dual-energy X-ray absorptiometry, which is more accurate.

5. Conclusions

Myokines play a crucial role in the communication between skeletal muscles and other distinct organs in order to adapt whole-body metabolism to nutritional changes. Our data suggest that changes in systemic osteonectin and musclin levels could contribute to improved muscle strength and endurance performance in healthy young males on RD undergoing regular resistance/aerobic training as compared to subjects on CKD, where no improvements in muscle strength and endurance performance were detected. Further studies with a longer duration and on a higher number of subjects are warranted to support the validity of these findings for other populations, such as professional athletes or patients with obesity and metabolic complications.

Author Contributions: Conceptualization, P.K., D.H., M.H. and Z.V.; methodology, P.K., I.P., K.Ř., T.H., I.L., Z.L., B.J.K., J.T., V.K., L.T., P.S., S.Š.H., M.M., D.H., Z.V. and M.H.; software, I.P., K.Ř., Z.L. and J.T.; validation, P.K., Z.L., D.H., Z.V. and M.H.; formal analysis, I.P., K.Ř., I.L. and Z.L.; investigation, P.K., M.M., S.Š.H., V.K., L.T., D.H., Z.V. and M.H.; resources, P.K., D.H., Z.V. and M.H.; data curation, P.K., D.H., Z.V. and M.H.; writing—original draft preparation, P.K. and M.H.; writing—review and editing, S.Š.H., M.M., P.S., D.H., Z.V. and M.H.; visualization, I.P., K.Ř., P.S. and S.Š.H.; supervision, D.H., Z.V. and M.H.; project administration, P.K., D.H., Z.V. and M.H.; funding acquisition, D.H., Z.V. and M.H. All authors have read and agreed to the published version of the manuscript.

Funding: Supported by the Ministry of Health, Czech Republic—conceptual development of a research organization (Institute for Clinical and Experimental Medicine—IKEM, IN 00023001), RVO VFN 64165, and by the project National Institute for Research of Metabolic and Cardiovascular Diseases (Programme EXCELES, Project No. LX22NPO5104)—Funded by the European Union—Next Generation EU.

Institutional Review Board Statement: This study was conducted in accordance with the World Medical Association Declaration of Helsinki—Ethical Principles for Medical Research Involving Human Subjects, as revised in 2013, and approved by the Human Ethics Review Board, First Faculty of Medicine and General University Hospital, Prague, Czech Republic (ethical approval code 764/18 S-IV).

Informed Consent Statement: Informed consent was obtained from all subjects involved in this study.

Data Availability Statement: The data presented in this study are available upon request from the corresponding author.

Conflicts of Interest: The authors declare no conflict of interest.

References

1. Mujika, I. Case Study: Long-Term Low-Carbohydrate, High-Fat Diet Impairs Performance and Subjective Well-Being in a World-Class Vegetarian Long-Distance Triathlete. *Int. J. Sport Nutr. Exerc. Metab.* **2019**, *29*, 339–344. [CrossRef]
2. Hector, A.J.; Phillips, S.M. Protein Recommendations for Weight Loss in Elite Athletes: A Focus on Body Composition and Performance. *Int. J. Sport Nutr. Exerc. Metab.* **2018**, *28*, 170–177. [CrossRef] [PubMed]
3. Freire, R. Scientific evidence of diets for weight loss: Different macronutrient composition, intermittent fasting, and popular diets. *Nutrition* **2020**, *69*, 110549. [CrossRef] [PubMed]
4. Yi, D.Y.; Kim, S.C.; Lee, J.H.; Lee, E.H.; Kim, J.Y.; Kim, Y.J.; Kang, K.S.; Hong, J.; Shim, J.O.; Lee, Y.; et al. Clinical practice guideline for the diagnosis and treatment of pediatric obesity: Recommendations from the Committee on Pediatric Obesity of the Korean Society of Pediatric Gastroenterology Hepatology and Nutrition. *Korean J. Pediatr.* **2019**, *62*, 3–21. [CrossRef] [PubMed]
5. Seo, M.H.; Lee, W.Y.; Kim, S.S.; Kang, J.H.; Kang, J.H.; Kim, K.K.; Kim, B.Y.; Kim, Y.H.; Kim, W.J.; Kim, E.M.; et al. 2018 Korean Society for the Study of Obesity Guideline for the Management of Obesity in Korea. *J. Obes. Metab. Syndr.* **2019**, *28*, 40–45. [CrossRef]
6. Kim, J.Y. Optimal Diet Strategies for Weight Loss and Weight Loss Maintenance. *J. Obes. Metab. Syndr.* **2021**, *30*, 20–31. [CrossRef]
7. Burke, L.M. Ketogenic low-CHO, high-fat diet: The future of elite endurance sport? *J. Physiol.* **2021**, *599*, 819–843. [CrossRef]
8. Ruiz-Castellano, C.; Espinar, S.; Contreras, C.; Mata, F.; Aragon, A.A.; Martinez-Sanz, J.M. Achieving an Optimal Fat Loss Phase in Resistance-Trained Athletes: A Narrative Review. *Nutrients* **2021**, *13*, 3255. [CrossRef]
9. Kirk, B.; Feehan, J.; Lombardi, G.; Duque, G. Muscle, Bone, and Fat Crosstalk: The Biological Role of Myokines, Osteokines, and Adipokines. *Curr. Osteoporos. Rep.* **2020**, *18*, 388–400. [CrossRef]
10. Severinsen, M.C.K.; Pedersen, B.K. Muscle-Organ Crosstalk: The Emerging Roles of Myokines. *Endocr. Rev.* **2020**, *41*, 594–609. [CrossRef]
11. Watanabe-Takano, H.; Ochi, H.; Chiba, A.; Matsuo, A.; Kanai, Y.; Fukuhara, S.; Ito, N.; Sako, K.; Miyazaki, T.; Tainaka, K.; et al. Mechanical load regulates bone growth via periosteal Osteocrin. *Cell Rep.* **2021**, *36*, 109380. [CrossRef] [PubMed]
12. Kanai, Y.; Yasoda, A.; Mori, K.P.; Watanabe-Takano, H.; Nagai-Okatani, C.; Yamashita, Y.; Hirota, K.; Ueda, Y.; Yamauchi, I.; Kondo, E.; et al. Circulating osteocrin stimulates bone growth by limiting C-type natriuretic peptide clearance. *J. Clin. Investig.* **2017**, *127*, 4136–4147. [CrossRef] [PubMed]
13. Bord, S.; Ireland, D.C.; Moffatt, P.; Thomas, G.P.; Compston, J.E. Characterization of osteocrin expression in human bone. *J. Histochem. Cytochem.* **2005**, *53*, 1181–1187. [CrossRef]
14. Hu, C.; Zhang, X.; Zhang, N.; Wei, W.Y.; Li, L.L.; Ma, Z.G.; Tang, Q.Z. Osteocrin attenuates inflammation, oxidative stress, apoptosis, and cardiac dysfunction in doxorubicin-induced cardiotoxicity. *Clin. Transl. Med.* **2020**, *10*, e124. [CrossRef]
15. Jorgensen, L.H.; Petersson, S.J.; Sellathurai, J.; Andersen, D.C.; Thayssen, S.; Sant, D.J.; Jensen, C.H.; Schroder, H.D. Secreted protein acidic and rich in cysteine (SPARC) in human skeletal muscle. *J. Histochem. Cytochem.* **2009**, *57*, 29–39. [CrossRef]
16. Jorgensen, L.H.; Jepsen, P.L.; Boysen, A.; Dalgaard, L.B.; Hvid, L.G.; Ortenblad, N.; Ravn, D.; Sellathurai, J.; Moller-Jensen, J.; Lochmuller, H.; et al. SPARC Interacts with Actin in Skeletal Muscle in Vitro and in Vivo. *Am. J. Pathol.* **2017**, *187*, 457–474. [CrossRef] [PubMed]
17. Morrissey, M.A.; Jayadev, R.; Miley, G.R.; Blebea, C.A.; Chi, Q.; Ihara, S.; Sherwood, D.R. SPARC Promotes Cell Invasion In Vivo by Decreasing Type IV Collagen Levels in the Basement Membrane. *PLoS Genet.* **2016**, *12*, e1005905. [CrossRef]
18. Oost, L.J.; Kustermann, M.; Armani, A.; Blaauw, B.; Romanello, V. Fibroblast growth factor 21 controls mitophagy and muscle mass. *J. Cachexia Sarcopenia Muscle* **2019**, *10*, 630–642. [CrossRef] [PubMed]
19. Uddin, M.S.; Kabir, M.T.; Tewari, D.; Al Mamun, A.; Barreto, G.E.; Bungau, S.G.; Bin-Jumah, M.N.; Abdel-Daim, M.M.; Ashraf, G.M. Emerging Therapeutic Promise of Ketogenic Diet to Attenuate Neuropathological Alterations in Alzheimer's Disease. *Mol. Neurobiol.* **2020**, *57*, 4961–4977. [CrossRef]
20. Tabaie, E.A.; Reddy, A.J.; Brahmbhatt, H. A narrative review on the effects of a ketogenic diet on patients with Alzheimer's disease. *AIMS Public Health* **2022**, *9*, 185–193. [CrossRef]
21. Abduljawad, A.A.; Elawad, M.A.; Elkhalifa, M.E.M.; Ahmed, A.; Hamdoon, A.A.E.; Salim, L.H.M.; Ashraf, M.; Ayaz, M.; Hassan, S.S.U.; Bungau, S. Alzheimer's Disease as a Major Public Health Concern: Role of Dietary Saponins in Mitigating Neurodegenerative Disorders and Their Underlying Mechanisms. *Molecules* **2022**, *27*, 6804. [CrossRef] [PubMed]
22. Li, Y.P.; Reid, M.B. Effect of tumor necrosis factor-alpha on skeletal muscle metabolism. *Curr. Opin. Rheumatol.* **2001**, *13*, 483–487. [CrossRef] [PubMed]
23. Kumar, S.; Behl, T.; Sachdeva, M.; Sehgal, A.; Kumari, S.; Kumar, A.; Kaur, G.; Yadav, H.N.; Bungau, S. Implicating the effect of ketogenic diet as a preventive measure to obesity and diabetes mellitus. *Life Sci.* **2021**, *264*, 118661. [CrossRef] [PubMed]
24. Ghitea, T.C.; Aleya, L.; Tit, D.M.; Behl, T.; Stoicescu, M.; Sava, C.; Iovan, C.; El-Kharoubi, A.; Uivarosan, D.; Pallag, A.; et al. Influence of diet and sport on the risk of sleep apnea in patients with metabolic syndrome associated with hypothyroidism—A 4-year survey. *Environ. Sci. Pollut. Res. Int.* **2022**, *29*, 23158–23168. [CrossRef] [PubMed]
25. Popescu-Spineni, D.M.; Guja, L.; Cristache, C.M.; Pop-Tudose, M.E.; Munteanu, A.M. The Influence of Endocannabinoid System on Women Reproduction. *Acta Endocrinol.* **2022**, *18*, 209–215. [CrossRef] [PubMed]

26. Karasu, T.; Marczylo, T.H.; Maccarrone, M.; Konje, J.C. The role of sex steroid hormones, cytokines and the endocannabinoid system in female fertility. *Hum. Reprod. Update* **2011**, *17*, 347–361. [CrossRef]
27. Leal, L.G.; Lopes, M.A.; Batista, M.L., Jr. Physical Exercise-Induced Myokines and Muscle-Adipose Tissue Crosstalk: A Review of Current Knowledge and the Implications for Health and Metabolic Diseases. *Front. Physiol.* **2018**, *9*, 1307. [CrossRef]
28. Kysel, P.; Haluzikova, D.; Dolezalova, R.P.; Lankova, I.; Lacinova, Z.; Kasperova, B.J.; Trnovska, J.; Hradkova, V.; Mraz, M.; Vilikus, Z.; et al. The Influence of Cyclical Ketogenic Reduction Diet vs. Nutritionally Balanced Reduction Diet on Body Composition, Strength, and Endurance Performance in Healthy Young Males: A Randomized Controlled Trial. *Nutrients* **2020**, *12*, 2832. [CrossRef]
29. World Medical Association Declaration of Helsinki: Ethical Principles for Medical Research Involving Human Subjects. Available online: https://www.wma.net/policies-post/wma-declaration-of-helsinki-ethical-principles-for-medical-research-involving-human-subjects/ (accessed on 2 January 2023).
30. FDA (U.S. Food and Drug Administration). *Guidance for the Clinical Evaluation of Weight-Control Drugs*; FDA: Rockville, MD, USA, 1996.
31. Lee, J.H.; Jun, H.S. Role of Myokines in Regulating Skeletal Muscle Mass and Function. *Front. Physiol.* **2019**, *10*, 42. [CrossRef]
32. Eckel, J. Myokines in metabolic homeostasis and diabetes. *Diabetologia* **2019**, *62*, 1523–1528. [CrossRef]
33. Subbotina, E.; Sierra, A.; Zhu, Z.; Gao, Z.; Koganti, S.R.; Reyes, S.; Stepniak, E.; Walsh, S.A.; Acevedo, M.R.; Perez-Terzic, C.M.; et al. Musclin is an activity-stimulated myokine that enhances physical endurance. *Proc. Natl. Acad. Sci. USA* **2015**, *112*, 16042–16047. [CrossRef]
34. Sierra, A.P.R.; Fontes-Junior, A.A.; Paz, I.A.; de Sousa, C.A.Z.; Manoel, L.; Menezes, D.C.; Rocha, V.A.; Barbeiro, H.V.; Souza, H.P.; Cury-Boaventura, M.F. Chronic Low or High Nutrient Intake and Myokine Levels. *Nutrients* **2022**, *15*, 153. [CrossRef]
35. Yan, Q.; Sage, E.H. SPARC, a matricellular glycoprotein with important biological functions. *J. Histochem. Cytochem.* **1999**, *47*, 1495–1506. [CrossRef]
36. Ghanemi, A.; Melouane, A.; Yoshioka, M.; St-Amand, J. Secreted Protein Acidic and Rich in Cysteine (Sparc) KO Leads to an Accelerated Ageing Phenotype Which Is Improved by Exercise Whereas SPARC Overexpression Mimics Exercise Effects in Mice. *Metabolites* **2022**, *12*, 125. [CrossRef] [PubMed]
37. Aoi, W.; Naito, Y.; Takagi, T.; Tanimura, Y.; Takanami, Y.; Kawai, Y.; Sakuma, K.; Hang, L.P.; Mizushima, K.; Hirai, Y.; et al. A novel myokine, secreted protein acidic and rich in cysteine (SPARC), suppresses colon tumorigenesis via regular exercise. *Gut* **2013**, *62*, 882–889. [CrossRef]
38. Velingkar, A.; Vuree, S.; Prabhakar, P.K.; Kalshikam, R.R.; Kondeti, S. Fibroblast Growth Factor 21 as a Potential Master Regulator in Metabolic Disorders. *Am. J. Physiol. Endocrinol. Metab.* **2023**. [CrossRef] [PubMed]
39. Potthoff, M.J.; Inagaki, T.; Satapati, S.; Ding, X.; He, T.; Goetz, R.; Mohammadi, M.; Finck, B.N.; Mangelsdorf, D.J.; Kliewer, S.A.; et al. FGF21 induces PGC-1alpha and regulates carbohydrate and fatty acid metabolism during the adaptive starvation response. *Proc. Natl. Acad. Sci. USA* **2009**, *106*, 10853–10858. [CrossRef] [PubMed]
40. Porflitt-Rodriguez, M.; Guzman-Arriagada, V.; Sandoval-Valderrama, R.; Tam, C.S.; Pavicic, F.; Ehrenfeld, P.; Martinez-Huenchullan, S. Effects of aerobic exercise on fibroblast growth factor 21 in overweight and obesity. A systematic review. *Metabolism* **2022**, *129*, 155137. [CrossRef]
41. Matsui, M.; Kosaki, K.; Tanahashi, K.; Akazawa, N.; Osuka, Y.; Tanaka, K.; Kuro, O.M.; Maeda, S. Relationship between physical activity and circulating fibroblast growth factor 21 in middle-aged and older adults. *Exp. Gerontol.* **2020**, *141*, 111081. [CrossRef]
42. Atakan, M.M.; Kosar, S.N.; Guzel, Y.; Tin, H.T.; Yan, X. The Role of Exercise, Diet, and Cytokines in Preventing Obesity and Improving Adipose Tissue. *Nutrients* **2021**, *13*, 1459. [CrossRef]
43. Pedersen, B.K.; Febbraio, M.A. Muscles, exercise and obesity: Skeletal muscle as a secretory organ. *Nat. Rev. Endocrinol.* **2012**, *8*, 457–465. [CrossRef] [PubMed]
44. Cipryan, L.; Dostal, T.; Plews, D.J.; Hofmann, P.; Laursen, P.B. Adiponectin/leptin ratio increases after a 12-week very low-carbohydrate, high-fat diet, and exercise training in healthy individuals: A non-randomized, parallel design study. *Nutr. Res.* **2021**, *87*, 22–30. [CrossRef] [PubMed]
45. Hansen, N.W.; Hansen, A.J.; Sams, A. The endothelial border to health: Mechanistic evidence of the hyperglycemic culprit of inflammatory disease acceleration. *IUBMB Life* **2017**, *69*, 148–161. [CrossRef] [PubMed]
46. Summer, S.S.; Brehm, B.J.; Benoit, S.C.; D'Alessio, D.A. Adiponectin changes in relation to the macronutrient composition of a weight-loss diet. *Obesity* **2011**, *19*, 2198–2204. [CrossRef] [PubMed]
47. Ruth, M.R.; Port, A.M.; Shah, M.; Bourland, A.C.; Istfan, N.W.; Nelson, K.P.; Gokce, N.; Apovian, C.M. Consuming a hypocaloric high fat low carbohydrate diet for 12 weeks lowers C-reactive protein, and raises serum adiponectin and high density lipoprotein-cholesterol in obese subjects. *Metabolism* **2013**, *62*, 1779–1787. [CrossRef] [PubMed]
48. Sajoux, I.; Lorenzo, P.M.; Gomez-Arbelaez, D.; Zulet, M.A.; Abete, I.; Castro, A.I.; Baltar, J.; Portillo, M.P.; Tinahones, F.J.; Martinez, J.A.; et al. Effect of a Very-Low-Calorie Ketogenic Diet on Circulating Myokine Levels Compared with the Effect of Bariatric Surgery or a Low-Calorie Diet in Patients with Obesity. *Nutrients* **2019**, *11*, 2368. [CrossRef]

49. Paoli, A.; Moro, T.; Bosco, G.; Bianco, A.; Grimaldi, K.A.; Camporesi, E.; Mangar, D. Effects of n-3 polyunsaturated fatty acids (omega-3) supplementation on some cardiovascular risk factors with a ketogenic Mediterranean diet. *Mar. Drugs* **2015**, *13*, 996–1009. [CrossRef]
50. Paoli, A.; Cenci, L.; Pompei, P.; Sahin, N.; Bianco, A.; Neri, M.; Caprio, M.; Moro, T. Effects of Two Months of Very Low Carbohydrate Ketogenic Diet on Body Composition, Muscle Strength, Muscle Area, and Blood Parameters in Competitive Natural Body Builders. *Nutrients* **2021**, *13*, 374. [CrossRef]

nutrients

Article

Fish Oil Supplementation Improves the Repeated-Bout Effect and Redox Balance in 20–30-Year-Old Men Submitted to Strength Training

Gustavo Barquilha [1], Cesar Miguel Momesso Dos Santos [1,2,3], Kim Guimaraes Caçula [1], Vinícius Coneglian Santos [4], Tatiana Geraldo Polotow [1], Cristina Vardaris Vasconcellos [1], José Alberto Fernandes Gomes-Santos [1], Luiz Eduardo Rodrigues [1], Rafael Herling Lambertucci [5], Tamires Duarte Afonso Serdan [1,6], Adriana Cristina Levada-Pires [1], Elaine Hatanaka [1], Maria Fernanda Cury-Boaventura [1], Paulo Barbosa de Freitas [1], Tania Cristina Pithon-Curi [1], Laureane Nunes Masi [1], Marcelo Paes Barros [1], Rui Curi [1,7], Renata Gorjão [1] and Sandro Massao Hirabara [1,*]

[1] Interdisciplinary Post-Graduate Program in Health Sciences, Institute of Physical Activity Sciences and Sports, Cruzeiro do Sul University, Sao Paulo 01506-000, Brazil
[2] ENAU Faculty, Ribeirão Pires 09424-130, Brazil
[3] United Metropolitan Colleges, Centro Universitário FMU, Sao Paulo 01503-001, Brazil
[4] Department of Physiology and Biophysics, Institute of Biomedical Sciences, University of Sao Paulo, Sao Paulo 05508-000, Brazil
[5] Institute of Health and Society, Federal University of Sao Paulo, Santos 11015-020, Brazil
[6] Department of Molecular Pathobiology, New York University, New York, NY 10010, USA
[7] Instituto Butantan, Sao Paulo 05503-900, Brazil
* Correspondence: sandromh@yahoo.com.br

Abstract: Herein, we investigated the effect of fish oil supplementation combined with a strength-training protocol, for 6 weeks, on muscle damage induced by a single bout of strength exercise in untrained young men. Sixteen men were divided into two groups, supplemented or not with fish oil, and they were evaluated at the pre-training period and post-training period. We investigated changes before and 0, 24, and 48 h after a single hypertrophic exercise session. Creatine kinase (CK) and lactate dehydrogenase (LDH) activities, plasma interleukin-6 (IL-6) and C-reactive protein (CRP) levels, and the redox imbalance were increased in response to the single-bout session of hypertrophic exercises at baseline (pre-training period) and decreased during the post-training period in the control group due to the repeated-bout effect (RBE). The fish oil supplementation exacerbated this reduction and improved the redox state. In summary, our findings demonstrate that, in untrained young men submitted to a strength-training protocol, fish oil supplementation is ideal for alleviating the muscle injury, inflammation, and redox imbalance induced by a single session of intense strength exercises, highlighting this supplementation as a beneficial strategy for young men that intend to engage in strength-training programs.

Keywords: non-linear strength training; *n-3* polyunsaturated fatty acids; inflammation; muscle damage; oxidative stress

Citation: Barquilha, G.; Dos Santos, C.M.M.; Caçula, K.G.; Santos, V.C.; Polotow, T.G.; Vasconcellos, C.V.; Gomes-Santos, J.A.F.; Rodrigues, L.E.; Lambertucci, R.H.; Serdan, T.D.A.; et al. Fish Oil Supplementation Improves the Repeated-Bout Effect and Redox Balance in 20–30-Year-Old Men Submitted to Strength Training. *Nutrients* **2023**, *15*, 1708. https://doi.org/10.3390/nu15071708

Academic Editors: Stephen Ives and Robert Wessells

Received: 16 February 2023
Revised: 23 March 2023
Accepted: 29 March 2023
Published: 31 March 2023

1. Introduction

Muscle damage induced by unusual eccentric exercises results in several skeletal muscle changes, including the release of muscle enzymes into the blood, a reduction in muscle strength, an increase in muscle soreness, and the activation of the inflammatory process and oxidative stress [1]. Nowadays, it is known that these alterations, when well-controlled, are required for adequate and complete muscle recovery [1]. However, when the inflammatory response and oxidative stress are exacerbated, an imbalance occurs in these processes, impairing or delaying muscle repair and regeneration [1,2]. This condition also causes high ATP generation via anaerobic metabolism, muscle inflammation, and oxidative stress. Oxidative stress leads to a change in iron homeostasis and antioxidant depletion, as observed by variations in the reduced: oxidized glutathione ratio [3,4]. Indirect markers of muscle damage—e.g., the plasma activities of creatine kinase (CK) and lactate dehydrogenase (LDH) [4–6]—are frequently evaluated to monitor the efficiency and risks of strength-training protocols in exercising subjects and athletes. Usually, the plasma activities of these enzymes increase within 6 to 8 h after a strength exercise session, peaking between 48 and 72 h and remaining elevated for up to 7 days [5,7]. Delayed-onset muscle soreness associated with muscle injury also peaks between 24 and 48 h post-exercise, and it is more pronounced in non-trained individuals and older people than in high-performance strength athletes [5,7].

After a muscle injury induced by eccentric contractions, the inflammatory response initiates tissue repair and regeneration [1,8]. This response involves the release of cytokines, including interleukin-6 (IL-6), interleukin-1b (IL-1b), and tumor necrosis factor-a (TNF-α). IL-6 is the main cytokine to increase after physical exercise [9,10]. These cytokines also increase in strenuous, high-intensity, and intermittent exercises [11,12]. These three pro-inflammatory cytokines act on the liver, stimulating the production and release of C-reactive protein (CRP), an indicator of systemic acute inflammation [13]. After a single extenuating aerobic or strength exercise session, CRP plasma levels increase [14,15].

The term the "repeated-bout effect" (RBE) commonly refers to the protective adaptation against muscle injury caused by an identical or a similar bout of eccentric exercises after a single bout of eccentric exercise or after a period of strength training [16–18]. This phenomenon has been observed in several animal and human models and usually lasts from weeks to months [19]. The mechanisms involved in the RBE are not entirely understood, but several theories have been proposed, including mechanical, cellular, and neural adaptations [18,19]. Potential interventions for increasing this effect can also help to decrease the impact of muscle damage in subjects during their training program.

Muscle disorders (e.g., lesions, oxidative stress, inflammation, and atrophy) often occur in several conditions, including in exercise-induced injuries and chronic diseases (e.g., obesity, diabetes, metabolic syndrome, and cardiovascular diseases). Cryotherapy [20–22] and the administration of antioxidant and anti-inflammatory agents [23,24], including omega-3 polyunsaturated fatty acids (*n-3* PUFAs) [25–28], have been proposed to provide a protective effect in these muscle disorders or exercise-induced muscle injury. The main anti-inflammatory *n-3* PUFAs comprise eicosapentaenoic acid (EPA) and docosahexaenoic acid (DHA), which have been demonstrated to reduce plasma lipids [3,29,30], oxidative stress conditions [31,32], insulin sensitivity [31,33,34], and inflammation [33,35]. Similar to regular and moderate aerobic exercise, *n-3* PUFAs reduce fat mass [36,37] and cardiovascular risks [38,39], but they do not raise physical capabilities [40] or performance [41,42]. Interestingly, when young endurance athletes are supplemented with highly purified *n-3* PUFAs (2.1 g DHA + 240 mg EPA per day for 10 wks), they present reduced muscle damage (plasma CK and LDH activities), inflammatory markers (plasma IL-6 and IL-1β), and muscle soreness after an eccentric-induced muscle damage exercise session [43].

It was observed that *n-3* PUFAs increase muscle strength gain in older women submitted to a strength-training program [44] or resistance exercise training [44–46]. Due to their anti-inflammatory effects, *n-3* PUFAs have beneficial effects in some diseases, such as neurodevelopmental disorders related to oxidative stress (Rett Syndrome) [47], multiple sclerosis [48], and depression [49]. The effects of fish oil supplementation on muscle damage induced by different physical exercise protocols have also been demonstrated in several studies. Acute supplementation (3 days, 3 g per day) with krill oil (a natural source of *n-3* PUFAs) was sufficient to reduce muscle damage induced by exercise (plasma CK activity) and malondialdehyde content, a stress oxidative marker, but it did not have a significant effect on inflammatory cytokines in resistance-trained young men [28]. In another study, it was observed that supplementation with fish oil for 6 wks in untrained young men induces a reduction in oxidative stress markers (thiobarbituric-acid-reactive substances and H_2O_2-induced DNA damage) after a single bout of eccentric exercise, but it did not have an effect on muscle damage markers or muscle soreness [50]. In highly trained athletes (power training or high-intensity interval training activities), krill oil supplementation (2.5 g per day), for 12 weeks, was associated with reduced oxidative stress after a high-intensity physical exercise session [51]. In untrained young men, fish oil supplementation (600 mg EPA and 260 mg DHA per day), for 8 weeks, was able to attenuate the muscle strength loss, range of motion, muscle soreness, and plasma IL-6 increase induced by a session of maximal voluntary eccentric contractions of the elbow flexors [52]. Some of these effects (the range of motion and serum CK activity) were also observed when subjects were supplemented for a shorter period (4 weeks) [53]. Previously, fish oil supplementation (3 g per day) for 4 weeks was also associated with reduced muscle soreness, an increase in plasma IL-6, and muscle peak power after downhill running at 65% VO_{2max} for 60 min [54]. In another study, it was observed that fish oil supplementation (6 g per day, for 7 weeks) improved muscle recovery and decreased muscle soreness after a damaging eccentric exercise session in recreationally active participants [55]. In summary, previous studies have evaluated the modulation of *n-3* PUFA supplementation on muscle injury induced by a single session of damaging exercises in untrained and trained participants, as well as in athletes. However, there are no studies addressing the combined effect of fish oil supplementation and strength exercise training on muscle damage in untrained participants. Thus, our study aimed to demonstrate the effect of *n-3* PUFA supplementation in combination with a strength-training protocol for 6 weeks on the muscle injury, inflammation, and redox balance induced by a single bout of intense strength exercises in untrained young men. For this purpose, we evaluated the plasma levels of cytokines and C-reactive protein, cortisol and testosterone, the activities of creatine kinase and lactate dehydrogenase, and redox state parameters (total iron, heme iron, reduced and oxidized glutathione, and Trolox equivalent antioxidant capacity—TEAC).

2. Material and Methods

2.1. Participants

All experimental procedures were carried out following the approval of the Ethical Committee for Research of the Cruzeiro do Sul University (Protocol Number: 0392009) and performed in compliance with the Helsinki Declaration. Initially, a total of 21 healthy men, between 20 and 30 years old, were eligible to participate in the study. All participants were classified as physically active using the International Questionnaire of Physical Activity, but they had not engaged in any aerobic or resistance training program in the last 12 months. In this study, we decided to investigate only young men to eliminate the influence of the hormonal variations observed in women due to the menstrual cycle, since female hormones have been associated with different leukocyte responses during exercise-induced muscle injury [56]. Individuals with muscle injury, endocrine disease, and hormonal or nutritional supplement usage were excluded from the study. The participants were randomly divided into two groups: a control group (n = 10) and a group supplemented with fish oil, a natural source of *n-3* PUFAs (n = 11). At the end of the experimental protocol, 2 participants from the control group and 3 from the fish oil group were excluded from the study for different reasons: withdrawal from participating in the study (1 from the fish oil group), an inability to attend the strength-training protocol (at least 85% of participation; 2 from the control group and 1 from the fish oil group), and inadequate supplementation (at least 90% adherence, as assessed by the capsule count at the end of the experimental protocol; 1 from the fish oil group). Thus, at the end, 8 participants of each group completed the experimental protocol, and their results were used in the analysis.

2.2. A Single Bout of a Strength Exercise Protocol

A single bout of strength exercises, consisting of 6 sets of 10 maximum repetitions, with intervals of 1 min between sets, was applied at baseline (the pre-training period) and after six weeks of training (the post-training period). The temporal responses (before and 0, 24, and 48 h after the single session) of muscle damage markers—the plasma activity of CK and LDH, and the circulating concentration of inflammatory cytokines (IL-6, TNF-α, and IL-1β) and CRP—were monitored, according to previous studies [5,57]. Plasma cortisol and testosterone levels were measured before and immediately after the single session of strength exercises, and the redox parameters were only measured after 24 h.

2.3. Strength-Training Protocol

All participants were supervised by a well-experienced professional in strength training, for the whole training protocol period (6 weeks), which was performed at the Cruzeiro do Sul University (Sao Paulo, Brazil). As an exclusion criterion, a minimum participation of 85% was required for the entire strength-training protocol. All participants were submitted to a strength-training protocol, which comprised a daily undulating periodization model [17] for six weeks, three times per week. Briefly, the participants performed the following training schedule: weeks 1, 3, and 6 (hypertrophy)—6 series of 10 repetitions with a 1 min interval (6 × 10 with 1 min interval); weeks 2 and 4 (strength)—5 × 5 with a 3 min interval; and week 5 (resistance)—2 × 20 with a 1 min interval.

2.4. Supplementation with Fish Oil

Fish oil capsules were provided by the Naturalis Nutricao & Farma LTDA (Sao Paulo, Brazil). The participants received 3 capsules of fish oil per day as recommended by the manufacturer. As demonstrated in previous studies, nutritional intervention or fish oil supplementation changes the fatty acid profile after a few weeks [53,58]. A high-performance liquid chromatography (HPLC) analysis for the determination of the fatty acid profile in the fish oil capsules demonstrated that each capsule contained 260 mg EPA and 202 mg DHA. Therefore, the daily doses of *n-3* PUFAs were 780 mg of EPA and 606 mg of DHA. The participants were supplemented for the six weeks of the daily undulating strength training. At the end of the experimental protocol, the remaining fish oil capsules were counted to determine the adherence of the participants to the fish oil supplementation. One participant was excluded because he had less than 90% adherence.

2.5. Blood Collection and Plasma Separation

The participants were instructed to not eat for at least four h before blood collection for a biochemical analysis of the plasma. The participants were instructed to have their regular breakfast after waking up (up to 07:00–08:00 a.m.), and blood collection was performed between 11:00 and 12:00 a.m.; therefore, all participants were in the same feeding state. Samples were collected before and 0, 24, and 48 h after a single session of a bout of strength exercise. After that, the blood samples were immediately processed for plasma separation, which was aliquoted and kept at $-80\ °C$ until analysis.

2.6. Measurements of Plasma Cytokines and C-Reactive Protein

IL-6, TNF-α, and IL-1β were measured using a quantitative immunoassay, an Enzyme-Linked Immunosorbent Assay (ELISA), with kits obtained from R&D System (Minneapolis, MN, USA). The concentration of plasma CRP was determined using a commercial kit from Bioclin (Belo Horizonte, Minas Gerais, Brazil) with immunoturbidimetry.

2.7. Plasma Activities of Creatine Kinase and Lactate Dehydrogenase

The activities of plasma CK and LDH were measured using a commercial kit from Bioclin (Belo Horizonte, Minas Gerais, Brazil). CK catalyzes the dephosphorylation of creatine phosphate with the production of adenosine triphosphate (ATP), which reacts with glucose in hexokinase, forming glucose-6-phosphate (G6P). Glucose-6-phosphate dehydrogenase oxidizes G6P to 6-phosphogluconate, reducing nicotinamide adenine dinucleotide (NAD^+) to NADH, which can be detected via spectrophotometry at 340 nm. LDH catalyzes the pyruvate reduction using NADH, producing lactate and NAD^+. The decomposition of NADH is proportional to the enzyme activity, and it can be measured at 340 nm.

2.8. Measurements of Cortisol and Testosterone

The plasma concentrations of testosterone and cortisol were determined using ELISA, following the specifications of the kits from Cayman Chemical Company (Ann Arbor, MI, USA), according to the manufactures' instructions.

2.9. Determination of Redox State Parameters

2.9.1. Total Iron Determination

The plasma total iron concentration was determined using a kit from Doles-Bioquímica Clínica (Goiania, Brazil). The Fe^{2+}:ferrozine complex formed after reducing the ferric ions (Fe^{3+}) released from several sources during exercise was measured at 560 nm. The reducing system comprises 0.36 M hydroxylamine chloride, 0.10 M glycine, 14 mM thiosemicarbazide, and 0.50 mM octylphenoxypolyetoxyethanol, at pH 2 [59]. The specific effects of exercise on redox parameters and background levels in rested subjects were normalized to 1.0, and post-exercise values are, thus, expressed as relative values (compared to pre-exercise values). Areas under curves were calculated between background levels (pre-exercise) and 24 h post-exercise levels (AUCpre-24 h).

2.9.2. Heme Iron Determination

Plasma heme iron (from hemoglobin, myoglobin, and other heme proteins) was assayed using a method based on heme iron oxidation by the ferricyanide anion contained in a solution of 0.10 M KH_2PO_4, 60 mM $K_3[Fe(CN)_6]$, 77 mM KCN, and 82 mM Triton X-100. Heme iron cyanide is stoichiometrically detected at 540 nm, using hemoglobin as a standard curve. The background levels in rested subjects were normalized to 1.0, and post-exercise values are, thus, expressed as relative values (compared to pre-exercise values). Areas under curves were calculated between background levels (pre-exercise) and 24 h post-exercise levels (AUCpre-24 h).

2.9.3. Plasma Trolox Equivalent Antioxidant Capacity (TEAC)

The Trolox equivalent antioxidant capacity in plasma was assayed as described by Van den Berg et al. [60]. Briefly, a 2,2′-azinobis-(3-ethylbenzthiazoline-6-sulfonate radical solution ($ABTS^-$) was prepared by mixing 2.5 mM 2,2′-azobis-(2-amidinopropane) and HCl (ABAP) with 20 mM ABTS stock solution in 100 mM phosphate buffer (pH 7.4), containing 150 mM NaCl (PBS). The solution was heated for 12 min at 60 °C, protected from light, and stored at room temperature, and absorbance at 734 nm should be 0.35–0.40 to ensure sufficient ABTS—formation. Since ABTS—gradually decomposes (approximately 2% per hour), regular blanks (in the absence of samples) were recorded for appropriate subtractions. The background levels in rested subjects were normalized to 1.0, and post-exercise values are, thus, expressed as relative values (compared to pre-exercise values). Areas under curves were calculated between background levels (pre-exercise) and 24 h post-exercise levels (AUCpre-24 h).

2.9.4. Reduced and Oxidized Glutathione Measurements

The reduced (GSH) and oxidized (GSSG) glutathione content in plasma was measured as described by Rahman et al. (Rahman et al., 2006) [61]. The method is based on the reaction of reduced thiol groups (such as in GSH) with 5,5′-dithiobis-2-nitrobenzoic acid (DTNB) to form 5-thio-2-nitrobenzoic acid (TNB), which is stoichiometrically detected via absorbance at 412 nm. Purified GSH and GSSG were used as standards. The background levels in rested subjects were normalized to 1.0, and post-exercise values are, thus, expressed as relative values (compared to pre-exercise values). Areas under curves were calculated between background levels (pre-exercise) and 24 h post-exercise levels (AUCpre-24 h).

2.10. Statistical Analysis

The results are presented as mean $\pm$ standard error of the mean (S. E. M.) and analyzed using Student's *t*-test when comparing AUC changes (pre- and post-training) between the fish oil and control groups and using two-way ANOVA, followed by Bonferroni post-test for multiple comparisons to evaluate the effect of training and/or supplementation (control pre-training vs. control post-training; fish oil pre-training vs. fish oil post-training; control pre-training vs. fish oil pre-training; and control post-training vs. fish oil post-training). The Cohen's *d* effect size values were determined based on the mean differences between the fish oil and control groups and pooled SD: Cohen's $d = (M_2 - M_1)/SD_{pooled}$; $SD_{pooled} = \sqrt{((SD_1{}^2 + SD_2{}^2)/2)}$ [62,63].

3. Results

3.1. Plasma Activity of Creatine Kinase and Lactate Dehydrogenase

A single bout of strength exercise increased the CK and LDH activities in the control and *n-3* PUFA-fed groups at baseline (the pre-training period), as shown in Figure 1A,C, respectively. During the post-training period, this increase was attenuated in the control group and reduced by the fish oil supplementation. We did not find any statistical difference using the two-way ANOVA test, but when the AUCs of the control group and the fish oil group were compared using Student's *t*-test, we observed a marked difference, as demonstrated in Figure 1B,D. The AUCs of the CK and LDH activities were also analyzed using Cohen's *d* effect size; the supplemented group exhibited a higher attenuation than the control group (effect sizes of −1.44 and −1.40, respectively). The intra-assay coefficient of variance (CV%) was 3.6–7.0% for CK activity and 4.4–9.0% for LDH activity.

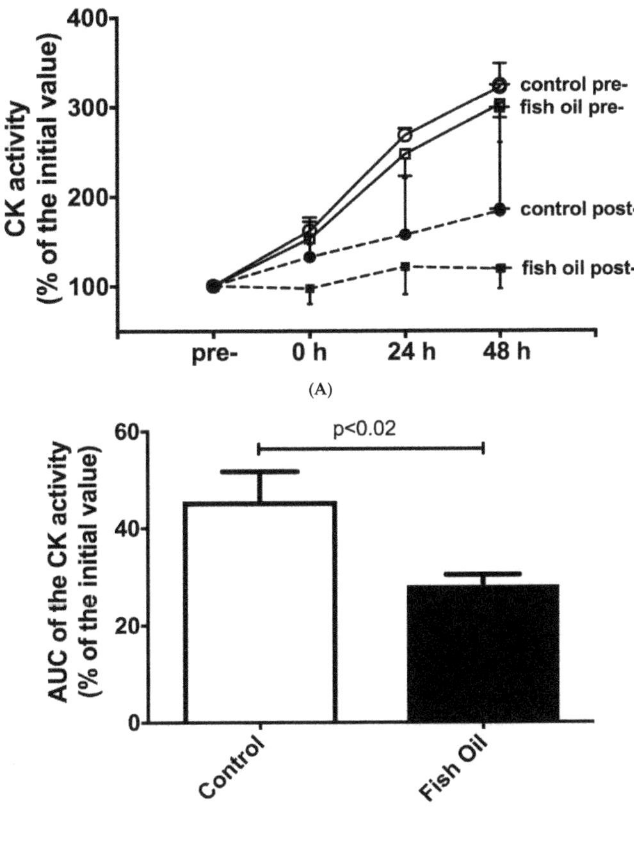

Figure 1. *Cont.*

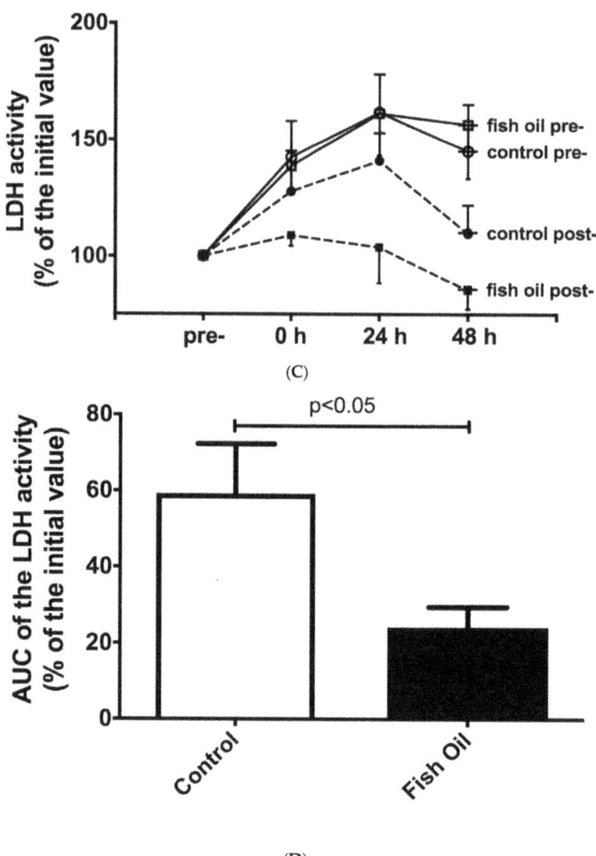

Figure 1. Effect of the fish oil supplementation on plasma activity of (**A**) creatine kinase (CK) and (**C**) lactate dehydrogenase (LDH), in response to a single bout of strength exercises, at baseline (pre-training period) and after 6 weeks of daily undulating strength training (post-training period). On the left, time-dependent plasma CK and LDH activities (before and 0, 24, and 48 h after a single bout of strength exercises). On the right, decrease in the area under curve (AUC) of the temporal plasma CK (**B**) and LDH (**D**) activities after 6 weeks of daily undulating strength training. Results presented as mean $\pm$ S.E.M. $p < 0.02$ for CK activity, and $p < 0.05$ for LDH activity, comparing control group with fish oil group.

3.2. Determination of Inflammation Markers

A single bout of strength exercises increased the plasma concentrations of IL-6 and CRP in both groups at baseline (the pre-training period). Following the strength-training protocol (the post-training period), this increase was significantly attenuated in the control group and exacerbated in the fish-oil-supplemented group (Figure 2A,C). No difference was found using the two-way ANOVA test, but a marked reduction was observed when the AUCs of the control group and the fish oil group were compared using Student's *t*-test, as demonstrated in Figure 2B,D. When the AUCs of the plasma IL-6 and CRP levels were compared using Cohen's *d* effect size, the supplemented group showed a higher reduction than the control group (effect sizes of -1.30 and -1.21, respectively). The linearity (r^2) for the IL-6 assay was 0.983. The intra-assay coefficient of variance (CV%) was 4.2–8.50% for IL-6 and 1.1–3.9% for CRP. We did not observe any significant alteration in the plasma IL-1b and TNF-α levels.

3.3. Plasma Testosterone: Cortisol Ratio

The testosterone: cortisol ratio was not significantly modified by a single bout of strength exercises before (pre-training) and after (post-training) six weeks of the strength-training protocol. Fish oil supplementation did not alter this response (Figure 3). The linearity (r^2) for the testosterone assay was 0.991, and for the cortisol assay, it was 0.966. The intra-assay coefficient of variance (CV%) was 4.1–6.2.0% for testosterone and 3.7–8.3% for cortisol.

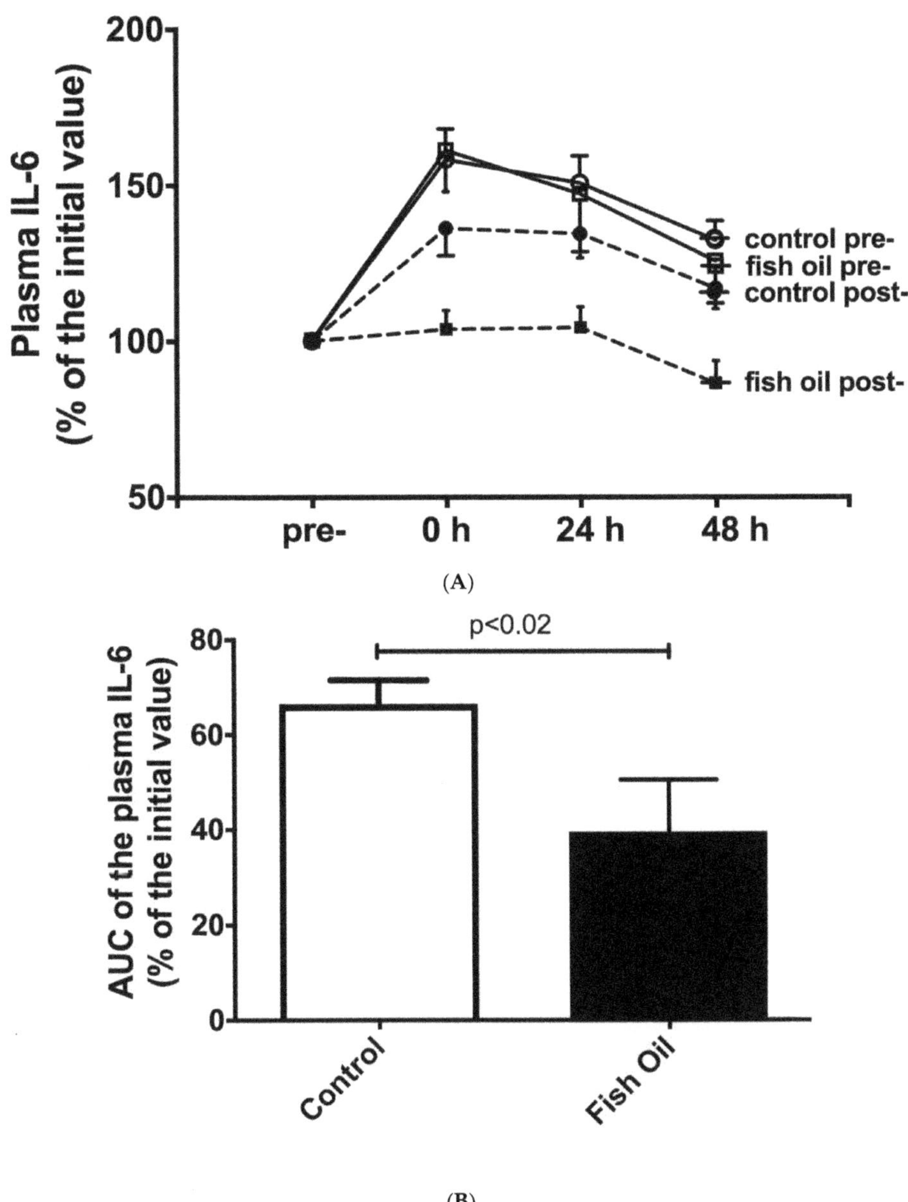

Figure 2. *Cont.*

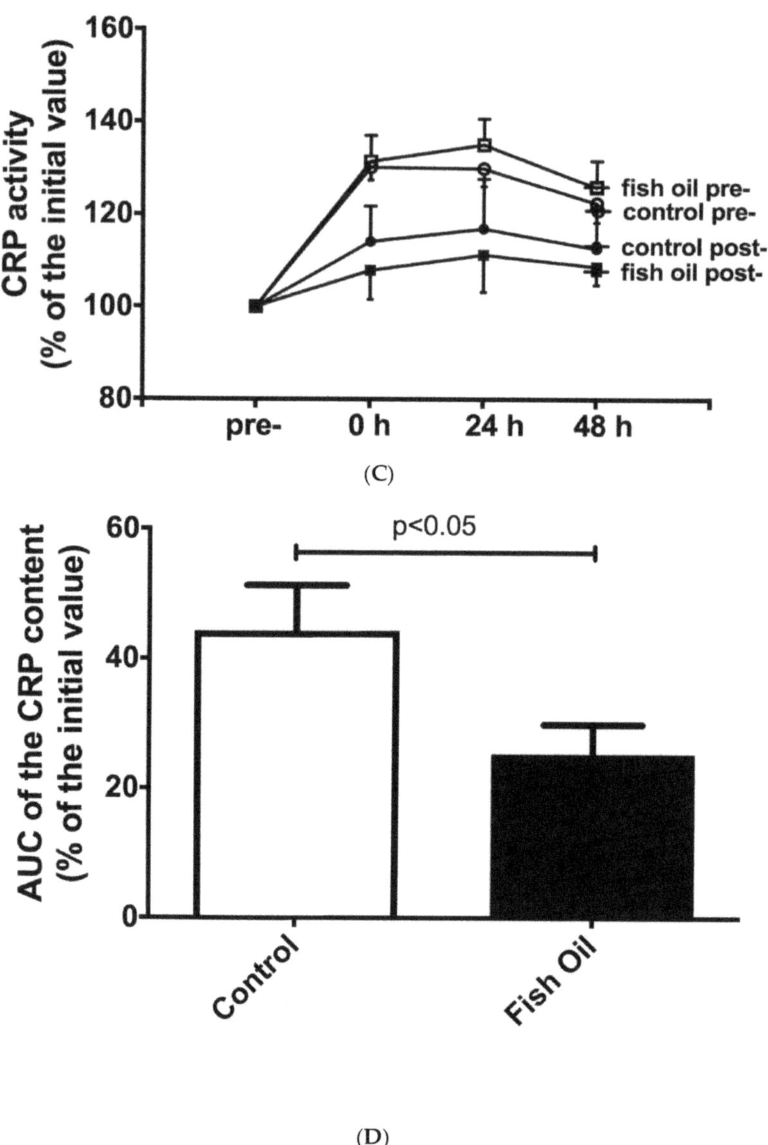

(D)

Figure 2. Effect of fish oil supplementation on plasma interlekin-6 (IL-6) (**A**) and C-reactive protein (CRP) (**C**) concentrations, in response to a single bout of strength exercises, at baseline (pre-training period) and after 6 weeks of non-linear strength training (post-training period). On the left, time-dependent plasma concentration (before and 0, 24, and 48 h after a single session of hypertrophic exercises). On the right, reduction in the area under curve (AUC) of the temporal plasma concentrations of IL-6 (**B**) and CRP (**D**) after 6 weeks of daily undulating training, associated or not with fish oil supplementation. Results presented as mean ± S.E.M. $p < 0.02$ for IL-6, and $p < 0.05$ for CRP, comparing control group with fish oil group.

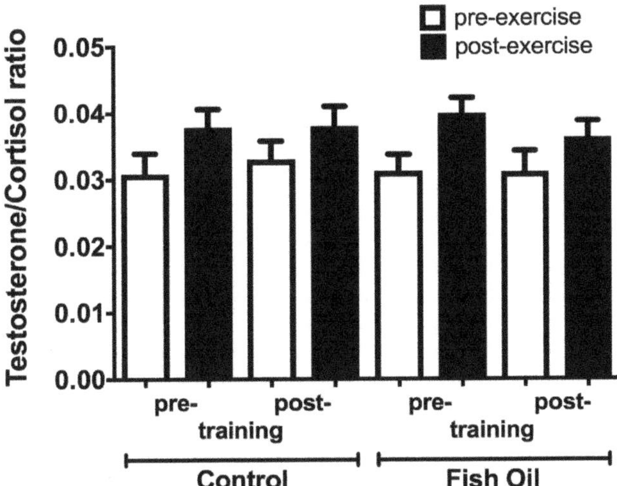

Figure 3. Effect of fish supplementation on plasma testosterone:cortisol ratio in response to a single bout of strength exercise before (pre-training) and after (post-training) 6 weeks of daily undulating strength training. Results presented as mean ± S.E.M.

3.4. Measurement of Plasma Redox Parameters

After six weeks of daily undulating strength training, it was found that fish oil supplementation did not modify the plasma concentrations of iron, heme iron, and TEAC after 24 h of a single bout of strength exercises when compared to those of the control group, using Student's *t*-test. However, it significantly increased GSH and decreased GSSG levels (effect sizes of +1.44 and −1.55, respectively) (Figure 4A). Consequently, the ratio of GSH/GSSH was increased by fish oil supplementation (an effect size of +5.97) (Figure 4B).

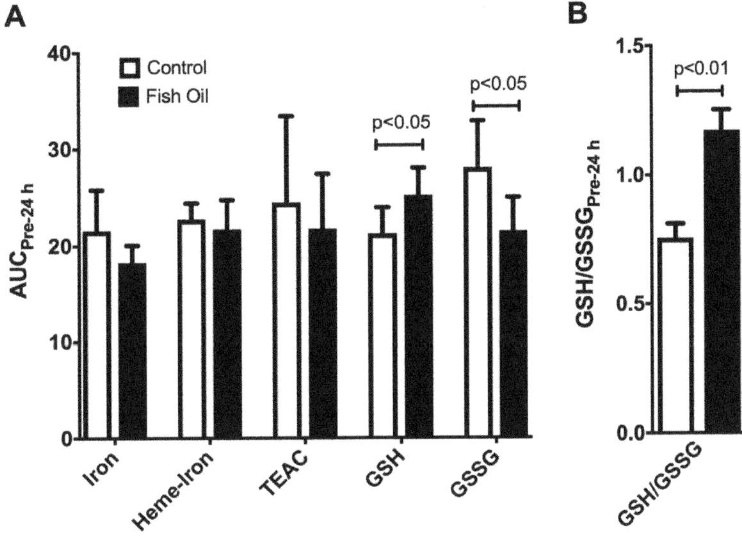

Figure 4. Effect of fish oil supplementation on redox parameters in response to a single bout of strength exercise after 6 weeks of daily undulating strength training (post-training period). Results presented as mean ± S.E.M. (**A**) Areas under curves (AUC) were calculated to express total concentrations of oxidative stress biomarkers in plasma pre- and 24 h post-exercise; (**B**) Reduced/oxidized glutathione ratios (GSH/GSSG) pre- and 24 h post-exercise.

4. Discussion

Various previous studies have demonstrated the beneficial effects of fish oil supplementation on markers of muscle injury (plasma CK and LDH activities), inflammation (plasma levels of pro-inflammatory cytokines), muscle soreness, and oxidative stress induced by different protocols of a single bout of damaging exercises, both in untrained and trained participants, including athletes [28,50–54]. In these studies, the participants were supplemented with fish oil prior to a protocol of a single bout of exercise that induces muscle damage. The main novelty of our study is that it addresses the effect of fish oil supplementation in combination with strength exercise training on exercise-induced muscle damage in untrained participants. Thus, our study is particularly important because it demonstrates that, in untrained young men submitted to strength exercise training, fish oil supplementation is ideal for alleviating the muscle injury, inflammation, and redox balance induced by a single bout of intense strength exercises.

At baseline (the pre-training period), a single bout of strength exercise increased plasma CK and LDH activities and IL-6 and CRP concentrations, classical markers of muscle damage and inflammation, respectively. However, these effects were significantly attenuated after six weeks of daily undulating training (the post-training period), demonstrating a protective muscle adaptation to the training. The RBE occurs when the individual presents attenuation in muscle injury, inflammation, and soreness after the same or similar bouts of physical exercise or training over time. Thus, the RBE is an important physiological adaptation to protect the skeletal muscle against excessive damage and inflammation, reducing the soreness and muscle recovery time after successive bouts of the same or similar physical exercise sessions or training [16–18]. Two components are mainly involved in the RBE in our study: (i) the first bout of strength exercise and (ii) the strength training. Regular physical training promotes anti-inflammatory and antioxidant responses [64,65], which additionally contribute to the RBE.

After muscle damage, an adequate and well-controlled inflammatory response is required to completely restore muscle homeostasis and for recovery [1]. This response involves the recruitment of leukocytes into injured tissue and the production of pro-inflammatory cytokines, consequently increasing these mediators' local and systemic concentrations [1,10]. However, an exacerbated inflammatory response after eccentric exercises can impair or delay muscle repair and regeneration. In our study, we observed an increased temporal plasma release of IL-6 and CRP after a single bout of strength exercises, but there were no differences in TNF-α or IL-1β plasma concentrations. Previous studies have also found no alterations in pro-inflammatory cytokines induced by physical activity [5,14]. The pro-inflammatory cytokines TNF-α, IL-1β, and IL-6, are essential for the acute inflammatory response, as they stimulate the production of acute-phase proteins, including CRP. This response depends on the characteristics of the physical exercise involved, including the intensity, volume, and intervals among series [15,66]. Other authors suggest that pro-inflammatory cytokines are locally produced by the exercised muscles and released into circulation but rapidly degrade, remaining stable in plasma for a short period [8]. These observations can explain, at least in part, our results concerning the pro-inflammatory cytokines IL-1β and TNF-α.

Supplementation with *n*-3 PUFA additionally increased the RBE, as demonstrated by the reduction in the plasma activities of CK and LDH and the circulating concentrations of IL-6 and CRP. Although the mechanisms involved in the RBE are not entirely known yet, cellular modifications may occur as a result of the fish oil supplementation, improving the protective adaptation against muscle damage induced by strength exercises. Some studies suggest that the increased recruitment of sarcomeres during contraction decreases mechanical stress, avoiding the rupture of myofibrils [67]. The reduced inflammatory process in the participants submitted to the fish oil supplementation could further attenuate the response induced by strength exercises.

The relationship between testosterone and cortisol in response to physical exercise indicates physical stress or an imbalance between anabolic and catabolic processes [68,69].

We did not observe any alteration in the testosterone/cortisol ratio as a result of the strength-training protocol or the fish oil supplementation, suggesting that our experimental protocol could not modify physical stress or the anabolic/catabolic balance. Uchida et al. [57] evaluated the influence of different intensities (50, 75, 90, and 110% of 1RM) of the bench press exercise on the same hormones, and they also did not find any changes in the plasma concentrations of both steroid hormones. The authors suggested that the possible cause of this effect was the low volume of exercise and muscle mass involved in the bench press exercise. Crewther et al. [70] studied the impact of three different sessions of squat exercises (45%, 75%, and 88% of 1RM) on plasma testosterone and cortisol concentrations. Interestingly, the session of 75% led to the highest increase compared to the other sessions. Thus, the modulation of the testosterone/cortisol ratio depends on the experimental protocol.

Several authors have used antioxidant compounds to reduce oxidative stress induced by physical exercise [71,72], including interventions with fish oil [25–28]. Regarding redox parameters at the end of the non-linear strength training in our study, no changes in the plasma concentrations of iron, heme iron, and TEAC were observed as a result of the fish oil supplementation. However, increased GSH, decreased GSSG, and consequently an increased GSH/GSSG ratio were found in the supplemented group, suggesting an improved antioxidant defense. GSH rapidly reacts nonenzymatically with reactive oxygen/nitrogen species (ROS/RNS), including the hydroxyl radical, dinitrogen trioxide (N2O3), and peroxynitrite [73]. Moreover, GSH also participates in enzymatic antioxidant defense, e.g., as a substrate of the GPx-mediated reduction of peroxides, resulting in the production of GSSG. The fish oil supplementation improved the GSH/GSSG antioxidant system. An elevated GSH/GSSG ratio is required to control the reducing environment [74]. The effects of *n-3* PUFAs and/or physical exercise might be effective under conditions of an impaired redox balance [26], including in older people [75] and in metabolic and inflammatory diseases [47,76].

The anti-inflammatory effect of *n-3* PUFA has been demonstrated by various research groups, and it has been related to the beneficial effects of these metabolites in different inflammatory diseases, obesity, diabetes mellitus, metabolic syndrome, cardiovascular diseases, fatty liver disease, and cancer [77–80]. The mechanisms of action of *n-3* PUFA involve several signaling pathways, including the activation of GPR120 [81], the generation of anti-inflammatory and/or pro-resolution lipid mediators (resolvins, protectins, and maresins) [82], and the reduction of pro-inflammatory lipid derivatives (prostaglandin and thromboxane 2 series, and leukotriene 4 series) [83]. Our group also demonstrated that *n-3* PUFA supplementation improves mitochondrial function in the skeletal muscle of an animal model of high-fat diet-induced obesity [34]. We propose herein that *n-3* PUFA could potentialize the repeated-bout effect induced by strength training through several actions, including (1) the anti-inflammatory effect, reducing the production of pro-inflammatory cytokines; (2) improved mitochondrial function in skeletal muscle, decreasing the generation of lipid derivatives and reactive oxygen species; and (3) decreased oxidative stress, resulting in diminished muscle damage. This proposition and the main findings of this work are summarized in Figure 5.

Our study is the first to demonstrate the beneficial effects of fish oil supplementation in combination with a strength-training protocol for 6 weeks on the muscle damage markers, inflammation, and redox imbalance induced by a single bout of strength exercises. It is important to describe some of the limitations of our study. First, we investigated the effects of *n-3* PUFAs and strength training for a short period (6 weeks); further studies are required to evaluate the effects for longer periods. Second, we assessed the adherence to the fish oil supplementation only by counting the remaining fish oil capsules at the end of the experimental protocol; a direct measurement (e.g., the determination of plasma fatty acid profiles) is lacking. Third, we analyzed only young men; further studies are required to analyze young women at different phases of the menstrual cycle and other groups of participants, including older people. Lastly, we used a small sample size (n = 8 per group), which could have reduced the statistical power of our analysis. However, the Cohen's *d* effect size values of our data ($d > 1.0$) suggest a large effect of the fish oil supplementation. In addition, the effects of fish oil supplementation on exercise-induced muscle injury were also observed in previous studies that used a similar number of participants (n = 7–11 per group) to demonstrate the effects of the supplementation [53–55]. Thus, based on the findings of previous studies, the well-controlled strength-training protocol that we used, and the Cohen's d effect size values that we found, our results seem to be statistically representative.

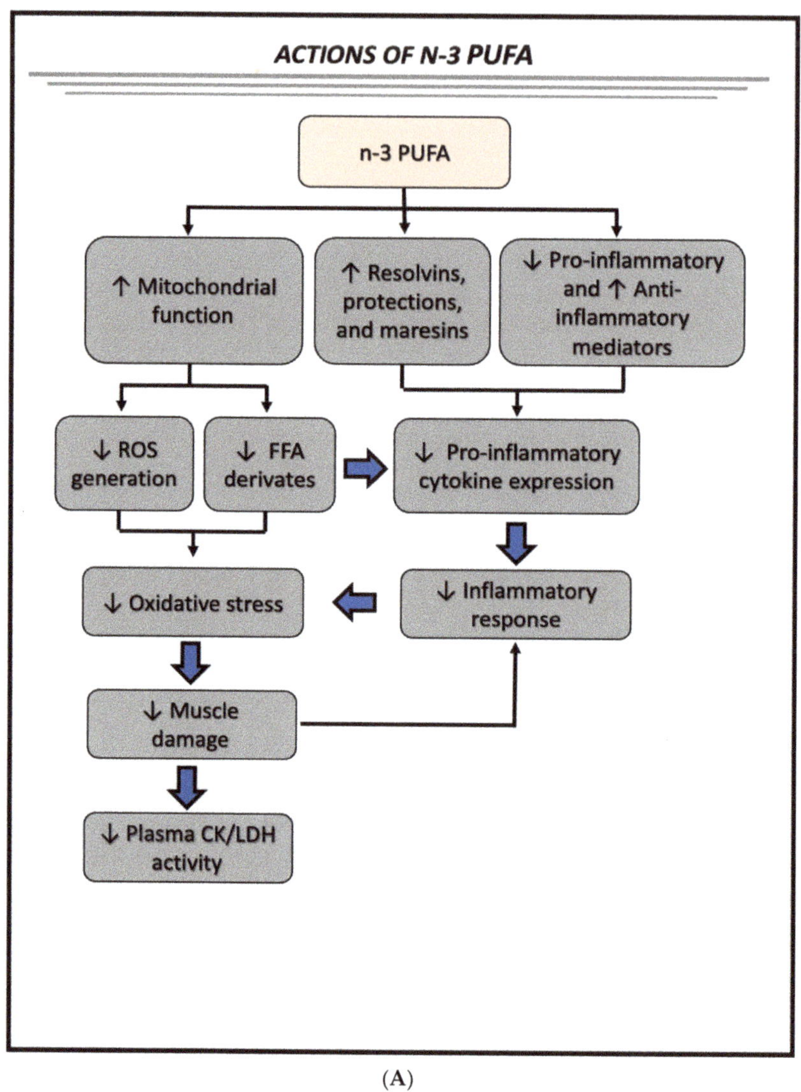

(A)

Figure 5. *Cont.*

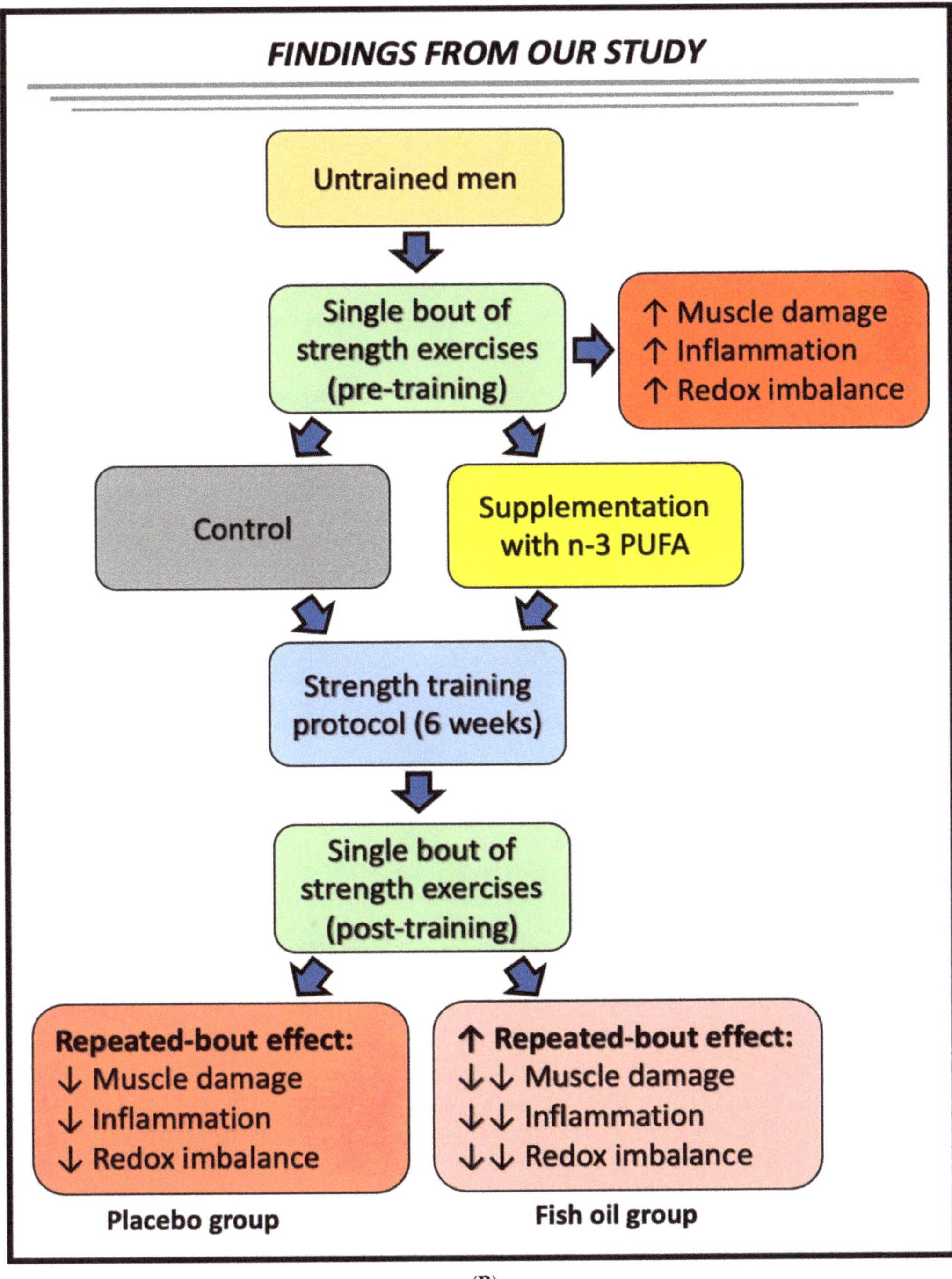

(**B**)

Figure 5. Actions of *n-3* PUFA (**A**) and effects of *n-3* PUFA supplementation on muscle damage induced by a single session of strength exercises after 6 weeks of non-linear strength training (post-training period) (**B**).

In summary, supplementation with *n-3* PUFAs improved the RBE and redox parameters in healthy young men submitted to daily undulating training for six weeks, as demonstrated by the decreased muscle damage (plasma activities of CK and LDH), pro-inflammatory markers (IL-6 and CRP), and redox biomarkers (increased GSH/GSSG ratio) after a bout of strength exercises. Thus, our study is of particular interest because it demonstrates that, in untrained young men submitted to a strength-training protocol, fish oil supplementation is ideal for alleviating the muscle injury, inflammation, and redox imbalance induced by a single session of intense strength exercise. Our findings highlight fish oil supplementation as an effective nutritional strategy to reduce the muscle damage, inflammation, and redox imbalance in untrained individuals who intend to engage in strength-training programs. Further studies are necessary to determine the persistence of this modulation for prolonged training periods and the effects of fish oil supplementation combined with strength exercise training in other groups of participants, including young women and older people.

Author Contributions: Design, conception, and supervision of the study: R.H.L., T.D.A.S., A.C.L.-P., E.H., M.F.C.-B., P.B.d.F., T.C.P.-C., L.N.M., R.G., M.P.B., R.C. and S.M.H. Literature search, collection, analysis, and interpretation of the findings: G.B., C.M.M.D.S., K.G.C., V.C.S., L.E.R., T.G.P., C.V.V., J.A.F.G.-S. and S.M.H. Manuscript preparation and first draft writing: G.B., C.M.M.D.S., V.C.S., T.G.P., J.A.F.G.-S. and S.M.H. Manuscript editing and review: G.B., R.H.L., T.D.A.S., A.C.L.-P., E.H., M.F.C.-B., P.B.d.F., T.C.P.-C., L.N.M., R.G., M.P.B., R.C. and S.M.H. Figure and table preparation: G.B., K.G.C., L.E.R., C.V.V. and S.M.H. All authors have read and agreed to the published version of the manuscript.

Funding: This research received no external funding.

Institutional Review Board Statement: The study was conducted in accordance with the Declaration of Helsinki, and approved by the Ethical Committee for Research of the Cruzeiro do Sul University (protocol number 0392009).

Informed Consent Statement: Informed consent was obtained from all subjects involved in the study.

Data Availability Statement: The data presented in this study are available in the article.

Acknowledgments: This study was supported by grants from the Sao Paulo Research Foundation (FAPESP: 2018/09868–7, 2010/08147-2), the Coordination for the Improvement of Higher-Level Personnel (CAPES), the National Council for Scientific and Technological Development (CNPq), and the Dean's Office for Post-Graduate Studies and Research of the Cruzeiro do Sul University. The authors are grateful to Naturalis Nutrition and Pharma Ltd. (Sao Paulo, Brazil) for donating the fish oil capsules.

Conflicts of Interest: The authors declare no conflict of interest.

References

1. Peake, J.M.; Neubauer, O.; Gatta, P.A.D.; Nosaka, K. Muscle damage and inflammation during recovery from exercise. *J. Appl. Physiol.* **2017**, *122*, 559–570. [CrossRef] [PubMed]
2. Armstrong, R.B. Initial events in exercise-induced muscular injury. *Med. Sci. Sport. Exerc.* **1990**, *22*, 429–435.
3. Margonis, K.; Fatouros, I.G.; Jamurtas, A.Z.; Nikolaidis, M.G.; Douroudos, I.; Chatzinikolaou, A.; Mitrakou, A.; Mastorakos, G.; Papassotiriou, I.; Taxildaris, K.; et al. Oxidative stress biomarkers responses to physical overtraining: Implications for diagnosis. *Free Radic. Biol. Med.* **2007**, *43*, 901–910. [CrossRef] [PubMed]
4. González-Hernández, J.M.; Jiménez-Reyes, P.; Cerón, J.J.; Tvarijonaviciute, A.; Llorente-Canterano, F.J.; Martínez-Aranda, L.M.; García-Ramos, A. Response of Muscle Damage Markers to an Accentuated Eccentric Training Protocol: Do Serum and Saliva Measurements Agree? *J. Strength Cond. Res.* **2022**, *36*, 2132–2138. [CrossRef]
5. Uchida, M.C.; Nosaka, K.; Ugrinowitsch, C.; Yamashita, A.; Martins, E., Jr.; Moriscot, A.S.; Aoki, M.S. Effect of bench press exercise intensity on muscle soreness and inflammatory mediators. *J. Sport. Sci.* **2009**, *27*, 499–507. [CrossRef]
6. Chen, T.C.; Yang, T.J.; Huang, M.J.; Wang, H.S.; Tseng, K.W.; Chen, H.L.; Nosaka, K. Damage and the repeated bout effect of arm, leg, and trunk muscles induced by eccentric resistance exercises. *Scand. J. Med. Sci. Sport.* **2019**, *29*, 725–735. [CrossRef]
7. Chen, T.C.; Chen, H.L.; Cheng, L.F.; Chou, T.Y.; Nosaka, K. Effect of Leg Eccentric Exercise on Muscle Damage of the Elbow Flexors after Maximal Eccentric Exercise. *Med. Sci. Sport. Exerc.* **2021**, *53*, 1473–1481. [CrossRef]

8. Paulsen, G.; Mikkelsen, U.R.; Raastad, T.; Peake, J.M. Leucocytes, cytokines and satellite cells: What role do they play in muscle damage and regeneration following eccentric exercise? *Exerc. Immunol. Rev.* **2012**, *18*, 42–97.

9. Kanda, K.; Sugama, K.; Hayashida, H.; Sakuma, J.; Kawakami, Y.; Miura, S.; Yoshioka, H.; Mori, Y.; Suzuki, K. Eccentric exercise-induced delayed-onset muscle soreness and changes in markers of muscle damage and inflammation. *Exerc. Immunol. Rev.* **2013**, *19*, 72–85.

10. Philippou, A.; Tryfonos, A.; Theos, A.; Nezos, A.; Halapas, A.; Maridaki, M.; Koutsilieris, M. Expression of tissue remodelling, inflammation- and angiogenesis-related factors after eccentric exercise in humans. *Mol. Biol. Rep.* **2021**, *48*, 4047–4054. [CrossRef]

11. Bruunsgaard, H.; Galbo, H.; Halkjaer-Kristensen, J.; Johansen, T.L.; MacLean, D.A.; Pedersen, B.K. Exercise-induced increase in serum interleukin-6 in humans is related to muscle damage. *J. Physiol.* **1997**, *499 Pt 3*, 833–841. [CrossRef] [PubMed]

12. Cannon, J.G.; Fielding, R.A.; Fiatarone, M.A.; Orencole, S.F.; Dinarello, C.A.; Evans, W.J. Increased interleukin 1 beta in human skeletal muscle after exercise. *Am. J. Physiol.* **1989**, *257*, R451–R455. [CrossRef]

13. Isaacs, A.W.; Macaluso, F.; Smith, C.; Myburgh, K.H. C-Reactive Protein Is Elevated Only in High Creatine Kinase Responders to Muscle Damaging Exercise. *Front. Physiol.* **2019**, *10*, 86. [CrossRef]

14. Barquilha, G.; Uchida, M.C.; Santos, V.C.; Moura, N.R.; Lambertucci, R.H.; Hatanaka, E.; Cury-Boaventura, M.F.; Pithon-Curi, T.C.; Gorjão, R.; Hirabara, S.M. Characterization of the Effects of One Maximal Repetition Test on Muscle Injury and Inflammation Markers. *WebmedCentral* **2011**, *2*, 1–8.

15. Bernat-Adell, M.D.; Collado-Boira, E.J.; Moles-Julio, P.; Panizo-González, N.; Martínez-Navarro, I.; Hernando-Fuster, B.; Hernando-Domingo, C. Recovery of Inflammation, Cardiac, and Muscle Damage Biomarkers after Running a Marathon. *J. Strength Cond. Res.* **2021**, *35*, 626–632. [CrossRef]

16. Ebbeling, C.B.; Clarkson, P.M. Exercise-induced muscle damage and adaptation. *Sport. Med.* **1989**, *7*, 207–234. [CrossRef] [PubMed]

17. Kraemer, W.J.; Fleck, S.J.; Evans, W.J. Strength and power training: Physiological mechanisms of adaptation. *Exerc. Sport. Sci. Rev.* **1996**, *24*, 363–397. [CrossRef]

18. Hyldahl, R.D.; Chen, T.C.; Nosaka, K. Mechanisms and Mediators of the Skeletal Muscle Repeated Bout Effect. *Exerc. Sport. Sci Rev.* **2017**, *45*, 24–33. [CrossRef] [PubMed]

19. McHugh, M.P.; Connolly, D.A.; Eston, R.G.; Gleim, G.W. Exercise-induced muscle damage and potential mechanisms for the repeated bout effect. *Sport. Med.* **1999**, *27*, 157–170. [CrossRef]

20. Ferreira-Junior, J.B.; Bottaro, M.; Vieira, A.; Siqueira, A.F.; Vieira, C.A.; Durigan, J.L.Q.; Cadore, E.L.; Coelho, L.G.M.; Simões, H.G.; Bemben, M.G. One session of partial-body cryotherapy (−110 °C) improves muscle damage recovery. *Scand. J. Med. Sci. Sport.* **2015**, *25*, e524–e530. [CrossRef]

21. Kwiecien, S.Y.; McHugh, M.P. The cold truth: The role of cryotherapy in the treatment of injury and recovery from exercise. *Eur. J. Appl. Physiol.* **2021**, *121*, 2125–2142. [CrossRef] [PubMed]

22. Haq, A.; Ribbans, W.J.; Hohenauer, E.; Baross, A.W. The Comparative Effect of Different Timings of Whole Body Cryotherapy Treatment with Cold Water Immersion for Post-Exercise Recovery. *Front. Sport. Act. Living* **2022**, *4*, 940516. [CrossRef] [PubMed]

23. Malaguti, M.; Angeloni, C.; Hrelia, S. Polyphenols in exercise performance and prevention of exercise-induced muscle damage. *Oxid. Med. Cell. Longev.* **2013**, *2013*, 825928. [CrossRef] [PubMed]

24. Calella, P.; Cerullo, G.; Di Dio, M.; Liguori, F.; Di Onofrio, V.; Gallè, F.; Liguori, G. Antioxidant, anti-inflammatory and immunomodulatory effects of spirulina in exercise and sport: A systematic review. *Front. Nutr.* **2022**, *9*, 1048258. [CrossRef]

25. Holdsworth, C.T.; Copp, S.W.; Hirai, D.M.; Ferguson, S.K.; Sims, G.E.; Hageman, K.S.; Stebbins, C.L.; Poole, D.C.; Musch, T.I. The effects of dietary fish oil on exercising skeletal muscle vascular and metabolic control in chronic heart failure rats. *Appl. Physiol. Nutr. Metab.* **2014**, *39*, 299–307. [CrossRef] [PubMed]

26. Marques, C.G.; Santos, V.C.; Levada-Pires, A.C.; Jacintho, T.M.; Gorjão, R.; Pithon-Curi, T.C.; Cury-Boaventura, M.F. Effects of DHA-rich fish oil supplementation on the lipid profile, markers of muscle damage, and neutrophil function in wheelchair basketball athletes before and after acute exercise. *Appl. Physiol. Nutr. Metab.* **2015**, *40*, 596–604. [CrossRef] [PubMed]

27. Visconti, L.M.; Cotter, J.A.; Schick, E.E.; Daniels, N.; Viray, F.E.; Purcell, C.A.; Brotman, C.B.R.; Ruhman, K.E.; Escobar, K.A. Impact of varying doses of omega-3 supplementation on muscle damage and recovery after eccentric resistance exercise. *Metabol. Open* **2021**, *12*, 100133. [CrossRef] [PubMed]

28. Yang, S.; He, Q.; Shi, L.; Wu, Y. Impact of *Antarctic krill* oil supplementation on skeletal muscle injury recovery after resistance exercise. *Eur. J. Nutr.* **2022**, *62*, 1345–1356. [CrossRef]

29. Yosefy, C.; Viskoper, J.R.; Laszt, A.; Priluk, R.; Guita, E.; Varon, D.; Illan, Z.; Berry, E.M.; Savion, N.; Adan, Y.; et al. The effect of fish oil on hypertension, plasma lipids and hemostasis in hypertensive, obese, dyslipidemic patients with and without diabetes mellitus. *Prostaglandins Leukot. Essent. Fat. Acids* **1999**, *61*, 83–87. [CrossRef]

30. Shibabaw, T. Omega-3 polyunsaturated fatty acids: Anti-inflammatory and anti-hypertriglyceridemia mechanisms in cardiovascular disease. *Mol. Cell. Biochem.* **2021**, *476*, 993–1003. [CrossRef]

31. Giordano, E.; Visioli, F. Long-chain omega 3 fatty acids: Molecular bases of potential antioxidant actions. *Prostaglandins Leukot. Essent. Fat. Acids* **2014**, *90*, 1–4. [CrossRef] [PubMed]

32. Dong, J.; Feng, X.; Zhang, J.; Zhang, Y.; Xia, F.; Liu, L.; Jin, Z.; Lu, C.; Xia, Y.; Papadimos, T.J.; et al. ω-3 Fish oil fat emulsion preconditioning mitigates myocardial oxidative damage in rats through aldehydes stress. *Biomed. Pharmacother.* **2019**, *118*, 109198. [CrossRef]

33. Tsitouras, P.D.; Gucciardo, F.; Salbe, A.D.; Heward, C.; Harman, S.M. High omega-3 fat intake improves insulin sensitivity and reduces CRP and IL6, but does not affect other endocrine axes in healthy older adults. *Horm. Metab. Res.* **2008**, *40*, 199–205. [CrossRef]

34. Martins, A.R.; Crisma, A.R.; Masi, L.N.; Amaral, C.L.; Marzuca-Nassr, G.N.; Bomfim, L.H.M.; Teodoro, B.G.; Queiroz, A.L.; Serdan, T.D.A.; Torres, R.P.; et al. Attenuation of obesity and insulin resistance by fish oil supplementation is associated with improved skeletal muscle mitochondrial function in mice fed a high-fat diet. *J. Nutr. Biochem.* **2018**, *55*, 76–88. [CrossRef]

35. Djuricic, I.; Calder, P.C. Beneficial Outcomes of Omega-6 and Omega-3 Polyunsaturated Fatty Acids on Human Health: An Update for 2021. *Nutrients* **2021**, *13*, 2421. [CrossRef] [PubMed]

36. Hill, A.M.; Worthley, C.; Murphy, K.J.; Buckley, J.D.; Ferrante, A.; Howe, P.R. *n-3* Fatty acid supplementation and regular moderate exercise: Differential effects of a combined intervention on neutrophil function. *Br. J. Nutr.* **2007**, *98*, 300–309. [CrossRef] [PubMed]

37. Monnard, C.R.; Dulloo, A.G. Polyunsaturated fatty acids as modulators of fat mass and lean mass in human body composition regulation and cardiometabolic health. *Obes. Rev.* **2021**, *22* (Suppl. 2), e13197. [CrossRef]

38. Warner, J.G., Jr.; Ullrich, I.H.; Albrink, M.J.; Yeater, R.A. Combined effects of aerobic exercise and omega-3 fatty acids in hyperlipidemic persons. *Med. Sci. Sport. Exerc.* **1989**, *21*, 498–505. [CrossRef]

39. Ma, T.; He, L.; Luo, Y.; Zhang, G.; Cheng, X.; Bai, Y. Use of fish oil and mortality of patients with cardiometabolic multimo rbidity: A prospective study of UK biobank. *Nutr. Metab. Cardiovasc. Dis.* **2022**, *32*, 2751–2759. [CrossRef]

40. Jost, Z.; Tomczyk, M.; Chroboczek, M.; Calder, P.C.; Laskowski, R. Improved Oxygen Uptake Efficiency Parameters Are Not Correlated with VO2peak or Running Economy and Are Not Affected by Omega-3 Fatty Acid Supplementation in Endurance Runners. *Int. J. Environ. Res. Public Health* **2022**, *19*, 14043. [CrossRef]

41. Oostenbrug, G.S.; Mensink, R.P.; Hardeman, M.R.; De Vries, T.; Brouns, F.; Hornstra, G. Exercise performance, red blood cell deformability, and lipid peroxidation: Effects of fish oil and vitamin E. *J. Appl. Physiol.* **1997**, *83*, 746–752. [CrossRef] [PubMed]

42. Philpott, J.D.; Witard, O.C.; Galloway, S.D.R. Applications of omega-3 polyunsaturated fatty acid supplementation for sport performance. *Res. Sport. Med.* **2019**, *27*, 219–237. [CrossRef] [PubMed]

43. Ramos-Campo, D.J.; Ávila-Gandía, V.; López-Román, F.J.; Miñarro, J.; Contreras, C.; Soto-Méndez, F.; Domingo Pedrol, J.C.; Luque-Rubia, A.J. Supplementation of Re-Esterified Docosahexaenoic and Eicosapentaenoic Acids Reduce Inflammatory and Muscle Damage Markers after Exercise in Endurance Athletes: A Randomized, Controlled Crossover Trial. *Nutrients* **2020**, *12*, 719. [CrossRef] [PubMed]

44. Rodacki, C.L.; Rodacki, A.L.; Pereira, G.; Naliwaiko, K.; Coelho, I.; Pequito, D.; Fernandes, L.C. Fish-oil supplementation enhances the effects of strength training in elderly women. *Am. J. Clin. Nutr.* **2012**, *95*, 428–436. [CrossRef] [PubMed]

45. Cornish, S.M.; Cordingley, D.M.; Shaw, K.A.; Forbes, S.C.; Leonhardt, T.; Bristol, A.; Candow, D.G.; Chilibeck, P.D. Effects of Omega-3 Supplementation Alone and Combined with Resistance Exercise on Skeletal Muscle in Older Adults: A Systematic Review and Meta-Analysis. *Nutrients* **2022**, *14*, 2221. [CrossRef]

46. da Cruz Alves, N.M.; Pfrimer, K.; Santos, P.C.; de Freitas, E.C.; Neves, T.; Pessini, R.A.; Junqueira-Franco, M.V.M.; Nogueira-Barbosa, M.H.; Greig, C.A.; Ferriolli, E. Randomised Controlled Trial of Fish Oil Supplementation on Responsiveness to Resistance Exercise Training in Sarcopenic Older Women. *Nutrients* **2022**, *14*, 2844. [CrossRef]

47. Leoncini, S.; De Felice, C.; Signorini, C.; Zollo, G.; Cortelazzo, A.; Durand, T.; Galano, J.M.; Guerranti, R.; Rossi, M.; Ciccoli, L.; et al. Cytokine Dysregulation in MECP2- and CDKL5-Related Rett Syndrome: Relationships with Aberrant Redox Homeostasis, Inflammation, and omega-3 PUFAs. *Oxid. Med. Cell. Longev.* **2015**, *2015*, 421624. [CrossRef]

48. Ramirez-Ramirez, V.; Macias-Islas, M.A.; Ortiz, G.G.; Pacheco-Moises, F.; Torres-Sanchez, E.D.; Sorto-Gomez, T.E.; Cruz-Ramos, J.A.; Orozco-Aviña, G.; Celis De La Rosa, A.J. Efficacy of fish oil on serum of TNF alpha, IL-1 beta, and IL-6 oxidative stress markers in multiple sclerosis treated with interferon beta-1b. *Oxid. Med. Cell. Longev.* **2013**, *2013*, 709493. [CrossRef]

49. Grosso, G.; Galvano, F.; Marventano, S.; Malaguarnera, M.; Bucolo, C.; Drago, F.; Caraci, F. Omega-3 fatty acids and depression: Scientific evidence and biological mechanisms. *Oxid. Med. Cell. Longev.* **2014**, *2014*, 313570. [CrossRef]

50. Gray, P.; Chappell, A.; Jenkinson, A.M.; Thies, F.; Gray, S.R. Fish oil supplementation reduces markers of oxidative stress but not muscle soreness after eccentric exercise. *Int. J. Sport. Nutr. Exerc. Metab.* **2014**, *24*, 206–214. [CrossRef]

51. Drobnic, F.; Storsve, A.B.; Burri, L.; Ding, Y.; Banquells, M.; Riera, J.; Björk, P.; Ferrer-Roca, V.; Domingo, J.C. Krill-Oil-Dependent Increases in HS-Omega-3 Index, Plasma Choline and Antioxidant Capacity in Well-Conditioned Power Training Athletes. *Nutrients* **2021**, *13*, 4237. [CrossRef]

52. Tsuchiya, Y.; Yanagimoto, K.; Nakazato, K.; Hayamizu, K.; Ochi, E. Eicosapentaenoic and docosahexaenoic acids-rich fish oil supplementation attenuates strength loss and limited joint range of motion after eccentric contr actions: A randomized, double-blind, placebo-controlled, parallel-group trial. *Eur. J. Appl. Physiol.* **2016**, *116*, 1179–1188. [CrossRef] [PubMed]

53. Tsuchiya, Y.; Ueda, H.; Yanagimoto, K.; Kato, A.; Ochi, E. 4-week eicosapentaenoic acid-rich fish oil supplementation partially protects muscular damage following eccentric contractions. *J. Int. Soc. Sport. Nutr.* **2021**, *18*, 8. [CrossRef] [PubMed]

54. Kyriakidou, Y.; Wood, C.; Ferrier, C.; Dolci, A.; Elliott, B. The effect of Omega-3 polyunsaturated fatty acid supplementation on exercise-induced muscle damage. *J. Int. Soc. Sport. Nutr.* **2021**, *18*, 9. [CrossRef] [PubMed]

55. VanDusseldorp, T.A.; Escobar, K.A.; Johnson, K.E.; Stratton, M.T.; Moriarty, T.; Kerksick, C.M.; Mangine, G.T.; Holmes, A.J.; Lee, M.; Endito, M.R.; et al. Impact of Varying Dosages of Fish Oil on Recovery and Soreness Following Eccentric Exercise. *Nutrients* **2020**, *12*, 2246. [CrossRef]

56. Funaki, A.; Gam, H.; Matsuda, T.; Ishikawa, A.; Yamada, M.; Ikegami, N.; Nishikawa, Y.; Sakamaki-Sunaga, M. Influence of Menstrual Cycle on Leukocyte Response Following Exercise-Induced Muscle Damage. *Int. J. Environ. Res. Public Health* **2022**, *19*, 9201. [CrossRef]

57. Uchida, M.C.; Crewther, B.T.; Ugrinowitsch, C.; Bacurau, R.F.; Moriscot, A.S.; Aoki, M.S. Hormonal responses to different resistance exercise schemes of similar total volume. *J. Strength Cond. Res.* **2009**, *23*, 2003–2008. [CrossRef]

58. Helge, J.W.; Ayre, K.J.; Hulbert, A.J.; Kiens, B.; Storlien, L.H. Regular exercise modulates muscle membrane phospholipid profile in rats. *J. Nutr.* **1999**, *129*, 1636–1642. [CrossRef]

59. Goodwin, J.F.; Murphy, B. The colorimetric determination of iron in biological material with reference to its measurement during chelation therapy. *Clin. Chem.* **1966**, *12*, 58–69. [CrossRef]

60. Van den Berg, R.; Haenen, G.; Van den Berg, H.; Bast, A. Applicability of an improved Trolox equivalent antioxidant capacity (TEAC) assay for evaluation of antioxidant capacity measurements of mixtures. *Food Chem.* **1999**, *66*, 511–517. [CrossRef]

61. Rahman, I.; Kode, A.; Biswas, S.K. Assay for quantitative determination of glutathione and glutathione disulfide levels using enzymatic recycling method. *Nat Protoc* **2006**, *1*, 3159–3165. [CrossRef] [PubMed]

62. Cohen, J. *Statistical Power Analysis for the Behavioral Sciences*; Routledge Academic: New York, NY, USA, 1988.

63. Lakens, D. Calculating and reporting effect sizes to facilitate cumulative science: A practical primer for *t*-tests and ANOVAs. *Front. Psychol.* **2013**, *4*, 863. [CrossRef]

64. Djordjevic, D.Z.; Cubrilo, D.G.; Puzovic, V.S.; Vuletic, M.S.; Zivkovic, V.I.; Barudzic, N.S.; Radovanovic, D.S.; Djuric, D.M.; Jakovljevic, V.L. Changes in athlete's redox state induced by habitual and unaccustomed exercise. *Oxid. Med. Cell. Longev.* **2012**, *2012*, 805850. [CrossRef]

65. El Assar, M.; Álvarez-Bustos, A.; Sosa, P.; Angulo, J.; Rodríguez-Mañas, L. Effect of Physical Activity/Exercise on Oxidative Stress and Inflammation in Muscle and Vascular Aging. *Int. J. Mol. Sci.* **2022**, *23*, 8713. [CrossRef] [PubMed]

66. Steensberg, A.; Fischer, C.P.; Keller, C.; Moller, K.; Pedersen, B.K. IL-6 enhances plasma IL-1ra, IL-10, and cortisol in humans. *Am. J. Physiol. Endocrinol. Metab.* **2003**, *285*, E433–E437. [CrossRef] [PubMed]

67. McHugh, M.P. Recent advances in the understanding of the repeated bout effect: The protective effect against muscle damage from a single bout of eccentric exercise. *Scand. J. Med. Sci. Sport.* **2003**, *13*, 88–97. [CrossRef]

68. Urhausen, A.; Gabriel, H.; Kindermann, W. Blood hormones as markers of training stress and overtraining. *Sport. Med.* **1995**, *20*, 251–276. [CrossRef]

69. Tait, J.L.; Bulmer, S.M.; Drake, J.M.; Drain, J.R.; Main, L.C. Impact of 12 weeks of basic military training on testosterone and cortisol responses. *BMJ Mil. Health* **2022**, e002179. [CrossRef]

70. Crewther, B.; Cronin, J.; Keogh, J.; Cook, C. The salivary testosterone and cortisol response to three loading schemes. *J. Strength Cond. Res.* **2008**, *22*, 250–255. [CrossRef]

71. Gomes, E.C.; Silva os de Oliveira, M.R. Oxidants, antioxidants, and the beneficial roles of exercise-induced production of reactive species. *Oxid. Med. Cell. Longev.* **2012**, *2012*, 756132. [CrossRef]

72. Kanzaki, K.; Watanabe, D.; Shi, J.; Wada, M. Mechanisms of eccentric contraction-induced muscle damage and nutritional supplementations for mitigating it. *J. Muscle Res. Cell Motil.* **2022**, *43*, 147–156. [CrossRef] [PubMed]

73. Kurozumi, R.; Kojima, S. Increase of intracellular glutathione by low-level NO mediated by transcription factor NF-kappaB in RAW 264.7 cells. *Biochim. Et Biophys. Acta* **2005**, *1744*, 58–67. [CrossRef] [PubMed]

74. Sies, H.; Moss, K.M. A role of mitochondrial glutathione peroxidase in modulating mitochondrial oxidations in liver. *Eur. J. Biochem.* **1978**, *84*, 377–383. [CrossRef] [PubMed]

75. Garrido, M.; Terron, M.P.; Rodriguez, A.B. Chrononutrition against oxidative stress in aging. *Oxid. Med. Cell. Longev.* **2013**, *2013*, 729804. [CrossRef]

76. Barbosa, A.M.; Francisco, P.D.C.; Motta, K.; Chagas, T.R.; dos Santos, C.; Rafacho, A.; Nunes, E.A. Fish oil supplementation attenuates changes in plasma lipids caused by dexamethasone treatment in rats. *Appl. Physiol. Nutr. Metab.* **2016**, *41*, 382–390. [CrossRef]

77. Gorjao, R.; Dos Santos, C.M.M.; Serdaos, D.A.; Diniz, V.L.S.; Alba-Loureiro, T.C.; Cury-Boaventura, M.F.; Hatanaka, E.; Levada-Pires, A.C.; Sato, F.T.; Pithon-Curi, T.C.; et al. New insights on the regulation of cancer cachexia by N-3 polyunsaturated fatty acids. *Pharmacol. Ther.* **2019**, *196*, 117–134. [CrossRef]

78. Liu, J.; Meng, Q.; Zheng, L.; Yu, P.; Hu, H.; Zhuang, R.; Ge, X.; Liu, Z.; Liang, X.; Zhou, X. Effect of omega-3 polyunsaturated fatty acids on left ventricular remodeling in chronic heart failure: A systematic review and meta-analysis. *Br. J. Nutr.* **2022**, 1–35. [CrossRef]

79. Siroma, T.K.; Machate, D.J.; Zorgetto-Pinheiro, V.A.; Figueiosdo, P.S.; Marcelino, G.; Hiane, P.A.; Bogo, D.; Pott, A.; Cury, E.R.J.; Guimarães, R.C.A.; et al. Polyphenols and ω-3 PUFAs: Beneficial Outcomes to Obesity and Its Related Metabolic Diseases. *Front. Nutr.* **2022**, *8*, 781622. [CrossRef]

80. Videla, L.A.; Hernandez-Rodas, M.C.; Metherel, A.H.; Valenzuela, R. Influence of the nutritional status and oxidative stress in the desaturation and elongation of n-3 and n-6 polyunsaturated fatty acids: Impact on non-alcoholic fatty liver disease. *Prostaglandins Leukot. Essent. Fat. Acids* **2022**, *181*, 102441. [CrossRef]

81. Talukdar, S.; Bae, E.J.; Imamura, T.; Morinaga, H.; Fan, W.; Li, P.; Lu, W.J.; Watkins, S.M.; Olefsky, J.M. GPR120 is an omega-3 fatty acid receptor mediating potent anti-inflammatory and insulin-sensitizing effects. *Cell* **2010**, *142*, 687–698.

82. Ferreira, I.; Falcato, F.; Bandarra, N.; Rauter, A.P. Resolvins, Protectins, and Maresins: DHA-Derived Specialized Pro-Resolving Mediators, Biosynthetic Pathways, Synthetic Approaches, and Their Role in Inflammation. *Molecules* **2022**, *27*, 1677. [CrossRef] [PubMed]
83. Schmitz, G.; Ecker, J. The opposing effects of n-3 and n-6 fatty acids. *Prog. Lipid Res.* **2008**, *47*, 147–155. [CrossRef] [PubMed]

 *nutrients* 

Article

Changes in Blood Markers of Oxidative Stress, Inflammation and Cardiometabolic Patients with COPD after Eccentric and Concentric Cycling Training

Mayalen Valero-Breton [1], Denisse Valladares-Ide [2], Cristian Álvarez [1], Reyna S. Peñailillo [3] and Luis Peñailillo [1,*]

[1] Exercise and Rehabilitation Sciences Institute, School of Physical Therapy, Faculty of Rehabilitation Sciences, Universidad Andres Bello, Santiago 7550196, Chile
[2] Long Active Life Laboratory, Instituto de Ciencias de la Salud, Universidad de O'Higgins, Rancagua 2841959, Chile
[3] Laboratory of Reproductive Biology, Center for Biomedical Research and Innovation (CIIB), Universidad de los Andes, Santiago 7620001, Chile
*   Correspondence: luis.penailillo@unab.cl

Abstract: Chronic obstructive pulmonary disease (COPD) patients manifest muscle dysfunction and impaired muscle oxidative capacity, which result in reduced exercise capacity and poor health status. This study examined the effects of 12-week eccentric (ECC) and concentric (CONC) cycling training on plasma markers of cardiometabolic health, oxidative stress, and inflammation in COPD patients. A randomized trial in which moderate COPD was allocated to ECC ($n = 10$; 68.2 ± 10.0 year) or CONC ($n = 10$; 71.1 ± 10.3 year) training groups. Participants performed 12-week ECC or CONC training, 2–3 sessions per week, 10 to 30 min per session. Before and after training, peak oxygen consumption, maximal power output (VO_{2peak} and PO_{max}), and time-to-exhaustion (TTE) tests were performed. Plasma antioxidant and oxidative markers, insulin resistance, lipid profile, and systemic inflammation markers were measured before and after training at rest. VO_{2peak}, PO_{max} and TTE remained unchanged after ECC and CONC. CONC induced an increase in antioxidants ($p = 0.01$), while ECC decreased antioxidant ($p = 0.02$) markers measured at rest. CONC induced lesser increase in oxidative stress following TTE ($p = 0.04$), and a decrease in insulin resistance ($p = 0.0006$) compared to baseline. These results suggest that CONC training induced an increase in insulin sensitivity, antioxidant capacity at rest, and lesser exercise-induced oxidative stress in patients with moderate COPD.

Keywords: pulmonary rehabilitation; chronic obstructive pulmonary disease; aerobic training; exercise prescription

Citation: Valero-Breton, M.; Valladares-Ide, D.; Álvarez, C.; Peñailillo, R.S.; Peñailillo, L. Changes in Blood Markers of Oxidative Stress, Inflammation and Cardiometabolic Patients with COPD after Eccentric and Concentric Cycling Training. *Nutrients* **2023**, *15*, 908. https://doi.org/10.3390/nu15040908

Academic Editor: Dariusz Nowak

Received: 19 December 2022
Revised: 2 February 2023
Accepted: 5 February 2023
Published: 11 February 2023

1. Introduction

Chronic obstructive pulmonary disease (COPD) is characterized by airflow limitation associated with inflammation of the airway [1]. A common extra-pulmonary manifestation of patients with COPD is skeletal muscle dysfunction, which is evidenced mainly by reduced muscle endurance and strength of the lower limbs [2]. Impaired muscle oxidative capacity, a shift toward a glycolytic muscle fiber type distribution, reduced capillarity, and reduced cross-sectional area of muscle fiber led to muscle dysfunction in COPD patients [2]. The reduced exercise capacity and increased fatigability are clinical manifestations that exacerbate poor health status and limit daily life activities in COPD patients [3]. The limited exercise capacity of these individuals also predisposes them to develop impaired metabolism of glucose and lipids, which may lead to metabolic syndrome [4]. Airway inflammation releases inflammatory molecules into the bloodstream, thus, COPD patients show evidence of an increased level of systemic inflammation chronically [2]. It has been shown that inflammation can increase muscle degradation, inhibiting muscle-specific protein expression, and increasing muscle cell apoptosis, worsening muscle dysfunction in COPD patients [5]. Thus, systemic inflammation has been proposed as a mechanism in the development of muscle dysfunction in COPD patients as interleukin-6 (IL-6), interleukin-1 beta (IL-1β), and tumor necrosis factor-alpha (TNF-α) have been shown to be elevated in these patients [6]. Based on the evidence, it is accepted that COPD patients also have elevated levels of oxidative stress due to the imbalance between the generation of reactive oxygen species (ROS) and the efficiency of the antioxidant mechanism [7]. All these manifestations affect skeletal muscle function and further worsen muscle dysfunction. Thus, searching for new rehabilitation strategies to reduce muscle dysfunction within the burden of this disease is necessary.

Several studies have shown that exercise training is the cornerstone of pulmonary rehabilitation as it markedly improves functional performance, muscle mass, and exercise capacity in COPD patients [8–10]. Endurance exercise training is the most common exercise modality prescribed in COPD [11–13], which has been shown to improve exercise tolerance and muscle oxidative capacity in these patients [12,14–16]. However, endurance training could also increase oxidative stress levels in severe COPD patients [17], which may not be desirable as it could impair skeletal muscle function. In contrast, another novel and less explored exercise training modality is eccentric cycling training. This type of training is characterized by eccentric contractions of the lower limb muscles that are performed when resisting the backward rotational movement of the cranks generated by an eccentric ergometer. Notably, eccentric cycling imposes lesser cardiopulmonary, metabolically, and perceptual demands (i.e., lower oxygen consumption, dyspnea, and blood pressure) than concentric cycling, making it safer for COPD patients [18]. Interestingly, eccentric cycling can produce a greater workload for the same metabolic demand than concentric cycling in COPD patients [18,19]. Furthermore, eccentric cycling training has been shown to increase muscle strength and mass to a greater extent than concentric cycling in moderate COPD patients [19,20], and a decrease (20%) in the homeostasis model assessment of insulin resistance index (HOMA-IR) after 12 weeks of continuous moderate-intensity eccentric cycling training [21]. Notably, these benefits were reported despite ~30-50% lesser cardiovascular strain and metabolic cost than conventional concentric cycling [21]. Interestingly, we have recently shown that oxidative stress levels (i.e., thiobarbituric acid reactive substances; TBARS) decreased after one acute bout of eccentric cycling at moderate intensity [22]. Hence, eccentric cycling training could be an ideal exercise modality for COPD patients. However, the long-term effects of eccentric cycling training on oxidative stress and inflammation in COPD patients have not been explored yet.

Interestingly, exercise training in COPD patients has been shown to produce positive [23], null [24], or negative changes in oxidative stress assessed at rest [25]. Discrepancies among studies may be due to differences in exercise prescription and heterogenicity of patients' severity. However, it is possible that in rest conditions, the oxidative molecules are not produced/released, and the antioxidant capacity is not fully displayed, hence, no changes have been observed. A single bout of exercise has been shown to induce exaggerated production of ROS and oxidative damage in COPD patients when exercise exceeds a certain intensity or duration [26]. Thus, it seems plausible to assess the oxidative and antioxidant response to an acute bout of exercise to dynamically assess oxidative stress handling in COPD patients. However, this has not been explored yet. There are tests that are used to evaluate exercise tolerance, such as time-to-exhaustion tests (TTE) [27]. Specifically, TTE consists of the patient sustaining a fixed workload for the longest time possible, which has been used in COPD patients [28]. However, to the best of our knowledge, no study has examined oxidative stress and antioxidant markers after TTE in COPD patients after an intervention of exercise training. Therefore, as oxidative stress has been involved in muscle mass loss and dysfunction in COPD patients [29,30], assessing oxidative stress changes after TTE in COPD patients could shed light on exercise-induced oxidative stress handling in these patients.

Thus, this study aimed to examine the effects of 12 weeks of eccentric (ECC) and conventional concentric (CONC) cycling training on markers of oxidative stress, inflammation, and cardiometabolic health in COPD patients measured at rest. Furthermore, as oxidative stress could be induced by acute exercise, we compared the changes in oxidative stress markers following a standardized submaximal exercise between ECC and CONC, before and after training in COPD patients.

2. Materials and Methods

2.1. Participants

This study is part of a larger study, further details of participants and training interventions have been published elsewhere [19]. In brief, it was a randomized prospective training study, in which twenty participants with moderate COPD (10 men and 10 women) volunteered to participate. Participants were diagnosed as moderate COPD patients according to the Global Initiative for Chronic Obstructive Lung Disease (GOLD; II) based on a pulmonologist assessment and spirometry tests. The sample size was estimated based on a reduction of 8.8% in total cholesterol levels after eccentric training (effect size of 1.5) reported in a previous study [31]. Considering an alpha level of 0.05 and a statistical power of 0.8, the sample size estimation revealed that eight participants per group would be sufficient (G * Power 3.1, Germany). Ten participants per group were included to account for a 20% dropout.

The exclusion criteria considered supplemental oxygen therapy, kidney, neurologic or cardiovascular disease, recent exacerbation with three or more days of steroid or antibiotic use, being part of strength or aerobic training in the last year; musculoskeletal injury of the legs during the past one year, and smoking in the last six months. Participants were instructed to refrain from other types of training and from any nutritional supplementation and to maintain regular daily routines during the study period. Written informed consent was obtained from all participants. The study was approved by the Institutional Ethics Committee and Eastern Metropolitan Health Service of the city (clinical trial registration number: DRKS00009755) and conducted according to the Helsinki declaration.

2.2. Study Design

Participants were allocated to ECC (men: 6, women: 4) or CONC (men: 4, women: 6), using a stratified randomized allocation scheme considering FEV_1 and age to be similar between groups. The physical characteristics of the participants are presented in the Results. All testing was completed in the laboratory at room temperature (20–22 °C) at the same time of day ($\pm$1 h). Participants were asked not to consume caffeine and alcohol for at least 3 h and 24 h, respectively.

Outcome measurements were collected 72 h before and after 12 weeks of ECC and CONC training. Participants were cited to the laboratory on two different days to complete all assessments. On day one, an incremental cycling test was performed to determine peak oxygen consumption (VO_{2peak}) and maximal concentric power output (PO_{max}). Forty-eight hours after, participants returned to the laboratory, and blood was withdrawn from the antecubital vein was collected at rest after 12 h of fasting. Furthermore, another blood sample was collected following a time-to-exhaustion cycling test (TTE). Participants received a light breakfast consisting of a cereal bar (78.4 kCal of energy; 1.1 g protein, 2.5 total fat, 13 g carbohydrate) and a milk box (40 kCal of energy; 0 g protein, 0 total fat, 8 g carbohydrate) after the first blood withdraw at rest. Immediately following the TTE a second blood sample was collected to determine the changes (Δ: Post-TTE—Pre-TTE) in plasma oxidative stress markers to determine exercise-induced oxidative stress.

2.3. Exercise Training

Participants attended four sessions before the commencement of the training to familiarize patients with exercise. During familiarization sessions, eccentric or concentric cycling was performed (5–20 min) at ~25% of their self-perceived maximum effort. After this, participants performed eccentric or concentric cycling training for 12 weeks with a total number of training sessions of 34. Training frequency started with two times per week, on weeks 1–2, and progressed to three times per week from weeks 3 to 12. Training intensity was moderate and increased from 11 to 13 on the rating of perceived exertion (RPE) from a 6-20 Borg's scale. Training time started from 10 min in week 1 to 30 min in week 12. Training total workload was different between CONC and ECC as for the same RPE, the ECC training group performed a much higher training average workload ECC = 226.6 ± 101.9 vs. CONC = 78.1 ± 62.7 kJ; $p \leq 0.05$). The CONC group trained in a conventional concentric recumbent cycle ergometer (Livestrong, LS 5.0R model, Austin, TX, USA). The ECC group trained in a recumbent eccentric cycle ergometer (Eccentric Trainer, Metitur, Finland) in which participants were instructed to resist the backward movements of the cranks, which is known to induce eccentric contractions of the knee extensor muscles mainly. All training sessions were performed in an airconditioned laboratory (temperature: 20–21 °C, relative humidity: 50%). Exercise training protocol induced changes in body composition and functional performance markers after training, which have been published previously elsewhere [19].

2.4. Measurements

2.4.1. Maximal Incremental Cycling Test

On day one, patients performed a maximal incremental test on an electromagnetically braked ergometer (Livestrong, LS 5.0 R model, Austin, TX, USA). The test started at 10 W for two minutes, with 10 W increase every minute until voluntary exhaustion or until the patients were unable to maintain 60 revolutions per minute of cadence [18]. Oxygen consumption (VO_2) and carbon dioxide production (VCO_2) were assessed using a breath-by-breath gas analyzer (Ergocard, Medisoft, Belgium). PO_{max} and VO_{2peak} were obtained as described in previous study [32].

2.4.2. Time-to-Exhaustion Test (TTE)

On day two, a constant power output time-to-exhaustion cycling test was performed at 75% of PO_{max} [33]. Participants were instructed to adopt their preferred cadence between 60 and 90 rpm and maintain the target power output for as long as possible. Verbal encouragement was provided; however, participants were not given feedback on their elapsed time or power output. The participants' TTE was reached when despite encouragement, their cadence fell 10 rpm below 60 rpm for 10 s or more. The TTE was recorded to the nearest second. Total work was calculated by multiplying time by average power output and expressed in kJ [27].

2.4.3. Blood Samples

As mentioned above, an initial blood sample was obtained from the antecubital vein following a 12 h overnight fast at rest before and 72 h after the last training session. Additionally, a second blood sample was collected following (~5 min) the TTE. Blood samples were collected in Vacutainer© EDTA tubes for plasma and tubes with pro-coagulant (10 mL) for serum. Tubes were centrifuged for 10 min at 4 °C and 4000 rpm. Plasma and serum were aliquoted in Eppendorf tubes with 500 μL of sample and stored at −80 °C until analyzed.

2.4.4. Oxidative Stress and Inflammatory Markers

Oxidative stress and inflammatory markers were analyzed by ELISA or colorimetric kits (Cayman Chemical Company, Ann Arbor, MI, USA) and read on a plate reader (Multiskan FC, Thermo Scientific, Beijing, China). All measurements were performed in plasma or serum samples according to the manufacturer's instructions. Interleukin 6 (IL-6), tumor necrosis factor α (TNF-α), interleukin 1 beta (IL-1β), total antioxidant capacity (TAC), superoxide dismutase (SOD), catalase (CAT), thiobarbituric acid reactive substances (TBARS) and glutathione peroxidase activity (GPx) were analyzed at rest (pre-TTE). Changes (Δ: Post-Pre) in these markers after TTE were also determined (TACTTE,

SODTTE, CATTTE, TBARSTTE, GPxTTE). All samples were analyzed in duplicate. Coefficient of variation of duplicates were: IL-6: $5.2 \pm 5.0\%$, TNF-α: $21.5 \pm 18.4\%$, IL-1β: $21.9 \pm 26.8\%$, TAC: $9.8 \pm 7.2\%$, SOD: $5.6 \pm 7.9\%$, CAT: $4.4 \pm 3.6\%$, TBARS: $3.1 \pm 3.7\%$, and GPx: $4.4 \pm 3.4\%$.

2.4.5. Cardiometabolic Health Markers

Insulin Sensitivity

Fasting plasma glucose (FPG) concentration was measured using an enzymatic method (Trinder, Genzyme Diagnostics, Charlottetown, Canada) and plasma insulin was assessed using a radioimmune assay (DPC, Houston, TX, USA). Homeostatic model assessment insulin resistance (HOMA-IR) was calculated according to Matthew's equation [34]. Whole-blood glycosylated hemoglobin (%Hb1ac) was measured using a latex immunoturbidimetric assay in a DCA analyzer (Siemens Medical, Devault, PA, USA).

Lipid Profile

Total cholesterol (TC) and triglycerides (TAG) concentrations were analyzed by enzymatic methods using standard kits (Wiener Lab Inc., Rosario, Argentina) in an automatic analyzer (Metrolab 2 300 Plus™, Metrolab Biomed Inc., Rosario, Argentina). Plasma high-density lipoprotein cholesterol (HDL) levels were measured by the same enzymatic method after phosphotungstate precipitation. Low-density lipoprotein cholesterol (LDL) levels were calculated using the Friedewald formula.

2.5. Statistical Analysis

The Shapiro–Wilk test was used to assess the distribution of the data as a whole deviates from a comparable normal distribution, and the analysis showed that all dependent variables were normally distributed. Baseline characteristics between groups were analyzed by Student t test. Changes in the dependent variables from the baseline to post-training were compared between the ECC and CONC groups by a two-way repeated measure analysis of variance. If a significant interaction (group $\times$ time) effect, was found a Fisher's least significance difference (LSD) post hoc test was used. The Mann–Whitney test was used to compare the percentage of change in oxidative stress variables and cardiometabolic health markers. Statistical significance was set at $p \leq 0.05$, and data are presented as mean and standard deviation (mean $\pm$ SD). All statistical analyses were performed using PRISM 8.0 (GraphPad, San Diego, CA, USA).

3. Results

3.1. Participants' Characteristics

Age (ECC = 68.2 ± 10.0 year, CONC = 71.1 ± 10.3 year; $p = 0.53$), body mass (ECC = 73.2 ± 11.7 kg, CONC = 69.7 ± 10.0 kg; $p = 0.47$), height (ECC = 161.2 ± 7.4 cm, CONC = 158.5 ± 12.0 cm; $p = 0.54$), body mass index (ECC = 29.1 ± 5.4 kg/m^2, CONC = 28.1 ± 5.9 kg/m^2; $p = 0.96$), forced expiratory volume in 1 second to forced vital capacity ratio (ECC = 63.1 ± 0.9, CONC = 66.8 ± 1.2; $p = 0.42$), and forced expiratory volume in 1-second post-bronchodilator (ECC = $68.7 \pm 15.1\%$ of predicted, CONC = $73.1 \pm 12.8\%$ of predicted; $p = 0.49$) were similar between groups.

3.2. Maximal Aerobic Capacity

The $\dot{V}O_{2peak}$ (1.2 ± 0.5 L/min and 1.2 ± 0.4 L/min; $p = 0.8$) and PO$_{max}$ (87.0 ± 54.1 W and 76.0 ± 37.1 W; $p = 0.6$) were similar at baseline between ECC and CONC, respectively. Moreover, as shown in Figure 1A, $\dot{V}O_{2peak}$ of both groups remained unchanged ($p > 0.05$) after training (ECC, Post: 1.3 ± 0.6, CONC, Post: 1.2 ± 0.6 L/min). PO$_{max}$ also remained unchanged after ECC (92.0 ± 55.5 W) and CONC (98.0 ± 54.3 W) training ($p > 0.05$).

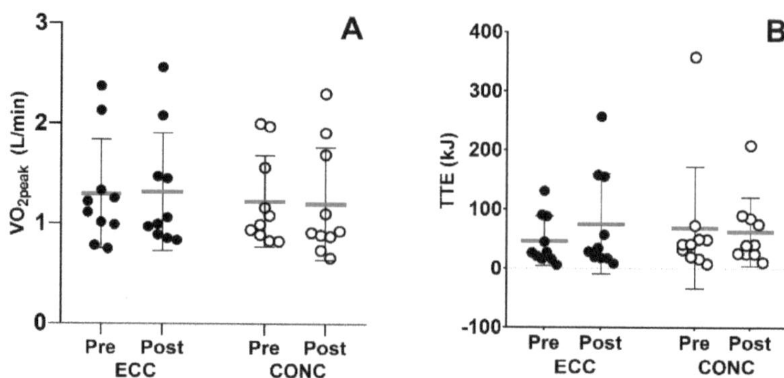

Figure 1. Maximal and submaximal aerobic performance. Peak oxygen consumption during the incremental cycling test (**A**) and workload performed in the time to exhaustion test (**B**). ECC: eccentric cycling, CONC: concentric cycling.

3.3. TTE

As shown in Figure 1B, total work performed in TTE in ECC (46.4 $\pm$ 41.7 kJ) and CONC (68.6 $\pm$ 103.3 kJ) were similar at baseline ($p = 0.2$). Moreover, both groups remained unchanged ($p > 0.05$) after training (ECC, Post: 75.1 $\pm$ 84.6, CONC, Post: 62.4 $\pm$ 58.1 kJ).

3.4. Systemic Oxidative Stress Markers

3.4.1. At Rest

As shown in Figure 2A, TAC concentration at rest showed a 137.9 $\pm$ 162.4% increase after CONC training (time effect, $p = 0.004$; training effect, $p = 0.6$; interaction effect, $p = 0.02$). Moreover, GPx activity decreased by 17.2 $\pm$ 20.8% after ECC training (time effect, $p = 0.3$; training effect, $p = 0.9$; interaction effect, $p = 0.03$) as shown in Figure 2C. In addition, the percentages of changes from baseline in both oxidative stress markers were statistically different as shown in Figure 2B (TAC, $p = 0.01$) and Figure 2D (Gpx activity, $p = 0.02$). Other oxidative stress markers measured at rest (Figure 2E–J) remained unchanged after ECC and CONC training: TBARS (time effect, $p = 0.6$; training effect, $p = 0.1$; interaction effect, $p = 0.8$), SOD (time effect, $p = 0.1$; training effect, $p = 0.2$; interaction effect, $p = 0.5$) and CAT (time effect, $p = 0.9$; training effect, $p = 0.1$; interaction effect, $p = 0.5$).

3.4.2. Submaximal Cycling

Changes in oxidative markers following TTE are shown in Figure 3. Oxidative stress values obtained were normalized by workload performance on the TTE of each patient in kJ. Absolute value changes in TAC_{TTE} were similar after ECC and CONC training (time effect, $p = 0.5$; training effect, $p = 0.9$; interaction effect, $p = 0.4$; Figure 3A) and no differences were found in the percentages of change from baseline in any group ($p = 0.5$; Figure 3B). Changes in absolute values of GPx_{TTE} were similar after ECC and CONC (time effect, $p = 0.5$; training effect, $p = 0.9$; interaction effect, $p = 0.4$; Figure 3C), similar to when the percentage of change from baseline were compared ($p = 0.6$; Figure 3D). The absolute values of $TBARS_{TTE}$ decreased 244.8 $\pm$ 566.5% only after CONC training ($p = 0.04$; Figure 3E), but no difference was found in the percentage of change from baseline between groups in this marker ($p = 0.5$; Figure 3F). No significant changes in absolute values were found in SOD_{TTE} (time effect, $p = 0.7$; training effect, $p = 0.5$; interaction effect, $p = 0.3$; Figure 3G) and CAT_{TTE} (time effect, $p = 0.3$; training effect, $p = 0.8$; interaction effect, $p = 0.4$; Figure 3I) after any training. Furthermore, percentages of change from baseline of SOD_{TTE} ($p = 0.4$) and CAT_{TTE} ($p = 0.4$) were similar between ECC and CONC (Figure 3).

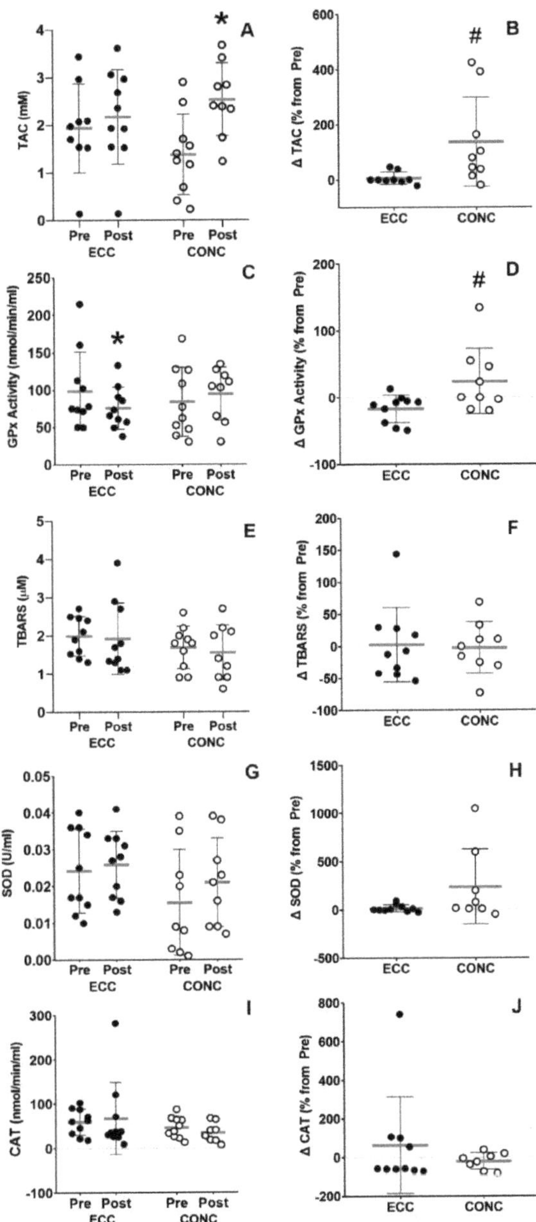

Figure 2. Oxidative stress at rest. Total antioxidant capacity (TAC) levels in absolute values (**A**) as percentage of change from pre-training values (**B**). Glutathione peroxidase activity (GPx) in absolute values (**C**) as percentage of change from pre-training values (**D**). Thiobarbituric acid reactive substances (TBARS) levels in absolute values (**E**) as percentage of change from pre-training values (**F**). Superoxide dismutase (SOD) levels in absolute values (**G**) as percentage of change from pre-training values (**H**). Catalase (CAT) levels in absolute values (**I**) as percentage of change from pre-training values (**J**). ECC: eccentric cycling, CONC: concentric cycling. *: $p < 0.05$ vs. Pre-. #: $p < 0.05$ ECC vs. CONC.

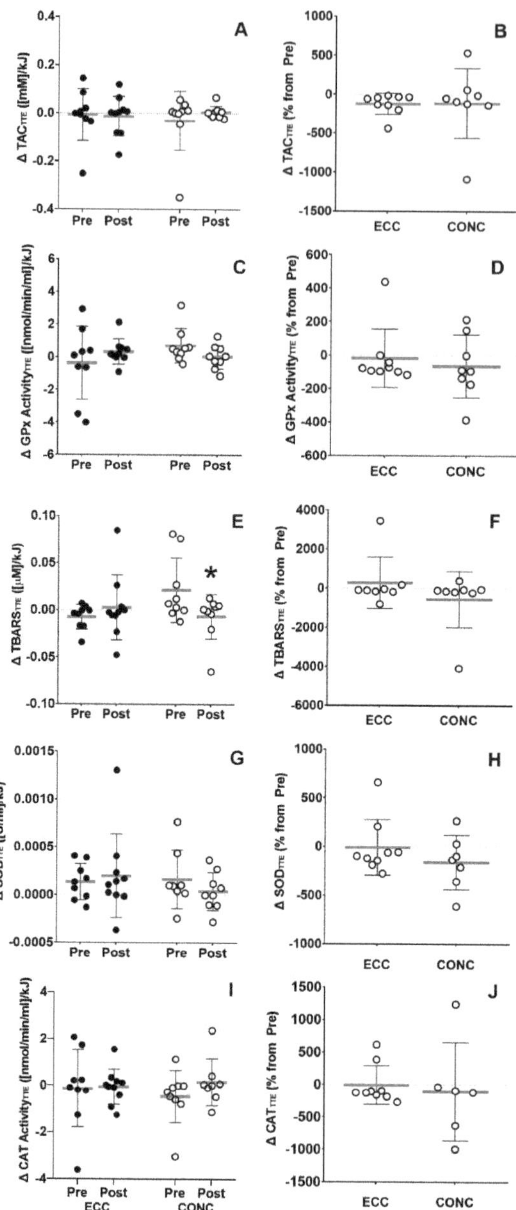

Figure 3. Oxidative stress markers change from rest to immediately following time to exhaustion test (TTE) normalized by the total workload during TTE. Absolute value changes in total antioxidant capacity (TAC$_{TTE}$) (**A**) and as a percentage of change from baseline (**B**). Absolute value changes in glutathione peroxidase activity (GPx$_{TTE}$) (**C**) and as a percentage of change from baseline (**D**). Absolute value changes in thiobarbituric acid reactive substances (TBARS$_{TTE}$) (**E**) and as a percentage of change from baseline (**F**). Absolute value changes in superoxide dismutase (SOD$_{TTE}$) (**G**) and as a percentage of change from baseline (**H**). Absolute value changes in catalase (CAT$_{TTE}$) (**I**) and as a percentage of change from baseline (**J**). ECC: eccentric cycling, CONC: concentric cycling. *: $p < 0.05$ vs. Pre-.

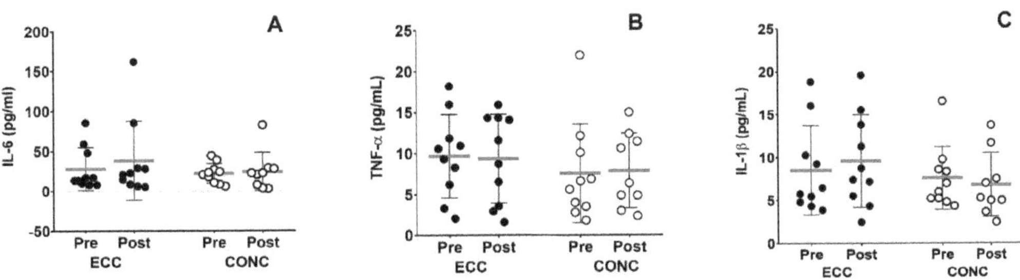

Figure 4. Concentration of systemic inflammation markers. Interleukin 6 (IL-6) (**A**), tumor necrosis factor α (TNF-α) (**B**), and interleukin 1 (IL-1) (**C**) concentrations. ECC: eccentric cycling, CONC: concentric cycling.

3.5. Systemic Inflammation Markers at Rest

The IL-6 (time effect, $p = 0.2$; training effect, $p = 0.4$; interaction effect, $p = 0.6$; Figure 4A), TNF-α (time effect, $p = 0.9$; training effect, $p = 0.3$; interaction effect, $p = 0.8$; Figure 4B) and IL-1β (time effect, $p = 0.5$; training effect, $p = 0.4$; interaction effect, $p = 0.5$; Figure 4C) concentrations remained unchanged after ECC and CONC training.

3.6. Cardiometabolic Health Markers

3.6.1. Insulin Sensitivity

The HOMA-IR index decreased ($-30.8 \pm 22.9\%$) after CONC (from 3.86 ± 2.27 to 2.7 ± 1.93), but not after ECC training (time effect, $p = 0.005$; training effect, $p = 0.3$; interaction effect, $p = 0.01$; Figure 5A). The percentage of change in HOMA-IR showed a greater decrease in CONC compared to ECC ($p = 0.03$; Figure 5B). No significant changes in %Hb1Ac were observed after ECC or CONC training (time effect, $p = 0.6$; training effect, $p = 0.1$; interaction effect, $p > 0.9$; Figure 5C). No differences in the percentage of change after training in %Hb1Ac were observed after ECC or CONC ($p = 0.7$; Figure 5D).

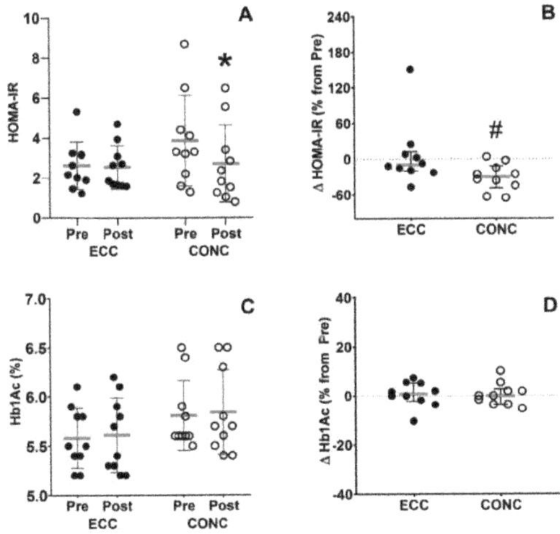

Figure 5. Insulin sensitivity. Homeostatic model assessment insulin resistance (HOMA-IR) in absolute values (**A**) and as a percentage of change from pre-training values (**B**). Whole-blood glycosylated hemoglobin (%Hb1ac) in absolute values (**C**) and as a percentage of change from pre-training values (**D**). ECC: eccentric cycling, CONC: concentric cycling. *: $p < 0.05$ Pre vs. Post, #: $p < 0.05$ ECC vs. CONC.

3.6.2. Lipid Profile

No significant differences after ECC and CONC training in TAG levels (time effect, $p = 0.4$; training effect, $p = 0.7$; interaction effect, $p > 0.9$; Figure 6A), TC (time effect, $p = 0.5$; training effect, $p = 0.6$; interaction effect, $p = 0.9$; Figure 6C), HDL (time effect, $p = 0.6$; training effect, $p = 0.6$; interaction effect, $p = 0.7$; Figure 6E) and LDL (time effect, $p = 0.6$; training effect, $p = 0.4$; interaction effect, $p = 0.7$; Figure 6G) levels were observed. No differences were shown in percentages of changes from baseline in TAG, TC, and LDL between groups (Figure 6B,D,G), while HDL levels showed a tendency to increase after CONC compared to ECC ($p = 0.06$; Figure 6E).

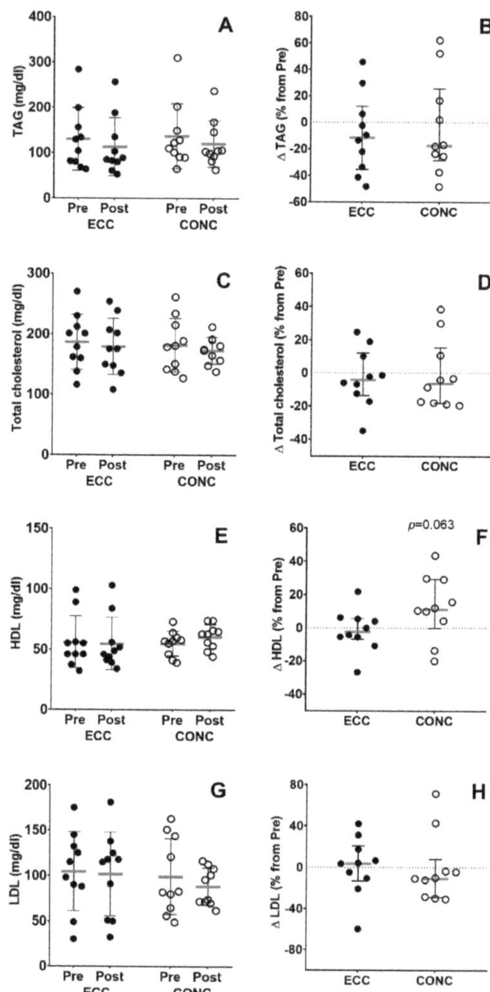

Figure 6. Lipid profile. Triglycerides (TAG) levels in absolute values (**A**) as a percentage of change from pre-training values (**B**). Total cholesterol levels in absolute values (**C**) as a percentage of change from pre-training values (**D**). High-density lipoprotein cholesterol (HDL) levels in absolute values (**E**) as a percentage of change from pre-training values (**F**). Low-density lipoprotein cholesterol (LDL) levels in absolute values (**G**) as a percentage of change from pre-training values (**H**). ECC: eccentric cycling, CONC: concentric cycling.

4. Discussion

This study aimed to examine the effects of eccentric cycling (ECC) and conventional concentric cycling (CONC) training on plasma markers of oxidative stress, systemic inflammation, and cardiometabolic health in patients with moderate chronic obstructive pulmonary disease (COPD). The main findings of this study were: (1) maximal aerobic capacity and time-to-exhaustion (TTE) performance were maintained after 12 weeks of ECC and CONC training; (2) CONC induced an increase in total antioxidant capacity (TAC) and ECC induced a decrease in GPx activity at rest; (3) exercise-induced oxidative stress (i.e., TBARS) showed a smaller increase after CONC training; (4) insulin sensitivity and HDL were improved after CONC training only. Thus, our initial hypothesis was supported as CONC training induced greater improvements in oxidative stress at rest and after exercise, and cardiometabolic health markers changed more favorably after CONC training compared with ECC, while inflammatory markers remained unchanged after both training interventions.

We found that neither ECC nor CONC training improved maximal aerobic capacity (VO_{2peak}) or time-to-exhaustion test (TTE) performance (Figure 1). COPD patients manifest pulmonary parenchyma fibrosis, airway remodeling, and dynamic hyperinflation, which may interfere with oxygen exchange [35]. Therefore, it is possible that no changes in oxygen consumption observed in these patients [20,36], were possibly due to the same reasons that the performance in the TTE test was also not changed after CONC or ECC. However, Porszasz et al. reported that endurance training (35 min per session at 75% of the peak work rate attained on incremental test) induced a 20% increase in PO_{max}, which was accompanied by an increase in TTE performance [37]. These equivocal results in endurance performance may be due to the large heterogeneity in the primary factors (i.e., ventilatory response, dynamic hyperinflation) affecting exercise tolerance in COPD patients [38]. It is also possible that our training interventions were not sufficiently intense stimuli as to induce aerobic adaptations compared with previous study [37]. Mechanisms underpinning the absence of aerobic adaptations in this population may warrant further research.

Increased serum TBARS levels and reduced GPx activity have been associated with the disease severity of COPD patients [39]. We found that CONC training induced $137.9 \pm 162.4\%$ increase in TAC concentration (Figure 2A), while ECC training induced $17.2 \pm 20.8\%$ decrease in GPx activity (Figure 2C,D) measured at rest in moderate COPD patients. Equivocal results have been published previously. For instance, Rabinovich et al. showed that COPD patients decreased their antioxidant capacity after 8 weeks of concentric high-intensity interval training (HIIT), while healthy sedentary individuals increased their muscle antioxidant markers (i.e., glutathione levels; GSH). Interestingly, Zarrindast et al. showed that an 8-week moderate endurance training reduced lipid peroxidation (i.e., 8-iso PGF 2α), and increased TAC levels in elderly women [40]. It has been speculated that after regular training, there is an upregulation of the antioxidant enzymatic systems as an adaptive response. Specifically, superoxide dismutase (SOD), catalase (CAT), and glutathione peroxidase (GPx) activity act synergistically with non-enzymatic antioxidants to reduce the total antioxidant capacity. However, we did not observe changes in SOD and CAT concentrations in plasma, which is similar to Shin et al. which showed maintenance in TBARS levels and SOD activities at rest after 6 months of moderate endurance training, while GPx activity increased by 12% after training [41]. Interestingly, the same study reported that TBARS levels tended to increase ($p = 0.052$) in the control group after 6 months of endurance training in middle age obese women. Equivocal results regarding changes in oxidant and antioxidant agents in plasma after training interventions could be due to the heterogeneity in the disease severity of COPD patients. Overall, our results reveal that 12 weeks of moderate-intensity continuous endurance concentric training (CONC) may have positively changed the non-enzymatic antioxidant capacity and have maintained the TBARS levels of COPD patients at rest.

In our study, the TTE test was used to assess endurance capacity, and to induce a standardized (75% of individual PO_{max}) physiological challenge to COPD patients to assess the response to exercise-induced oxidative stress before and after training interventions. Our results showed that the increase in TBARS levels following TTE was significantly smaller after CONC training (Figure 3E). Mercken et al. reported that COPD patients have elevated systemic and pulmonary oxidative stress at rest and in response to acute exercise compared with age-matched healthy control individuals [42]. This may suggest that healthy individuals can tolerate exercise-induced oxidative stress more effectively than COPD patients. Mercken et al. also showed for the first time that intensive supervised pulmonary rehabilitation (8-week training program) decreased exercise-induced oxidative stress following submaximal exercise [42]. This improved redox handling after exercise could be attributable to adaptive responses involving more efficient muscle oxidative metabolism (e.g., lesser mitochondrial ROS production) or a better capacity of endogenous antioxidant systems to handle oxidative stress. However, it is still not known if the systemic increase in oxidative stress markers in COPD patients is a reversible process and whether other exercise interventions could also modify this scenario as it has been shown to occur in other metabolic and cardiovascular diseases [43].

Systemic inflammatory markers were unchanged after ECC and CONC training (Figure 4). These results are similar to previous reports, in which circulating pro-inflammatory cytokines were unchanged in COPD patients regardless of endurance or strength-based training interventions [15,44]. Greater intensity and longer interventions may be necessary to modify inflammatory markers in this population.

We found a $30.8 \pm 22.9\%$ decrease in HOMA-IR ($p = 0.0006$) only after CONC training (Figure 5A,B). This result becomes relevant considering that our COPD patients had elevated baseline HOMA-IR index (ECC: 2.6 ± 1.2; CONC: 4.1 ± 2.3), suggesting insulin resistance [34]. Changes observed after CONC training were similar to the 20.4% decrease in HOMA-IR reported by Matos et al. after eight weeks of HIIT (at 80–110% of PO_{max}) in obese patients with insulin resistance [45]. We have also previously shown similar decreases in HOMA-IR (–50%) after 12 weeks of concentric HIIT (at 70–100% maximum heart rate) in the same patients [46]. These positive changes induced by CONC training in the present study may be attributable to increases in skeletal muscle oxidative enzymes, mitochondrial biogenesis and glucose transporters induced by endurance exercise training [47]. The lack of changes after ECC training may be related to lesser metabolic demand imposed by eccentric contractions, which did not stimulate the skeletal muscle oxidative pathways and thus, no changes in oxidative capacity and cardiometabolic health markers were induced.

Although reductions in circulating lipids have been previously reported after eccentric training [31,48–53], we found no change in lipid profile markers after ECC training (Figure 6). An association has been observed between the magnitude of muscle damage induced by the eccentric contractions and changes in circulating lipids after eccentric training [54], so we speculate that the familiarization period and gradual increase in training workload in ECC may have avoided muscle damage. This could explain the lack of changes in lipid profile markers in the present study in comparison to previous eccentric training studies [31,48–53]. Interestingly, we found a tendency to increase HDL ($p = 0.06$) after CONC training (Figure 6F). For instance, Sillanpää et al. reported that endurance training (21 weeks) decreased total cholesterol (–5.6%), LDL (–5.4%), and TAG (–10.0%), without changes in HDL in healthy middle-aged men [55]. More recently, Boukabous et al. showed that endurance training (45 min per session at 55% heart rate reserve, 8 weeks) decreased LDL (–17.6%), and increased HDL (+7.6%) in older women [56]. Therefore, although both training programs (CONC and ECC) were performed at similar perceptual intensity (RPE of 11–13) and ECC performed ~3-fold greater workload than CONC, CONC showed larger improvements in markers of cardiometabolic health compared to ECC. However, lipid profile and insulin sensitivity changes reported in this study seem smaller compared to previous studies, which may be due to the lower training intensity (RPE = 13) of both interventions. Further studies should implement higher intensity and prolonged (more than 12 weeks) training interventions in order to induce greater cardiometabolic risk marker changes in COPD patients.

We acknowledge the limitations of our study. First, high variability in the outcome parameters may be due to differences in the severity of the disease of our patients; as we used patients with moderate COPD, this category may be too wide in symptoms and muscle dysfunction degrees, which increased variability. However, a stratified randomization was performed to distribute the participant in order to minimize this effect. Second, a small sample size was also recruited for this study, which also could have affected the variability of the results. Third, although all participants were diagnosed with moderate COPD, different clinical phenotypes exist [57]. It has been observed that emphysema and bronchiolitis can largely differ among patients of the same severity, which is evidenced by different levels of dyspnea while spirometric values are similar. The heterogeneity of the sample could have been reduced if more comprehensive inclusion criteria were used, e.g., exacerbation/hospitalizations per year. Finally, other factors such as daily physical activity and nutrition were not controlled and could have affected the results. Thus, our results need to be analyzed with caution as larger sample size clinical trials are needed to extrapolate these results to the general population.

5. Conclusions

In conclusion, moderate-intensity continuous concentric cycling training was more efficient to increase antioxidant capacity (enzymatic and non-enzymatic components) at rest and to improve exercise-induced oxidative stress control. Furthermore, it improved insulin sensitivity and HDL levels in moderate COPD patients to a greater extent than eccentric cycling training. Thus, moderate-intensity continuous concentric cycling training could be prescribed to reduce oxidative stress and improve markers of cardiometabolic health in comparison to eccentric cycling, which may be more indicated to increase muscle mass and functional performance [19]. Further research should focus on a combination of both concentric and eccentric cycling to target muscle dysfunction in COPD patients.

Author Contributions: Conceptualization, L.P.; methodology, M.V.-B.; software, M.V.-B.; validation, M.V.-B. and D.V.-I.; formal analysis, M.V.-B.; investigation, M.V.-B.; resources, L.P.; data curation, M.V.-B.; writing—original draft preparation, L.P. and M.V.-B.; writing—review and editing, L.P., M.V.-B., R.S.P., C.Á. and D.V.-I.; visualization, L.P. and M.V.-B.; supervision, L.P.; project administration, L.P.; funding acquisition, L.P. All authors have read and agreed to the published version of the manuscript.

Funding: This work was funded by a research grant awarded to L.P. (#1211962) and D.V.I. (#11190949) by ANID of Chile.

Institutional Review Board Statement: The study was conducted according to the guidelines of the Declaration of Helsinki and approved by the Institutional Review Board of Universidad Finis Terrae and Eastern Metropolitan Health Service of the city (clinical trial registration number: DRKS00009755).

Informed Consent Statement: Informed consent was obtained from all subjects involved in the study.

Data Availability Statement: Data are contained within the article.

Conflicts of Interest: All authors do not have any conflicts of interest to declare. The results of the study are presented clearly, honestly, and without fabrication, falsification, or inappropriate data manipulation.

References

1. Vogelmeier, C.F.; Criner, G.J.; Martinez, F.J.; Anzueto, A.; Barnes, P.J.; Bourbeau, J.; Celli, B.R.; Chen, R.; Decramer, M.; Fabbri, L.M.; et al. Global Strategy for the Diagnosis, Management and Prevention of Chronic Obstructive Lung Disease 2017 Report: GOLD Executive Summary. *Respirology* **2017**, *22*, 575–601. [CrossRef]
2. Jaitovich, A.; Barreiro, E. Skeletal Muscle Dysfunction in Chronic Obstructive Pulmonary Disease. What We Know and Can Do for Our Patients. *Am. J. Respir. Crit. Care Med.* **2018**, *198*, 175–186. [CrossRef] [PubMed]
3. Donaldson, A.V.; Maddocks, M.; Martolini, D.; Polkey, M.; Man, W.D.-C. Muscle function in COPD: A complex interplay. *Int. J. Chronic Obstr. Pulm. Dis.* **2012**, *7*, 523–535. [CrossRef]
4. James, B.D.; Jones, A.V.; Trethewey, R.E.; Evans, R. Obesity and metabolic syndrome in COPD: Is exercise the answer? *Chronic Respir. Dis.* **2018**, *15*, 173–181. [CrossRef]

5. Zhao, H.; Li, P.; Wang, J. The role of muscle-specific MicroRNAs in patients with chronic obstructive pulmonary disease and skeletal muscle dysfunction. *Front. Physiol.* **2022**, *13*, 954364. [CrossRef]
6. Singh, S.; Verma, S.; Kumar, S.; Ahmad, M.; Nischal, A.; Singh, S.K.; Dixit, R. Correlation of severity of chronic obstructive pulmonary disease with potential biomarkers. *Immunol. Lett.* **2018**, *196*, 1–10. [CrossRef]
7. Barreiro, E.; Fermoselle, C.; Mateu-Jimenez, M.; Sánchez-Font, A.; Pijuan, L.; Gea, J.; Curull, V. Oxidative stress and inflammation in the normal airways and blood of patients with lung cancer and COPD. *Free Radic. Biol. Med.* **2013**, *65*, 859–871. [CrossRef]
8. De Brandt, J.; Spruit, M.A.; Derave, W.; Hansen, D.; Vanfleteren, L.E.G.W.; Burtin, C. Changes in structural and metabolic muscle characteristics following exercise-based interventions in patients with COPD: A systematic review. *Expert Rev. Respir. Med.* **2016**, *10*, 521–545. [CrossRef] [PubMed]
9. Iepsen, U.W.; Jørgensen, K.J.; Ringbaek, T.; Hansen, H.; Skrubbeltrang, C.; Lange, P. A Systematic Review of Resistance Training Versus Endurance Training in COPD. *J. Cardiopulm. Rehabil. Prev.* **2015**, *35*, 163–172. [CrossRef]
10. Zeng, Y.; Jiang, F.; Chen, Y.; Chen, P.; Cai, S. Exercise assessments and trainings of pulmonary rehabilitation in COPD: A literature review. *Int. J. Chronic Obstr. Pulm. Dis.* **2018**, *13*, 2013–2023. [CrossRef]
11. Spruit, M.; Gosselink, R.; Troosters, T.; De Paepe, K.; Decramer, M. Resistance versus endurance training in patients with COPD and peripheral muscle weakness. *Eur. Respir. J.* **2002**, *19*, 1072–1078. [CrossRef] [PubMed]
12. Vogiatzis, I.; Simoes, D.C.M.; Stratakos, G.; Kourepini, E.; Terzis, G.; Manta, P.; Athanasopoulos, D.; Roussos, C.; Wagner, P.D.; Zakynthinos, S. Effect of pulmonary rehabilitation on muscle remodelling in cachectic patients with COPD. *Eur. Respir. J.* **2010**, *36*, 301–310. [CrossRef] [PubMed]
13. Li, Y.; Wu, W.; Wang, X.; Chen, L. Effect of Endurance Training in COPD Patients Undergoing Pulmonary Rehabilitation: A Meta-Analysis. *Comput. Math. Methods Med.* **2022**, *2022*, 4671419. [CrossRef]
14. Nolan, C.M.; Rochester, C.L. Exercise Training Modalities for People with Chronic Obstructive Pulmonary Disease. *COPD J. Chronic Obstr. Pulm. Dis.* **2019**, *16*, 378–389. [CrossRef] [PubMed]
15. Vogiatzis, I.; Stratakos, G.; Simoes, D.C.M.; Terzis, G.; Georgiadou, O.; Roussos, C.; Zakynthinos, S. Effects of rehabilitative exercise on peripheral muscle TNF, IL-6, IGF-I and MyoD expression in patients with COPD. *Thorax* **2007**, *62*, 950–956. [CrossRef]
16. Whittom, F.; Jobin, J.; Simard, P.M.; Leblanc, P.; Simard, C.; Bernard, S.; Belleau, R.; Maltais, F. Histochemical and morphological characteristics of the vastus lateralis muscle in patients with chronic obstructive pulmonary disease. *Med. Sci. Sports Exerc.* **1998**, *30*, 1467–1474. [CrossRef]
17. Barreiro, E.; Rabinovich, R.; Marin-Corral, J.; Barberà, J.A.; Gea, J.; Roca, J. Chronic endurance exercise induces quadriceps nitrosative stress in patients with severe COPD. *Thorax* **2009**, *64*, 13–19. [CrossRef]
18. Nickel, R.; Troncoso, F.; Flores, O.; Gonzalez-Bartholin, R.; Mackay, K.; Diaz, O.; Jalon, M.; Peñailillo, L. Physiological response to eccentric and concentric cycling in patients with chronic obstructive pulmonary disease. *Appl. Physiol. Nutr. Metab.* **2020**, *45*, 1232–1237. [CrossRef]
19. Inostroza, M.; Valdés, O.; Tapia, G.; Núñez, O.; Kompen, M.J.; Nosaka, K.; Peñailillo, L. Effects of eccentric vs concentric cycling training on patients with moderate COPD. *Eur. J. Appl. Physiol.* **2022**, *122*, 489–502. [CrossRef]
20. MacMillan, N.J.; Kapchinsky, S.; Konokhova, Y.; Gouspillou, G.; Sena, R.D.S.; Jagoe, R.T.; Baril, J.; Carver, T.E.; Andersen, R.E.; Richard, R.; et al. Eccentric Ergometer Training Promotes Locomotor Muscle Strength but Not Mitochondrial Adaptation in Patients with Severe Chronic Obstructive Pulmonary Disease. *Front. Physiol.* **2017**, *8*, 114. [CrossRef]
21. Julian, V.; Thivel, D.; Miguet, M.; Pereira, B.; Costes, F.; Coudeyre, E.; Duclos, M.; Richard, R. Eccentric cycling is more efficient in reducing fat mass than concentric cycling in adolescents with obesity. *Scand. J. Med. Sci. Sports* **2019**, *29*, 4–15. [CrossRef]
22. Valladares-Ide, D.; Bravo, M.J.; Carvajal, A.; Araneda, O.F.; Tuesta, M.; Reyes, A.; Peñailillo, R.; Peñailillo, L. Changes in pulmonary and plasma oxidative stress and inflammation following eccentric and concentric cycling in stable COPD patients. *Eur. J. Appl. Physiol.* **2021**, *121*, 1677–1688. [CrossRef]
23. Alcazar, J.; Losa-Reyna, J.; Lopez, C.R.; Navarro-Cruz, R.; Alfaro-Acha, A.; Ara, I.; García-García, F.J.; Alegre, L.M.; Guadalupe-Grau, A. Effects of concurrent exercise training on muscle dysfunction and systemic oxidative stress in older people with COPD. *Scand. J. Med. Sci. Sports* **2019**, *29*, 1591–1603. [CrossRef]
24. Domaszewska, K.; Górna, S.; Pietrzak, M.; Podgórski, T. Oxidative Stress and Total Phenolics Concentration in COPD Patients—The Effect of Exercises: A Randomized Controlled Trial. *Nutrients* **2022**, *14*, 1947. [CrossRef] [PubMed]
25. Pinho, R.; Chiesa, D.; Mezzomo, K.; Andrades, M.; Bonatto, F.; Gelain, D.; Pizzol, F.D.; Knorst, M.; Moreira, J. Oxidative stress in chronic obstructive pulmonary disease patients submitted to a rehabilitation program. *Respir. Med.* **2007**, *101*, 1830–1835. [CrossRef] [PubMed]
26. Couillard, A.; Maltais, F.; Saey, D.; Debigaré, R.; Michaud, A.; Koechlin, C.; LeBlanc, P.; Préfaut, C. Exercise-induced Quadriceps Oxidative Stress and Peripheral Muscle Dysfunction in Patients with Chronic Obstructive Pulmonary Disease. *Am. J. Respir. Crit. Care Med.* **2003**, *167*, 1664–1669. [CrossRef]
27. Nicolò, A.; Sacchetti, M.; Girardi, M.; McCormick, A.; Angius, L.; Bazzucchi, I.; Marcora, S.M. A comparison of different methods to analyse data collected during time-to-exhaustion tests. *Sport Sci. Health* **2019**, *15*, 667–679. [CrossRef]
28. Saey, D.; Debigaré, R.; LeBlanc, P.; Mador, M.J.; Côté, C.H.; Jobin, J.; Maltais, F. Contractile Leg Fatigue after Cycle Exercise: A factor limiting exercise in patients with chronic obstructive pulmonary disease. *Am. J. Respir. Crit. Care Med.* **2003**, *168*, 425–430. [CrossRef] [PubMed]

29. Barreiro, E.; Gea, J.; Matar, G.; Hussain, S.N. Expression and Carbonylation of Creatine Kinase in the Quadriceps Femoris Muscles of Patients with Chronic Obstructive Pulmonary Disease. *Am. J. Respir. Cell Mol. Biol.* **2005**, *33*, 636–642. [CrossRef] [PubMed]
30. Barreiro, E. Role of Protein Carbonylation in Skeletal Muscle Mass Loss Associated with Chronic Conditions. *Proteomes* **2016**, *4*, 18. [CrossRef]
31. Paschalis, V.; Nikolaidis, M.G.; Theodorou, A.A.; Panayiotou, G.; Fatouros, I.G.; Koutedakis, Y.; Jamurtas, T. A Weekly Bout of Eccentric Exercise Is Sufficient to Induce Health-Promoting Effects. *Med. Sci. Sports Exerc.* **2011**, *43*, 64–73. [CrossRef]
32. Peñailillo, L.; Blazevich, A.; Numazawa, H.; Nosaka, K. Metabolic and Muscle Damage Profiles of Concentric versus Repeated Eccentric Cycling. *Med. Sci. Sports Exerc.* **2013**, *45*, 1773–1781. [CrossRef] [PubMed]
33. Whipp, B.J.; Ward, S.A. Quantifying intervention-related improvements in exercise tolerance. *Eur. Respir. J.* **2009**, *33*, 1254–1260. [CrossRef] [PubMed]
34. Matthews, D.R.; Hosker, J.P.; Rudenski, A.S.; Naylor, B.A.; Treacher, D.F.; Turner, R.C. Homeostasis model assessment: Insulin resistance and β-cell function from fasting plasma glucose and insulin concentrations in man. *Diabetologia* **1985**, *28*, 412–419. [CrossRef]
35. McNamara, R.J.; Houben-Wilke, S.; Franssen, F.M.; Smid, D.E.; Vanfleteren, L.E.; Groenen, M.T.; Uszko-Lencer, N.H.; Wouters, E.F.; Alison, J.A.; Spruit, M.A. Determinants of functional, peak and endurance exercise capacity in people with chronic obstructive pulmonary disease. *Respir. Med.* **2018**, *138*, 81–87. [CrossRef]
36. Rooyackers, J.; Berkeljon, D.; Folgering, H. Eccentric exercise training in patients with chronic obstructive pulmonary disease. *Int. J. Rehabil. Res.* **2003**, *26*, 47–49. [CrossRef]
37. Porszasz, J.; Emtner, M.; Goto, S.; Somfay, A.; Whipp, B.J.; Casaburi, R. Exercise Training Decreases Ventilatory Requirements and Exercise-Induced Hyperinflation at Submaximal Intensities in Patients With COPD. *Chest* **2005**, *128*, 2025–2034. [CrossRef] [PubMed]
38. Bauerle, O.; Chrusch, C.A.; Younes, M. Mechanisms by Which COPD Affects Exercise Tolerance. *Am. J. Respir. Crit. Care Med.* **1998**, *157*, 57–68. [CrossRef]
39. Kluchová, Z.; Petrášová, D.; Joppa, P.; Dorková, Z.; Tkáčová, R. The association between oxidative stress and obstructive lung impairment in patients with COPD. *Physiol. Res.* **2007**, *56*, 51–56. [CrossRef]
40. Zarrindast, S.; Ramezanpour, M.; Moghaddam, M. Effects of eight weeks of moderate intensity aerobic training and training in water on DNA damage, lipid peroxidation and total antioxidant capacity in sixty years sedentary women. *Sci. Sports* **2021**, *36*, e81–e85. [CrossRef]
41. Shin, Y.-A.; Lee, J.-H.; Song, W.; Jun, T.-W. Exercise training improves the antioxidant enzyme activity with no changes of telomere length. *Mech. Ageing Dev.* **2008**, *129*, 254–260. [CrossRef]
42. Mercken, E.M.; Hageman, G.J.; Schols, A.M.W.J.; Akkermans, M.A.; Bast, A.; Wouters, E.F.M. Rehabilitation Decreases Exercise-induced Oxidative Stress in Chronic Obstructive Pulmonary Disease. *Am. J. Respir. Crit. Care Med.* **2005**, *172*, 994–1001. [CrossRef] [PubMed]
43. Pedersen, B.K. The anti-inflammatory effect of exercise: Its role in diabetes and cardiovascular disease control. *Essays Biochem.* **2006**, *42*, 105–117. [CrossRef]
44. Ryrsø, C.K.; Thaning, P.; Siebenmann, C.; Lundby, C.; Lange, P.; Pedersen, B.K.; Hellsten, Y.; Iepsen, U.W. Effect of endurance versus resistance training on local muscle and systemic inflammation and oxidative stress in COPD. *Scand. J. Med. Sci. Sports* **2018**, *28*, 2339–2348. [CrossRef]
45. De Matos, M.A.; Vieira, D.V.; Pinhal, K.C.; Lopes, J.F.; Dias-Peixoto, M.F.; Pauli, J.R.; Magalhães, F.D.C.; Little, J.P.; Rocha-Vieira, E.; Amorim, F.T. High-Intensity Interval Training Improves Markers of Oxidative Metabolism in Skeletal Muscle of Individuals With Obesity and Insulin Resistance. *Front. Physiol.* **2018**, *9*, 1451. [CrossRef]
46. Álvarez, C.; Ramírez-Campillo, R.; Ramírez-Vélez, R.; Izquierdo, M. Effects and prevalence of nonresponders after 12 weeks of high-intensity interval or resistance training in women with insulin resistance: A randomized trial. *J. Appl. Physiol.* **2017**, *122*, 985–996. [CrossRef] [PubMed]
47. Evans, P.L.; McMillin, S.L.; Weyrauch, L.A.; Witczak, C.A. Regulation of Skeletal Muscle Glucose Transport and Glucose Metabolism by Exercise Training. *Nutrients* **2019**, *11*, 2432. [CrossRef] [PubMed]
48. Chen, T.C.; Hsieh, C.-C.; Tseng, K.-W.; Ho, C.-C.; Nosaka, K. Effects of Descending Stair Walking on Health and Fitness of Elderly Obese Women. *Med. Sci. Sports Exerc.* **2017**, *49*, 1614–1622. [CrossRef] [PubMed]
49. Chen, T.C.-C.; Tseng, W.-C.; Huang, G.-L.; Chen, H.-L.; Tseng, K.-W.; Nosaka, K. Superior Effects of Eccentric to Concentric Knee Extensor Resistance Training on Physical Fitness, Insulin Sensitivity and Lipid Profiles of Elderly Men. *Front. Physiol.* **2017**, *8*, 209. [CrossRef] [PubMed]
50. Drexel, H.; Saely, C.H.; Langer, P.; Loruenser, G.; Marte, T.; Risch, L.; Hoefle, G.; Aczel, S. Metabolic and anti-inflammatory benefits of eccentric endurance exercise—A pilot study. *Eur. J. Clin. Investig.* **2008**, *38*, 218–226. [CrossRef] [PubMed]
51. Yfanti, C.; Tsiokanos, A.; Fatouros, I.G.; A Theodorou, A.; Deli, C.K.; Koutedakis, Y.; Jamurtas, A.Z. Chronic Eccentric Exercise and Antioxidant Supplementation: Effects on Lipid Profile and Insulin Sensitivity. *J. Sports Sci. Med.* **2017**, *16*, 375–382. [PubMed]
52. Zeppetzauer, M.; Drexel, H.; Vonbank, A.; Rein, P.; Aczel, S.; Saely, C.H. Eccentric endurance exercise economically improves metabolic and inflammatory risk factors. *Eur. J. Prev. Cardiol.* **2013**, *20*, 577–584. [CrossRef] [PubMed]

53. Nikolaidis, M.G.; Paschalis, V.; Giakas, G.; Fatouros, I.G.; Sakellariou, G.K.; Theodorou, A.A.; Koutedakis, Y.; Jamurtas, A.Z. Favorable and Prolonged Changes in Blood Lipid Profile after Muscle-Damaging Exercise. *Med. Sci. Sports Exerc.* **2008**, *40*, 1483–1489. [CrossRef]

54. Paschalis, V.; Nikolaidis, M.G.; Giakas, G.; Theodorou, A.A.; Sakellariou, G.K.; Fatouros, I.G.; Koutedakis, Y.; Jamurtas, A.Z.; Jamurtas, T. Beneficial changes in energy expenditure and lipid profile after eccentric exercise in overweight and lean women. *Scand. J. Med. Sci. Sports* **2010**, *20*, e103–e111. [CrossRef] [PubMed]

55. Sillanpaa, E.; Hakkinen, A.; Punnonen, K.; Laaksonen, D.E. Effects of strength and endurance training on metabolic risk factors in healthy 40-65-year-old men. *Scand. J. Med. Sci. Sports* **2009**, *19*, 885–895. [CrossRef]

56. Boukabous, I.; Marcotte-Chénard, A.; Amamou, T.; Boulay, P.; Brochu, M.; Tessier, D.; Dionne, I.; Riesco, E. Low-Volume High-Intensity Interval Training Versus Moderate-Intensity Continuous Training on Body Composition, Cardiometabolic Profile, and Physical Capacity in Older Women. *J. Aging Phys. Act.* **2019**, *27*, 879–889. [CrossRef]

57. Barnes, P.J. COPD 2020: New directions needed. *Am. J. Physiol. Cell Mol. Physiol.* **2020**, *319*, L884–L886. [CrossRef]

Article

Chronic Low or High Nutrient Intake and Myokine Levels

Ana Paula Renno Sierra [1], Antônio Alves Fontes-Junior [2], Inês Assis Paz [2], Cesar Augustus Zocoler de Sousa [2], Leticia Aparecida da Silva Manoel [2], Duane Cardoso de Menezes [2], Vinicius Alves Rocha [2], Hermes Vieira Barbeiro [3], Heraldo Possolo de Souza [3] and Maria Fernanda Cury-Boaventura [2,*]

[1] School of Physical Education and Sport, University of São Paulo, Sao Paulo 05508-030, Brazil
[2] Interdisciplinary Post-Graduate Program in Health Sciences, Institute of Physical Activity and Sports Sciences, Cruzeiro do Sul University, Sao Paulo 01506-000, Brazil
[3] Emergency Medicine Department, LIM-51, University of São Paulo, Sao Paulo 01246-903, Brazil
* Correspondence: maria.boaventura@cruzeirodosul.edu.br

Abstract: Inadequate nutrient availability has been demonstrated to be one of the main factors related to endocrine and metabolic dysfunction. We investigated the role of inadequate nutrient intakes in the myokine levels of runners. Sixty-one amateur runners participated in this study. The myokine levels were determined using the Human Magnetic Bead Panel from plasma samples collected before and after the marathon. Dietary intake was determined using a prospective method of three food records. The runners with lower carbohydrate and calcium intakes had higher percentages of fat mass ($p < 0.01$). The runners with a sucrose intake comprising above 10% of their energy intake and an adequate sodium intake had higher levels of BDNF ($p = 0.027$ and $p = 0.031$). After the race and in the recovery period, the runners with adequate carbohydrate intakes (g/kg) (>5 g/kg/day) had higher levels of myostatin and musclin ($p < 0.05$). The runners with less than 45% of carbohydrate of EI had lower levels of IL-15 ($p = 0.015$) and BNDF ($p = 0.013$). The runners with higher cholesterol intakes had lower levels of irisin ($p = 0.011$) and apelin ($p = 0.020$), and those with a low fiber intake had lower levels of irisin ($p = 0.005$) and BDNF ($p = 0.049$). The inadequate intake influenced myokine levels, which promoted cardiometabolic tissue repair and adaptations to exercise.

Keywords: nutrition; exercise; endurance; musculoskeletal; physiology

Citation: Sierra, A.P.R.; Fontes-Junior, A.A.; Paz, I.A.; de Sousa, C.A.Z.; Manoel, L.A.d.S.; Menezes, D.C.d.; Rocha, V.A.; Barbeiro, H.V.; Souza, H.P.d.; Cury-Boaventura, M.F. Chronic Low or High Nutrient Intake and Myokine Levels. *Nutrients* 2023, 15, 153. https://doi.org/ 10.3390/nu15010153

Academic Editor: Olivier Bruyère

Received: 7 December 2022
Revised: 20 December 2022
Accepted: 22 December 2022
Published: 29 December 2022

1. Introduction

Exercise promotes the release of chemical messengers as a result of skeletal muscle contraction, called myokines and/or exerkines. The myokines modulate muscle mass, function, and regeneration by acting on protein synthesis, insulin sensitivity, fat oxidation, myogenesis, mitochondrial biogenesis, autophagy, mitophagy, and the remodeling of the extracellular matrix [1–4].

In addition, many myokines seem to contribute to cardiometabolic adaptations to exercise as a result of crosstalk with adipose tissues, the liver and heart contributing to glucose homeostasis, the browning of white adipose tissue (WAT), and cardioprotection [1,5–7].

More than 650 myokines have been described in response to exercise, and many researchers have been investigating their biological effects on different tissues. The most studied myokines include IL-6, IL-15, myostatin, leukemia inhibitory factor (LIF), secreted protein acidic rich in cysteine (SPARC), myonectin, monocyte chemoattractant protein-1 (MCP1), irisin, apelin, decorin, musclin, growth differentiation factor 15 (GDF-15), brain-derived neurotrophic factor (BDNF), fibroblast growth factor (FGF)-21, follistatin (FSTL), meteorin-like (Metrnl), fractalkine, and angiopoietin-like protein 4 [2,3,8]. The myokine levels are dependent on various forms of exercise and training [9].

Nutritional interventions such as caloric restriction or supplementation seem to modulate cytokines, adipokines, myokines, and cardiomiokynes [4,10,11]. Previous studies have reported a poor daily intake with low carbohydrate, dietary fiber source, fruit, dairy beverage, and vegetable intake in endurance runners [12–15]. Chronic low or higher nutrient availability promotes endocrine and cardiometabolic dysfunction [5,10].

Inadequate daily intake (DI) may influence the myokine response induced by endurance exercise, affecting the dynamics of muscle repair and cardiometabolic adaptations. The maladaptive response after endurance exercise may impair muscle function, performance, and health, or even increase the risk of acute cardiovascular events [16]

The aim of this study to investigate the effects of chronic low or high nutrient intakes on myokine levels before and after endurance exercise.

2. Material and Methods

2.1. Subjects

Seventy-four amateur Brazilian male marathon finishers (aged 30 to 55 years) participated in this study. The volunteer recruitment was performed by e-mail to all marathon runners registered in the São Paulo International Marathon in 2017 or 2018. Researchers randomly contacted volunteers to confirm their interest and availability to participate in all steps of the study (before the race, immediately after the race, and in the recovery period). Inclusion criteria were training more than 30 km per week and having previously participated in a half marathon or marathon, as well as not having cardiovascular, pulmonary, or kidney injury, and/or liver, kidney, inflammatory, or neoplastic diseases, or use alcohol or drugs.

The Ethics Committee of Cruzeiro do Sul University, Brazil (Permit Number: 3.895.058) approved this study in accordance with the Declaration of Helsinki. All volunteers read and signed the written informed consent document before starting to participate in the study.

Body composition, cardiopulmonary function, and DI were evaluated before the race. Of the seventy-four marathon runners, sixty filled in three food records before the marathon race (one week) for DI analyses. Therefore, we excluded fourteen runners from the analysis of the association between myokines and DI.

The São Paulo International Marathon began at 07:30 a.m. on 9 April 2017 and 8 April 2018. Fluid ingestion was provided during the race (water every 2 to 3 km; sports drinks at 12 km, 21.7 km, 33 km, and 42 km; and a carbohydrate at 28.8 km). The weather during the race was temperate (average temperature and humidity 19.8 °C, 72.8% in 2017, and 19.9 °C, 87.7% in 2018) (National Institute of Meteorology, Ministry of Agriculture, Livestock, and Supply).

An electronic digital scale platform (marte®, Sao Paulo, SP, Brazil) was used to measure body mass (kg) and height (cm). Body mass index (BMI) was calculated according to the International Society for Advancement of Kinanthropometry (ISAK) standard [weight (kg)/height (m^2)]. The body composition was determined by bioimpedance analysis (Biodynamics Corporation, Shoreline (WA), USA, 310e) 24 h before the marathon race in the fasting state.

2.2. Cardiopulmonary Exercise Test

After the medical history data collection, the cardiopulmonary exercise test (CPET) was realized between three and one week before the São Paulo International Marathon by a progressive treadmill test with a constant incline of 1%, and an initial speed of 8 km·h^{-1}, with an elevation of 1 km·h^{-1} every 1 min until voluntary exhaustion (TEB Apex 200, TEB, São Paulo, Brazil, speed 0–24 km/h, grade 0–35%). The volunteers were monitored with a standard 12-lead computerized electrocardiogram during the test (TEB®, ECG São Paulo, Brazil) to rule out any cardiac dysfunction at rest and during exertion. The respiratory gas exchange was measured through open-circuit and automatic, indirect calorimetry (Quark CPET, COSMED®, Rome, Italy). The VO2 max of the subjects was determined according to the American College of Sports Medicine [17]. All volunteers finished the International Marathon of São Paulo in 2017 (40 runners) and 2018 (34 runners).

2.3. Blood Sampling

Blood samples (10 mL) were collected 24 h before, and 24 h and 72 h after the race from the antecubital vein at the Cruzeiro do Sul University with at least 12 h without physical activity and from fasting runners. To obtain plasma samples, vacuum tubes containing ethylenediaminetetraacetic acid (10 mL, EDTA, 1 mg/mL) samples were immediately centrifuged at 4 °C, $400 \times g$ for 10 min and then stored at -80 °C for the later analysis of myokines at University of São Paulo. Immediately after the race, blood samples (10 mL) from the fed runners were maintained on ice for approximately 2 h at the International Marathon of São Paulo (competition venue, close to finish line) and then sent to Cruzeiro do Sul University (10 mL) for plasma collection as described above.

2.4. Determination of Myokines

Apelin, BDNF, FSTL, FGF-21, IL-6, IL-15, irisin, myostatin, and musclin plasmatic levels were evaluated using the MILLIPLEX® Human Myokine Magnetic Bead (MagPlex®-C microspheres) Panel protocol (HCYTOMAG-56K, EMD Millipore Corporation, MA, USA). The fluorescent-coded magnetic beads contain a specific capture antibody of each myokine on the surface with internally color-coded microspheres with two fluorescent dyes detected by Luminex® xMAP® technology. After the capture antibody incubation, a biotinylated detection antibody is added on the assay, followed by a reaction with a Streptavidin-PE conjugate. The high-speed digital signal processors of capture and detection components were analyzed by a Luminex® analyzer (MAGPIX®). The intra-assay precision (mean coefficient variation percentage) described by the MILLIPLEX® Human Myokine Magnetic Bead Panel instructions is <10%.

2.5. Dietary Intake

A prospective method of three food records was used to estimate DI during the week before the marathon race (3rd to 8th April in 2017, and 2nd to 7th April in 2018). Dietetics instructed the runners to fill in the meal time and all food and drinks that were ingested, including portion size and food brand, on two days of the week and one day of the weekend. In an interview with the runners one day before the race, the food records were checked by a trained nutrition undergraduate student to elucidate or complete missing food data. The energy intake (kcal), macronutrients (g or g/kg), and micronutrients (mg) were estimated by the professional Dietbox (http:/dietbox.me) website/app. The United States Department of Agriculture–Agricultural Research Service (USDA) food composition database and Brazilian Table of Food Composition database (TACO, University of Campinas, São Paulo, SP, Brazil) were used in the professional Dietbox website/app to provide the nutrient composition of foods.

2.6. Statistical Analyses

Data of general characteristics, DI, and myokines are reported as mean $\pm$ SEM of sixty endurance runners. Statistical analyses were performed using GraphPad Prism (GraphPad Prism version 9). The myokines were used as the independent variable. The normality of the data distribution was determined using the Kolmogorov–Smirnov test and the normality was rejected. Statistical analyses of myokines were evaluated using the Kruskal–Wallis test and Dunn's test for multiple comparison. Correlations between myokines and DI (macronutrients and micronutrients) were performed by the Spearman test. Statistical significance was accepted at the level of $p < 0.05$ in all analyses. Statistical analyses of myokine levels in runners with an adequate and inadequate intake of the percentage of sucrose and carbohydrate in the energy intake (EI), with fiber, calcium, sodium, and potassium intakes evaluated using the Mann–Whitney test, and carbohydrate, cholesterol, selenium, vitamin B3, and phosphorus intakes evaluated using the Kruskal–Wallis test. Cohen's d was calculated to estimate the effect size of the significant differences of myokine levels between runners with adequate and inadequate intake.

The number of runners with lower, adequate, higher, or very higher intake are described in Table 1.

Table 1. Number of runners with lower, adequate, higher, or very higher intake.

	Lower	Adequate	Higher	Very Higher
Sucrose (% of EI)		<10	>10	
		40	20	
Carbohydrate (% of EI)	<45	45–65	>65	
	19	38	3	
Carbohydrate (g/kg/day)	<3	3–5	>5	
	23	23	14	
Protein (g/kg/day)	<1.2	1.2–2.0	>2	
	24	21	15	
Cholesterol (mg)		<300	300–600	>600
		23	27	10
Fiber (g)	<25	>25		
	40	20		
Calcium (mg)	<1000	>1000		
	50	10		
Sodium (mg)		<2300	>2300	
		29	31	
Selenium (mcg)	<55	55–110	>110	
	7	20	33	
Vitamin B3 (mg)	<16	16–32	>32	
	16	29	15	
Phosphorus (mg)	<700	700–1400	>1400	
	5	30	25	
Potassium (mg)		<2000	>2000	
		21	39	

EI, Energy Intake.

3. Results

3.1. General Characteristics

The general characteristics of sixty runners are described as follows: age, 40.9 ± 1.0 years; body mass, 74.6 ± 1.3 kg; height, 1.73 ± 0.01 m; BMI, 24.9 ± 0.4 kg/m^2; percentage of fat mass, $21.6 \pm 0.6\%$; free fat mass, 58.3 ± 0.9 kg; race time, 258.7 ± 6.0 min, training experience, 6.9 ± 0.5 years; time in 10 km race, 46.7 ± 0.77 min; frequency of training, 4 (3.75–5.0) times/week; and training volume, 51.8 ± 2.7 km/week. The CPET parameters are: time of exhaustion, 11.5 ± 0.3 min; maximum speed of runners, 18.4 ± 0.3 km/h; anaerobic threshold oxygen consumption (VO$_2$ AT), 33.6 ± 1.0 mL/kg/min; respiratory compensation point oxygen consumption (VO$_2$ RCP), 51.6 ± 1.2 mL/kg/min; and peak oxygen consumption (VO$_2$ peak), 54.4 ± 1.3 mL/kg/min.

3.2. Dietary Intake

The energy and macronutrient intake are summarized in Table 2. We observed an adequate protein, total fat, and sucrose intake; however, a low energy, carbohydrate, and fiber intake were observed, as well as a higher cholesterol intake.

Table 2. Energy, macronutrient, cholesterol, and fiber intake.

	Daily Intake	**DV ***
Energy intake (kcal)	2319 ± 117	2907 ± 36
Energy availability (kcal/kg of FFM)	40 ± 1.98	>45
Carbohydrate (g/kg)	3.9 ± 0.3	* 5–12
Protein (g/kg)	1.6 ± 0.1	* 1.2–1.7
Total fat (% of EI)	29 ± 1	<30%
Sucrose (% of EI)	8 ± 1	<10%
Cholesterol (mg)	391 ± 28	<300
Fiber (g)	22 ± 1	>25

* Reference daily values (DV) based on Dietary Reference Intake (DRI) of the American Dietetic Association, Dietitians of Canada, and the American College of Sports Medicine (ADA/ACSM). The values are presented as mean ± mean standard error of 60 runners.

The micronutrient daily intake is summarized in Table 3. We observed low folic acid, vitamin D, calcium, and magnesium intakes.

Table 3. Micronutrient daily intakes in marathon runners.

Vitamins	**Daily Intake**	**DV ***	**Minerals**	**Daily Intake**	**DV ***
Vitamin A (mcg)	994 ± 183	900	Calcium (mg)	715 ± 57	1000
Vitamin B1 (mg)	1.73 ± 0.14	1.2	Iron (mg)	15.7 ± 1.5	8
Vitamin B2 (mg)	1.94 ± 0.15	1.3	Mn (mg)	2.8 ± 0.30	2.3
Vitamin B3 (mg)	28 ± 3	16	Se (mcg)	163 ± 18	55
Vitamin B6 (mg)	2.3 ± 0.2	1.7	Zinc (mg)	12.7 ± 0.9	11
Folic acid (mg)	286 ± 27	400	Mg (mg)	289 ± 16	420
Vitamin B12 (mcg)	6.4 ± 1.6	2.4	P (mg)	1340 ± 73	700
Vitamin C (mg)	141 ± 28	90	Potassium (g)	2.6 ± 130	4.7
Vitamin D (mcg)	3.7 ± 0.53	15	Sodium (g)	2.5 ± 1.6	1.5
Vitamin E (mg)	12.34 ± 1.5	15			

* Reference daily values (DV) based on Dietary Reference Intake (DRI) of the American Dietetic Association, Dietitians of Canada, and the American College of Sports Medicine (ADA/ACSM). Mn, manganese; Se, selenium; Mg magnesium; P, phosphorus. The values are presented as mean ± mean standard error of 60 runners.

3.3. Correlation: Dietary Intake and General Characteristics

Percentage of fat mass was negatively correlated with EI, % of adequate EI, carbohydrate, protein, sucrose, vitamin B2, calcium, manganese, and phosphorus intake (Table 3). Free fat mass was negatively correlated with the percentage of carbohydrate and protein of EI ($r = -31, p = 0.016$; $r = -26, p = 0.042$), and carbohydrate intake ($r = -27, p = 0.032$). Race time was negatively correlated with protein, cholesterol, vitamin B3, selenium, magnesium, potassium, and phosphorus (Table 4).

The runners with lower carbohydrate (Cohen's d = 1.37), phosphorus (Cohen's d = 1.48), and calcium intakes (Cohen's d = 1.16) had a higher percentage of fat mass (Figure 1A–C).

Table 4. Correlation of dietary intake with % of fat mass and race time.

% of Fat Mass	r	p	Race Time (min)	r	p
EI (kcal/kg of FFM)	−0.28	0.025	Protein (g/kg)	−0.35	0.012
% of adequate EI	−0.34	0.007	Cholesterol (mg)	−0.40	0.043
Carbohydrate (g/kg)	−0.41	0.0009	Vitamin B3 (mg)	−0.30	0.033
Protein (g/kg)	−0.32	0.013	Se (mcg)	−0.30	0.032
Sucrose (g)	−0.34	0.007	Mg (mg)	−0.36	0.011
Vitamin B2 (mg)	−0.27	0.033	K (mg)	−0.29	0.043
Calcium (mg)	−0.49	<p.0001	P (mg)	−0.38	0.006
Mn (mg)	−0.31	0.016			
P (mg)	−0.35	0.007			

EI, energy intake; Mn, manganese; Se, selenium; Mg, magnesium; P, phosphorus. The values are presented as mean ± mean standard error of 60 runners.

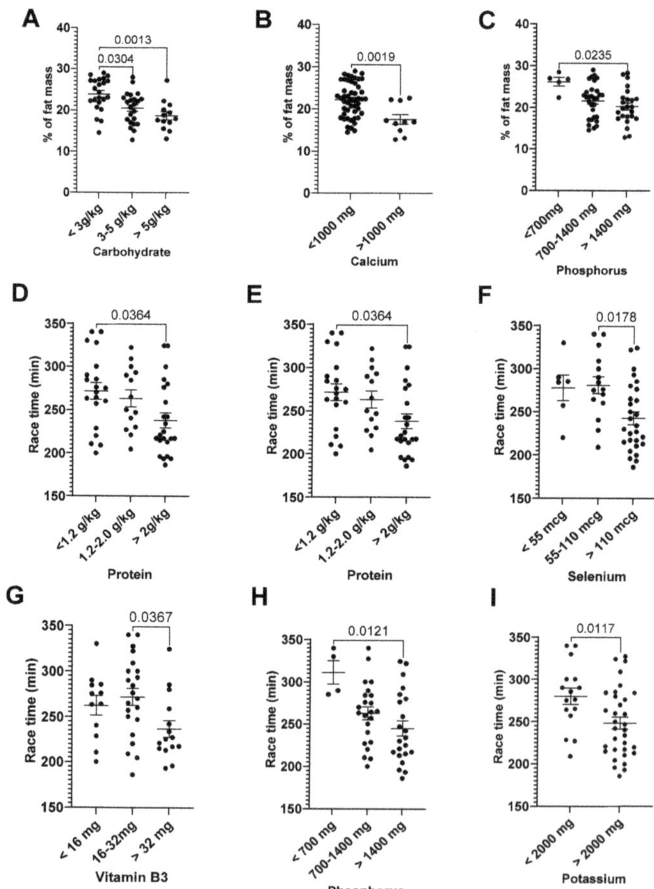

Figure 1. General characteristic and dietary intake. Percentages of fat mass and carbohydrate (**A**), calcium (**B**), and phosphorus (**C**) intake; race time and protein (**D**), cholesterol (**E**), selenium (**F**), vitamin B3 (**G**), phosphorus (**H**), and potassium (**I**) intake. The percentages of fat mass and race time were presented as mean ± EPM, as well as individuals' values.

Runners with higher intakes of protein (>2 g/kg/day, Cohen's d = 0.79), cholesterol (>600 mg, Cohen's d = 0.98), selenium (>110 mg, 2-fold RDA, Cohen's d = 0.99), vitamin B3 (>32 mg, 4-fold RDA, Cohen's d = 0.86), phosphorus (>1400 mg, 2-fold RDA, Cohen's d = 1.62), and potassium (>2000 mg, 100% RDA, Cohen's d = 0.79) had lower race times (Figure 1D–I).

3.3.1. Myokine Analyses

Running the marathon elevated BDNF (2-fold,), FSTL (2-fold), FGF-21 (4-fold), and IL-6 (5-fold) plasma levels (Figure 2). The IL-6 concentration was slightly reduced 72 h after the marathon (Figure 2D).

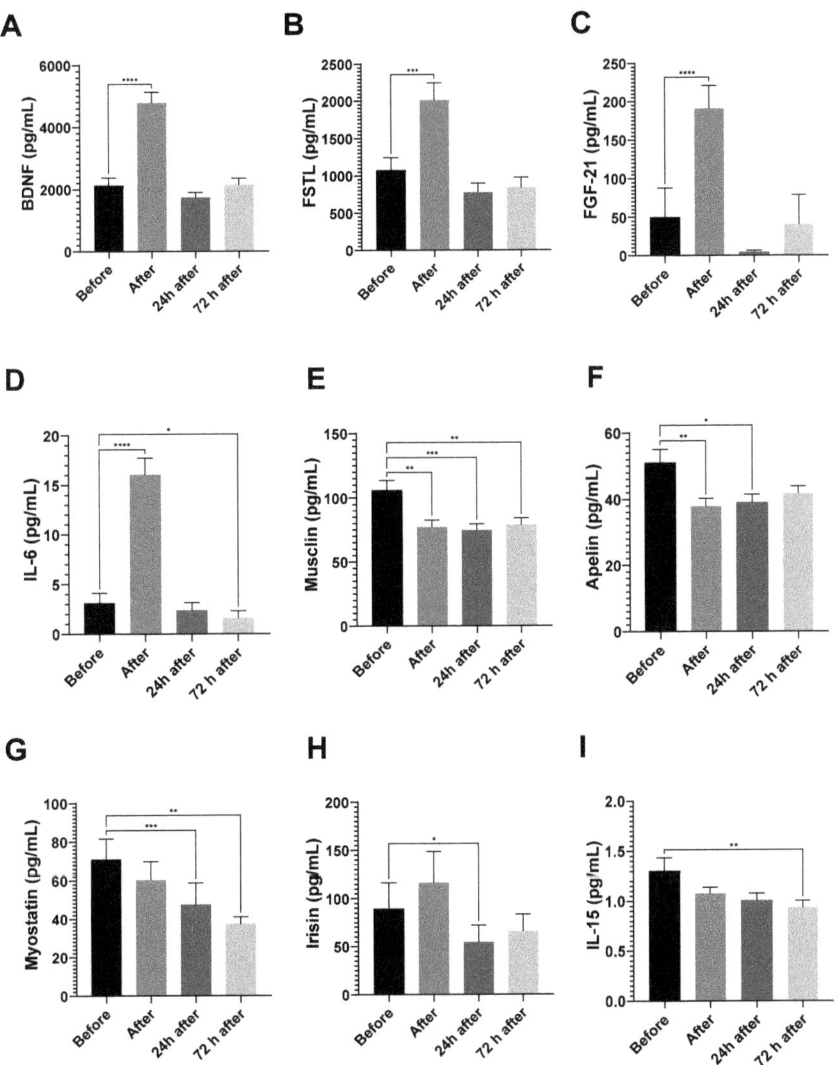

Figure 2. Myokine levels after the marathon race. The BDNF (**A**), FSTL (**B**), FGF-21 (**C**), IL-6 (**D**), musclin (**E**), apelin (**F**), myostatin (**G**), irisin (**H**), and IL-15 (**I**) levels are presented as mean ± EPM of 60 runners before the race, and 24 and 72 h after the race. * vs. before $p < 0.05$; ** for $p < 0.01$; *** vs. before $p < 0.001$; and **** vs. before $p < 0.0001$.

We also demonstrated a decrease in musclin and apelin after the race (Figure 2E,F); musclin, apelin, myostatin, and irisin levels 24 h after the marathon (Figure 2E–H); and musclin, myostatin, and IL-15 levels 72 h after the race (Figure 2E,G,I).

3.3.2. Myokines and DI

Before the race, BDNF was positively correlated with the percentage of carbohydrate and sucrose in the EI, as well as fiber intake (r = 0.36, p = 0.01; r = 0.38, p = 0.0053; r = 0.30, p = 0.033, respectively), and negatively correlated with sodium (r = −0.27, p = 0.049). The percentage of carbohydrate also had a positive correlation with FSTL (r = 0.32, p = 0.025). Musclin, myostatin, IL-15, irisin, and apelin were not associated with dietary intake before the race (data not shown).

Runners with a sucrose intake above 10% of their EI and adequate levels of sodium intake (<2300 mg/day) had higher levels of BDNF (Figure 3A,B).

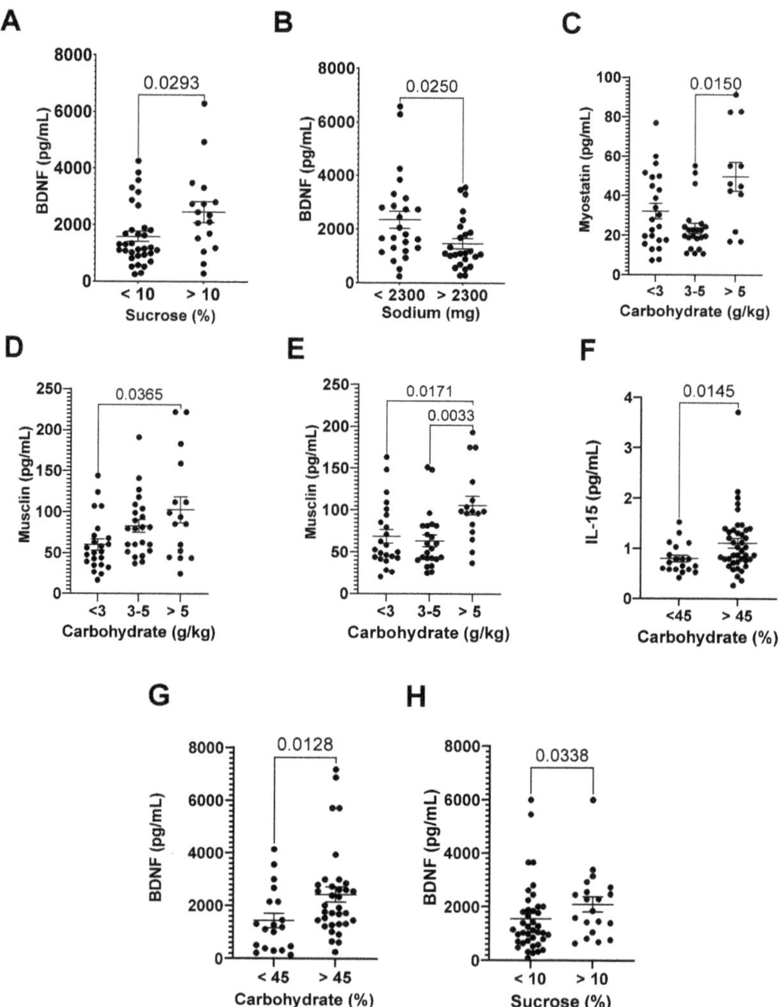

Figure 3. Dietary intake and myokine levels after the race and in the recovery period. The BDNF levels before the race (**A**,**B**), myostatin levels 72 h after the race (**C**), and musclin levels 24 h after (**D**) the race (**E**); IL-15 levels 24 h after the race (**F**); and BDNF levels 72 h after the race (**G**) and 24 h after the race (**H**). The myokine levels are presented as mean ± EPM, as well as individuals' values.

3.4. Myokines Induced by Exercise and Dietary Intake

After the race, carbohydrate intake (g/kg) was correlated with musclin and myostatin levels (r = 0.29, *p* = 0.027; r = 0.32, *p* = 0.014), and the percentage of carbohydrate of EI was correlated positively with musclin and FGF-21 levels (r = 0.26, *p* = 0.047; r = 0.33, *p* = 0.012).

Runners with adequate carbohydrate intake (>5 g/kg/day) had higher levels of myostatin 72 h after the race, with Cohen's d = 1.46 (Figure 3C), and higher levels of musclin after the race and 24 h after the race, with Cohen's d = 0.90 and 0.79 (Figure 3D,E).

In the recovery period, the percentage of carbohydrate of EI was correlated positively with IL-15 and BDNF (r = 0.34, *p* = 0.01; r = 0.30, *p* = 0.024), and sucrose intake had a positive correlation with BDNF and FSTL levels (r = 0.36, *p* = 0.0049; r = 0.30, *p* = 0.022).

Runners with less than 45% of carbohydrate of EI had lower levels of IL-15 (Cohen's d = 0.59, 24 h after race) and BNDF (72 h after the race, Cohen's d = 0.65) (Figure 3F,G), and those with less than 10% of sucrose had lower levels of BDNF (24 h after the race, Cohen's d = 0.42) (Figure 3H).

Protein intake (g/kg or % of EI) was not correlated with the myokines determined in this study in all periods (before and after the race, and in the recovery period) (data not shown).

Cholesterol intake negatively correlated with apelin and irisin levels (r = −0.26, *p* = 0.037; r = −0.31, *p* = 0.015, respectively) after the race. Runners with a higher cholesterol intake (>600 mg/day) had lower levels of irisin (Cohen's d = 0.71) and apelin (Cohen's d = 0.96) compared to runners with adequate cholesterol intake after the race (Figure 4A,B). Cholesterol intake also showed a negative correlation with apelin 24 h after the race (r = −0.26, *p* = 0.44).

Fiber intake had a correlation with irisin and BDNF levels 72 h after the race (r = 0.34, *p* = 0.00067; r = 0.28, *p* = 0.028). Runners with a low fiber intake (<25 g/day) had lower levels of irisin (Cohen's d = 0.49) and BDNF (Cohen's d = 0.60) (Figure 4C,D).

After the race and in the recovery period, vitamin C was positively correlated with IL-15, musclin, FSTL, myostatin, IL-6, and FGF-21 (Table 5).

Table 5. Correlation of vitamin C with myokines after the race.

	r	*p*
IL-15 24 h after	0.33	0.010
Musclin after	0.34	0.0075
Musclin 24 h after	0.27	0.038
FSTL 24 h after	0.26	0.049
Myostatin after	0.26	0.037
Myostatin 24 h after	0.27	0.033
IL-6 24 h after	0.30	0.019
FGF-21 24 h after	0.28	0.031

IL, interleukin; FSTL, follistatin; FGF, fibroblast growth factor. Correlations between vitamin C and DDI were determined in 60 runners via the Spearman test.

Thiamine (B1) was correlated with myostatin levels after the race (r = 0.31, *p* = 0.014), vitamin D was negatively correlated with IL-6 levels after the race (r = −0.26, *p* = 0.044), and musclin 72 h after the race (r = −0.26, *p* = 0.037). Manganese was correlated with musclin after the race (r = 0.30, *p* = 0.019), and selenium was negatively correlated with apelin 24 h after the race (r = −0.28, *p* = 0.030).

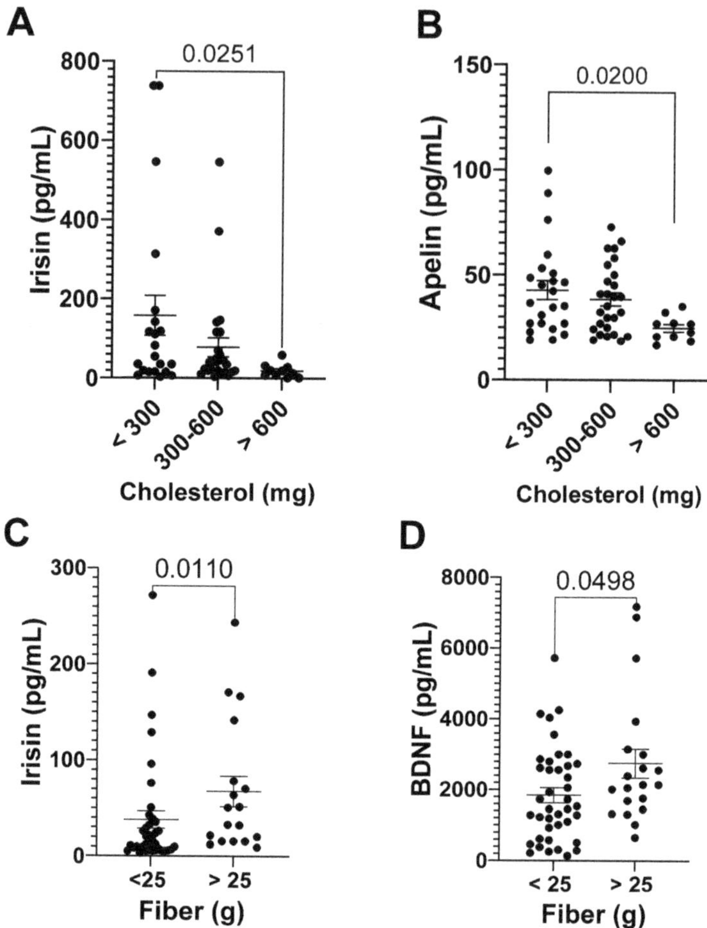

Figure 4. Myokine levels and cholesterol and fiber intake. Irisin (**A**) and apelin (**B**) levels after the race, and irisin (**C**) and BDNF (**D**) levels 72 h after the race. The myokine levels are presented as mean ± EPM, as well as individuals' values.

4. Discussion

Our study highlights the importance of carbohydrate, energy and vitamin C intake for the percentage of fat mass, race time, and myokine levels before (BDNF) and after the race (myostatin, musclin, irisin, apelin), and in the recovery period (BDNF, IL-15, FSTL, FGF-21, myostatin, and musclin). Chronic higher sodium and cholesterol, or low fiber consumption were demonstrated to reduce some myokine levels (BNDF, irisin, and apelin). Our results contribute to elucidating the mechanisms involved in the effects of low or higher nutrient intakes on tissue recovery and exercise adaptations.

In this study, amateur long-distance runners had low energy, carbohydrate, and fiber intakes and higher cholesterol intakes in accordance with scientific literature, which described low energy availability in elite and amateur athletes [12,18]. Relative Energy Deficiency in Sport (RED-S) has been reported in male and female elite athletes and impairs endocrine response (i.e., cortisol, insulin, IGF-1, adipokines, incretins), contributing to metabolic and immune dysfunction [19–22]. Previously, we suggested that pro-inflammatory cytokines induced by endurance exercise are higher in runners with a low energy, carbohydrate, and fiber intake, and the adequate carbohydrate intake tended to promote higher IL-10 levels in the recovery period of intense exercise [12]. RED (< 30 kcal·kg^{-1}

FFM·day^{-1}), accomplished by a low carbohydrate intake, also reduces the mobilization of fat stores, protein synthesis, metabolic rate, and glucose utilization, and the production of growth hormones [21,22]. Herein, we have demonstrated higher percentages of fat mass in runners with lower carbohydrate, phosphorus, and calcium intakes. There is crosstalk between adipose tissue, and calcium and phosphorus homeostasis, which involves some hormones such as leptin and parathormone [23].

We also observed higher percentages of runners with low energy (87.6%, <45 kcal/kg/FFM) and carbohydrate (76.7%, <5 g/kg/day) intakes. After the race and in the recovery period, we observed a positive correlation of myostatin and musclin with carbohydrate intake, and higher levels of these myokines with an adequate carbohydrate intake (5 to 8 g/kg/day). Myostatin acts in the process of protein degradation in muscle tissue via the activation of activin receptors (type I and II) leading to the phosphorylation and activation of SMAD-2 and SMAD-3, which forms a complex with SMAD-4, promoting the transcription of catabolic genes, and through the ubiquitin–proteosome system and autophagy activation [1,3,8], myostatin inhibition seems to increase the browning of WAT by activating the AMPK/PGC1-alpha/FNDC-5 pathway [24,25]. Activin receptors are also distributed in other tissues, and myostatin seems to impair the growth hormone (GH)/IGF1 axis in the liver [26]. The positive correlation of myostatin with carbohydrate intake does not contribute to our understanding of the role of carbohydrate intake in muscle mass repair after endurance exercise. However, it may elucidate the role of restricted caloric and low-carbohydrate diets, such as the Dietary Approaches to Stop Hypertension (DASH) diet, to improve muscle mass and cardiometabolic health [27].

Musclin activates PPAR-gamma and promotes mitochondrial biogenesis in WAT and skeletal muscle [28,29], and it also reduces glucose uptake in skeletal muscle through the inhibition of Akt/PKB and PPARγ and liver × receptor a (LXRα) activation [30]. Runners with adequate carbohydrate intake had higher musclin levels, which may contribute to improved exercise adaptations, such as the improvement of glucose homeostasis and the browning of adipose tissue.

After or in the recovery period, the percentage of carbohydrate of EI or sucrose intake was correlated positively with IL-15, FGF-21, BDNF, and FSTL levels, myokines that are responsible for muscle repair, whose functions include myogenesis (BDNF, FSTL, and IL-15), mitochondrial biogenesis (BDNF, IL-15), mytophagia (FGF-21, IL-15), autophagy (IL-15), satellite cell activation (BDNF), anti-inflammatory response (IL-15), vascular smooth muscle cell proliferation (IL-15), intramuscular fat oxidation, insulin sensitivity, and glucose intake (BDNF, FGF-21, IL-15) [1–3,5,8,31–34]. Many of these myokines act in several signaling pathways that will culminate in the activation of the transcription coactivator peroxisome proliferator-activated receptor-gamma coactivator (PGC)-1alpha and transcription factor PPAR-gamma, which modulate genes related to the muscle autocrine/paracrine effects cited above [25].

Moreover, these myokines promote crosstalk between skeletal muscle and other cardiometabolic tissues, which improves glucose homeostasis (IL-15, BDNF, and FSTL), enhances fuel utilization of glucose and lipids (FGF-21), and enacts a cardioprotective role (FGF-21 and BDNF) [1,5,9,33–35]. BDNF is a neurotrophin stimulated by metabolic stress and higher intracellular calcium levels, and it acts on myocardial contraction, decreasing the cardiomyocyte apoptosis and mitochondrial dysfunction, increasing motor neuron excitability and cardiomyocyte contraction, and improving lipid and glucose metabolism via p-AMPK and PGC-1α activation [31]. Interestingly, we also observed that runners with unhealthy behaviors, such as higher intakes of sodium or lower percentages of carbohydrate (<45%) or fiber intake, had lower BDNF levels.

Cholesterol intake was negatively associated with apelin and irisin levels, which improve carbohydrate and lipid metabolism, and have cardioprotection properties [8,32,36,37]. Runners with higher cholesterol intakes (>600 mg/day) had lower levels of irisin and apelin; moreover, runners with a lower fiber intake had lower levels of irisin. These myokines include a portion of the cell membrane protein, FNDC5, composed of 94-amino-acid residue fibronectin III (FNIII)-2 domains, and through the intracellular signaling pathway p38 and ERK1/2, they induce the browning of WAT, the upregulation of UCP1, and the improvement of glucose intake on skeletal muscle and insulin sensitivity [37]. Apelin seems to have a cardioprotective role, binding to the G-protein-coupled receptor (GPCR) and acting on the PI3K-Akt-NO signaling pathway, the reperfusion injury salvage kinase (RISK) pathway, the extracellular signal-related kinase 1/2 (ERK 1/2), protein kinase B/Akt, and eNOS [38]. Lipolysis on adipose tissue, lipid oxidation, and mitochondrial biogenesis on skeletal muscle are upregulated by apelin levels [39].

Muscle cells also require multiple protective enzyme systems involved in muscle function. In this study, we observed that after the race and in the recovery period, vitamin C was positively correlated with IL-15, musclin, FSTL, myostatin, IL-6, and FGF-21. Vitamin C has many physiological functions, including anti-inflammatory and anti-oxidative properties. However, the supplementation of vitamin C seems to be beneficial to vitamin deficiency, but leads to controversial responses in muscle mass and function, leading to the inhibition of protein synthesis pathways [40].

Inadequate dietary intake may influence some myokine levels responsible for the maintenance of muscle mass and function, and for the recovery of cardiometabolic tissues after endurance exercise, as well as enhancing adaptation to exercise. Professionals in the field of nutrition and sports medicine should emphasize the importance of adequate nutrition for performance improvement, the recovery of body tissues, and for the beneficial adaptations induced by exercise.

Author Contributions: Conceptualization, M.F.C.-B., A.P.R.S. and H.P.d.S.; methodology, A.P.R.S., A.A.F.-J., H.V.B., H.P.d.S. and M.F.C.-B.; software, A.P.R.S., A.A.F.-J. and I.A.P.; validation, I.A.P., C.A.Z.d.S., L.A.d.S.M., D.C.d.M. and V.A.R.; formal analysis, A.P.R.S., A.A.F.-J., C.A.Z.d.S., H.V.B. and M.F.C.-B.; investigation, A.P.R.S., A.A.F.-J., C.A.Z.d.S., L.A.d.S.M., D.C.d.M., V.A.R., H.V.B., H.P.d.S. and M.F.C.-B.; resources, H.P.d.S. and M.F.C.-B.; data curation, A.P.R.S., A.A.F.-J., C.A.Z.d.S., H.V.B. and M.F.C.-B.; writing—original draft preparation, A.P.R.S., A.A.F.-J., H.V.B., H.P.d.S. and M.F.C.-B.; writing—review and editing, A.P.R.S., A.A.F.-J., H.V.B., H.P.d.S. and M.F.C.-B.; visualization, A.P.R.S., A.A.F.-J., C.A.Z.d.S., L.A.d.S.M., D.C.d.M., V.A.R., H.V.B., H.P.d.S. and M.F.C.-B.; supervision, M.F.C.-B.; project administration, M.F.C.-B.; funding acquisition, M.F.C.-B. All authors have read and agreed to the published version of the manuscript.

Funding: This research was funded by Fundação de Amparo à Pesquisa do Estado de São Paulo (FAPESP) grant number [2018/26269].

Informed Consent Statement: Informed consent was obtained from all subjects involved in the study.

Data Availability Statement: The data presented in this study are available in the article.

Acknowledgments: The authors are grateful to the Sao Paulo International Marathon Organization (Yescom and IDEEA Institute) for their assistance and support in the São Paulo International Marathon and Dietbox Software for providing a free license.

Conflicts of Interest: The authors declare no conflict of interest.

References

1. Hoffmann, C.; Weigert, C. Skeletal Muscle as an Endocrine Organ: The Role of Myokines in Exercise Adaptations. *Cold Spring Harb. Perspect. Med.* **2017**, *7*, a029793. [CrossRef] [PubMed]
2. Laurens, C.; Bergouignan, A.; Moro, C. Exercise-Released Myokines in the Control of Energy Metabolism. *Front. Physiol.* **2020**, *11*, 91. [CrossRef] [PubMed]
3. Piccirillo, R. Exercise-Induced Myokines with Therapeutic Potential for Muscle Wasting. *Front. Physiol.* **2019**, *10*, 287. [CrossRef]
4. Sabaratnam, R.; Wojtaszewski, J.F.P.; Højlund, K. Factors mediating exercise-induced organ crosstalk. *Acta Physiol.* **2022**, *234*, e13766. [CrossRef]
5. Chow, L.S.; Gerszten, R.E.; Taylor, J.M.; Pedersen, B.K.; van Praag, H.; Trappe, S.; Febbraio, M.A.; Galis, Z.S.; Gao, Y.; Haus, J.M.; et al. Exerkines in health, resilience and disease. *Nat. Rev. Endocrinol.* **2022**, *18*, 273–289. [CrossRef] [PubMed]

5.   de Sousa, C.A.Z.; Sierra, A.P.R.; Galán, B.S.M.; Maciel, J.F.D.S.; Manoel, R.; Barbeiro, H.V.; de Souza, H.P.; Cury-Boaventura, M.F. Time Course and Role of Exercise-Induced Cytokines in Muscle Damage and Repair After a Marathon Race. *Front. Physiol.* **2021**, *12*, 752144. [CrossRef]

7.   Bay, M.L.; Pedersen, B.K. Muscle-Organ Crosstalk: Focus on Immunometabolism. *Front. Physiol.* **2020**, *11*, 567881. [CrossRef]

8.   Lee, J.H.; Jun, H.-S. Role of Myokines in Regulating Skeletal Muscle Mass and Function. *Front. Physiol.* **2019**, *10*, 42. [CrossRef]

9.   Domin, R.; Dadej, D.; Pytka, M.; Zybek-Kocik, A.; Ruchała, M.; Guzik, P. Effect of Various Exercise Regimens on Selected Exercise-Induced Cytokines in Healthy People. *Int. J. Environ. Res. Public Health* **2021**, *18*, 1261. [CrossRef]

10. Senesi, P.; Luzi, L.; Terruzzi, I. Adipokines, Myokines, and Cardiokines: The Role of Nutritional Interventions. *Int. J. Mol. Sci.* **2020**, *21*, 8372. [CrossRef]

11. Hennigar, S.R.; McClung, J.P.; Pasiakos, S.M. Nutritional interventions and the IL-6 response to exercise. *FASEB J.* **2017**, *31*, 3719–3728. [CrossRef]

12. Passos, B.N.; Lima, M.C.; Sierra, A.P.R.; Oliveira, R.A.; Maciel, J.F.S.; Manoel, R.; Rogante, J.I.; Pesquero, J.B.; Cury-Boaventura, M.F. Association of Daily Dietary Intake and Inflammation Induced by Marathon Race. *Mediat. Inflamm.* **2019**, *2019*, 1537274–1537278. [CrossRef] [PubMed]

13. Casazza, G.A.; Tovar, A.P.; Richardson, C.E.; Cortez, A.N.; Davis, B.A. Energy Availability, Macronutrient Intake, and Nutritional Supplementation for Improving Exercise Performance in Endurance Athletes. *Curr. Sports Med. Rep.* **2018**, *17*, 215–223. [CrossRef] [PubMed]

14. Bronkowska, M.; Kosendiak, A.; Orzeł, D. Assessment of the frequency of intake of selected sources of dietary fibre among persons competing in marathons. *Rocz. Panstw. Zakl. Hig.* **2018**, *69*, 347–351. [CrossRef] [PubMed]

15. Głąbska, D.; Jusińska, M. Analysis of the choice of food products and the energy value of diets of female middle- and long-distance runners depending on the self-assessment of their nutritional habits. *Rocz. Panstw. Zakl. Hig.* **2018**, *69*, 155–163. [PubMed]

16. Franklin, B.A.; Thompson, P.D.; Al-Zaiti, S.S.; Albert, C.M.; Hivert, M.F.; Levine, B.D.; Lobelo, F.; Madan, K.; Sharrief, A.Z.; Eijsvogels, T.M.; et al. Exercise-Related Acute Cardiovascular Events and Potential Deleterious Adaptations Following Long-Term Exercise Training: Placing the Risks Into Perspective–An Update: A Scientific Statement From the American Heart Association. *Circulation* **2020**, *141*, e705–e736. [CrossRef]

17. Thompson, P.D.; Arena, R.; Riebe, D.; Pescatello, L.S. ACSM's New Preparticipation Health Screening Recommendations from ACSM's Guidelines for Exercise Testing and Prescription, Ninth Edition. *Curr. Sports Med. Rep.* **2013**, *12*, 215–217. [CrossRef]

18. Shirley, M.K.; Longman, D.P.; Elliott-Sale, K.J.; Hackney, A.C.; Sale, C.; Dolan, E. A Life History Perspective on Athletes with Low Energy Availability. *Sports Med.* **2022**, *52*, 1223–1234. [CrossRef]

19. Mountjoy, M.; Sundgot-Borgen, J.; Burke, L.; Ackerman, K.E.; Blauwet, C.; Constantini, N.; Lebrun, C.; Lundy, B.; Melin, A.; Meyer, N.; et al. International Olympic Committee (IOC) Consensus Statement on Relative Energy Deficiency in Sport (RED-S): 2018 Update. *Int. J. Sport Nutr. Exerc. Metab.* **2018**, *28*, 316–331. [CrossRef]

20. Mountjoy, M.L.; Burke, L.M.; Stellingwerff, T.; Sundgot-Borgen, J. Relative Energy Deficiency in Sport: The Tip of an Iceberg. *Int. J. Sport Nutr. Exerc. Metab.* **2018**, *28*, 313–315. [CrossRef]

21. Hackney, A.C.; Elliott-Sale, K.J. Exercise Endocrinology: "What Comes Next?". *Endocrines* **2021**, *2*, 167–170. [CrossRef] [PubMed]

22. Elliott-Sale, K.J.; Tenforde, A.S.; Parziale, A.L.; Holtzman, B.; Ackerman, K.E. Endocrine Effects of Relative Energy Deficiency in Sport. *Int. J. Sport Nutr. Exerc. Metab.* **2018**, *28*, 335–349. [CrossRef] [PubMed]

23. Karava, V.; Christoforidis, A.; Kondou, A.; Dotis, J.; Printza, N. Update on the Crosstalk Between Adipose Tissue and Mineral Balance in General Population and Chronic Kidney Disease. *Front. Pediatr.* **2021**, *9*, 696942. [CrossRef] [PubMed]

24. Shan, T.; Liang, X.; Bi, P.; Kuang, S. Myostatin knockout drives browning of white adipose tissue through activating the AMPK-PGC1α-Fndc5 pathway in muscle. *FASEB J.* **2013**, *27*, 1981–1989. [CrossRef] [PubMed]

25. Cheng, C.-F.; Ku, H.-C.; Lin, H. PGC-1α as a Pivotal Factor in Lipid and Metabolic Regulation. *Int. J. Mol. Sci.* **2018**, *19*, 3447. [CrossRef] [PubMed]

26. Czaja, W.; Nakamura, Y.K.; Li, N.; Eldridge, J.A.; DeAvila, D.M.; Thompson, T.B.; Rodgers, B.D. Myostatin regulates pituitary development and hepatic IGF1. *Am. J. Physiol. Endocrinol.* **2019**, *316*, E1036–E1049. [CrossRef]

27. Perry, C.A.; Van Guilder, G.P.; Butterick, T.A. Decreased myostatin in response to a controlled DASH diet is associated with improved body composition and cardiometabolic biomarkers in older adults: Results from a controlled-feeding diet intervention study. *BMC Nutr.* **2022**, *8*, 1–11. [CrossRef]

28. Jeremic, N.; Chaturvedi, P.; Tyagi, S.C. Browning of White Fat: Novel Insight Into Factors, Mechanisms, and Therapeutics. *J. Cell Physiol.* **2017**, *232*, 61–68. [CrossRef]

29. Subbotina, E.; Sierra, A.; Zhu, Z.; Gao, Z.; Koganti, S.R.K.; Reyes, S.; Stepniak, E.; Walsh, S.A.; Acevedo, M.R.; Perez-Terzic, C.M.; et al. Musclin is an activity-stimulated myokine that enhances physical endurance. *Proc. Natl. Acad. Sci. USA* **2015**, *112*, 16042–16047. [CrossRef]

30. Liu, Y.; Huo, X.; Pang, X.F.; Zong, Z.H.; Meng, X.; Liu, G.L. Musclin Inhibits Insulin Activation of Akt/Protein Kinase B in Rat Skeletal Muscle. *J. Int. Med. Res.* **2008**, *36*, 496–504. [CrossRef]

31. Rentería, I.; García-Suárez, P.C.; Fry, A.C.; Moncada-Jiménez, J.; Machado-Parra, J.P.; Antunes, B.M.; Jiménez-Maldonado, A. The Molecular Effects of BDNF Synthesis on Skeletal Muscle: A Mini-Review. *Front. Physiol.* **2022**, *13*, 934714. [CrossRef] [PubMed]

32. Mughal, A.; O'Rourke, S.T. Vascular effects of apelin: Mechanisms and therapeutic potential. *Pharmacol. Ther.* **2018**, *190*, 139–147. [CrossRef] [PubMed]

33. Tanimura, Y.; Aoi, W.; Takanami, Y.; Kawai, Y.; Mizushima, K.; Naito, Y.; Yoshikawa, T. Acute exercise increases fibroblast growth factor 21 in metabolic organs and circulation. *Physiol. Rep.* **2016**, *4*, e12828. [CrossRef]
34. Tezze, C.; Romanello, V.; Sandri, M. FGF21 as Modulator of Metabolism in Health and Disease. *Front. Physiol.* **2019**, *10*, 419. [CrossRef] [PubMed]
35. Fujimoto, T.; Sugimoto, K.; Takahashi, T.; Yasunobe, Y.; Xie, K.; Tanaka, M.; Ohnishi, Y.; Yoshida, S.; Kurinami, H.; Akasaka, H.; et al. Overexpression of Interleukin-15 exhibits improved glucose tolerance and promotes GLUT4 translocation via AMP-Activated protein kinase pathway in skeletal muscle. *Biochem. Biophys. Res. Commun.* **2019**, *509*, 994–1000. [CrossRef]
36. Akbari, H.; Hosseini-Bensenjan, M.; Salahi, S.; Moazzen, F.; Aria, H.; Manafi, A.; Hosseini, S.; Niknam, M.; Asadikaram, G. Apelin and its ratio to lipid factors are associated with cardiovascular diseases: A systematic review and meta-analysis. *PLoS ONE* **2022**, *17*, e0271899. [CrossRef]
37. Waseem, R.; Shamsi, A.; Mohammad, T.; Hassan, M.I.; Kazim, S.N.; Chaudhary, A.A.; Rudayni, H.A.; Al-Zharani, M.; Ahmad, F.; Islam, A. FNDC5/Irisin: Physiology and Pathophysiology. *Molecules* **2022**, *27*, 1118. [CrossRef]
38. Folino, A.; Accomasso, L.; Giachino, C.; Montarolo, P.G.; Losano, G.; Pagliaro, P.; Rastaldo, R. Apelin-induced cardioprotection against ischaemia/reperfusion injury: Roles of epidermal growth factor and Src. *Acta Physiol.* **2018**, *222*, e12924. [CrossRef]
39. Esmaeili, S.; Bandarian, F.; Esmaeili, B.; Nasli-Esfahani, E. Apelin and stem cells: The role played in the cardiovascular system and energy metabolism. *Cell Biol. Int.* **2019**, *43*, 1332–1345. [CrossRef]
40. Valenzuela, P.L.; Morales, J.S.; Emanuele, E.; Pareja-Galeano, H.; Lucia, A. Supplements with purported effects on muscle mass and strength. *Eur. J. Nutr.* **2019**, *58*, 2983–3008. [CrossRef] [PubMed]

 *nutrients* 

Review

⌐ndothelial Glycocalyx Preservation—Impact of Nutrition and Lifestyle

Paula Franceković and Lasse Gliemann *

Department of Nutrition, Exercise and Sports, University of Copenhagen, Universitetsparken 13, DK-2100 Copenhagen, Denmark
* Correspondence: gliemann@nexs.ku.dk; Tel.: +45-35-321-632

Abstract: The endothelial glycocalyx (eGC) is a dynamic hair-like layer expressed on the apical surface of endothelial cells throughout the vascular system. This layer serves as an endothelial cell gatekeeper by controlling the permeability and adhesion properties of endothelial cells, as well as by controlling vascular resistance through the mediation of vasodilation. Pathogenic destruction of the eGC could be linked to impaired vascular function, as well as several acute and chronic cardiovascular conditions. Defining the precise functions and mechanisms of the eGC is perhaps the limiting factor of the missing link in finding novel treatments for lifestyle-related diseases such as atherosclerosis, type 2 diabetes, hypertension, and metabolic syndrome. However, the relationship between diet, lifestyle, and the preservation of the eGC is an unexplored territory. This article provides an overview of the eGC's importance for health and disease and describes perspectives of nutritional therapy for the prevention of the eGC's pathogenic destruction. It is concluded that vitamin D and omega-3 fatty acid supplementation, as well as healthy dietary patterns such as the Mediterranean diet and the time management of eating, might show promise for preserving eGC health and, thus, the health of the cardiovascular system.

Keywords: endothelial glycocalyx; lifestyle diseases; cardiovascular health; obesity; diabetes; hypertension; nutrition therapy; Mediterranean diet; vitamin D; dietary sulfur; dietary nitrates; mechanotransduction; endotoxemia; intermittent fasting

Citation: Franceković, P.; Gliemann, L. Endothelial Glycocalyx Preservation—Impact of Nutrition and Lifestyle. *Nutrients* **2023**, *15*, 2573. https://doi.org/10.3390/nu15112573

Academic Editor: Jose Lara

Received: 17 March 2023
Revised: 8 May 2023
Accepted: 11 May 2023
Published: 31 May 2023

1. Introduction

The endothelial glycocalyx (eGC), sometimes referred to as the endothelial surface layer, is a protective layer of glycoconjugates (e.g., carbohydrates covalently linked to other molecules such as amino acids, proteins, lipids, etc.) that covers the luminal side of the endothelial cells (see Figure 1). In homeostatic conditions, this layer serves as an endothelial cell gatekeeper by controlling the permeability of substances from the blood to the interstitium and initiating the adhesion of blood-carried molecules to the cell surface [1]. In addition, the eGC is crucial for proper endothelial nitric oxide production and vasodilatation. Thus, the eGC serves an important regulatory function for the cardiovascular system [2]. Pathogenic destruction of the eGC is detected in a large number of cardiovascular conditions, including atherosclerosis [3,4], hypertension [5], diabetes mellitus [6], chronic kidney disease [7], ischemia reperfusion syndrome [8], and septic shock [9]. Most of the current research regarding the eGC is aimed towards acute care surgery and perioperative care, because derangement of the eGC is proposed to increase the severity of sepsis. Presently, little is known about the regeneration or prevention of its pathogenic destruction in randomized controlled clinical trials [10,11]. It is suggested that the main reason behind this is a lack of reliable detection techniques is due to the ex vivo instability of the eGC [12]. In addition, this important structure of the cardiovascular system has been overseen by researchers, funding bodies, and pharmaceutical companies [13]. Nevertheless, newly developed methods, such as side-stream dark-field imaging, orthogonal polarization spectral imaging, and improved fixation techniques, have shed new light on the eGC and its role in health and disease [14,15]. The in-depth eGC knowledge is a missing link in primary, secondary, and tertiary cardiovascular disease prevention and treatment. The eGC is, therefore, a novel target for various healthcare professionals such as nutritionists, dietitians, clinicians, surgeons, oncologists, researchers, and many others. The relationship between diet and eGC destruction and/or restoration is still an unexplored territory, but it might be particularly important in cardiovascular disease prevention and treatment. The aims of the present review are the following: (1) to provide an overview of current eGC knowledge—including factors that influence eGC structure and function in microcirculation—in healthy conditions and in chronic vasculature-related inflammatory pathologies; and (2) to describe perspectives of nutritional therapy, as well as diet- and lifestyle-related behaviors for eGC destruction prevention and regeneration.

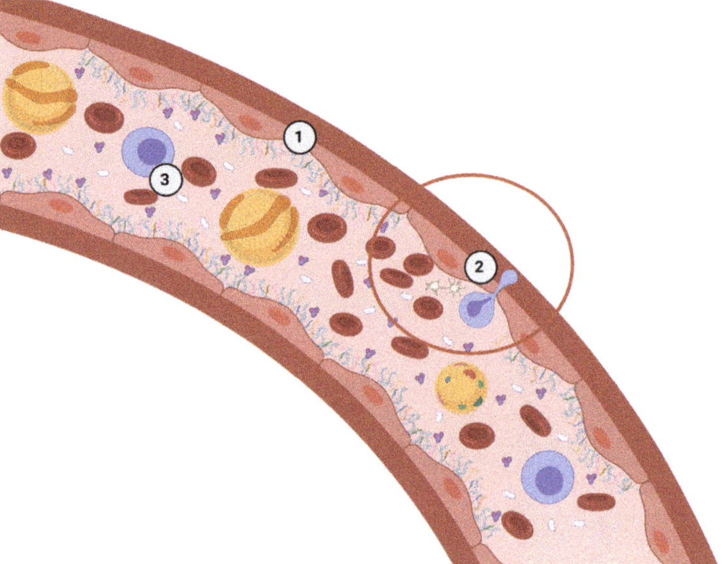

Figure 1. A model of the position of endothelial glycocalyx, which covers the apical surface of the endothelial cells (1). The circled area presents a part of the vessel without the glycocalyx (2). In this area, leukocytes, platelets, and other blood components can reach the endothelial cell receptors more easily and start blood clot formation, leukocyte migration, and other processes. This quicker adhesion is important in wound healing but might contribute to chronic disease progression as well. (3) Various blood components passing through a blood vessel. Created with BioRender.com.

2. The eGC in Healthy Conditions—Extended State of eGC

The eGC is a dynamic hair-like layer expressed on the apical surface of the endothelial cells, facing the lumen, throughout the vascular system. It is described as a "hair-like layer" because the structures of eGC resemble microscopic hair when observed under the microscope [16] (Figure 2). Additionally, the eGC is described as a "dynamic layer" because it undergoes continuous metabolic turnover that is dependent on the local environment; as such, in healthy, homeostatic conditions, the eGC is in a state of dynamic equilibrium with plasma proteins and exists in an 'extended state'. Most commonly, the components consist of a proteoglycan backbone with many glycosaminoglycan attachment sites. Syndecans, glypican-1, biglycans, and perlecans are the typical proteoglycans of the glycocalyx. These proteoglycans mainly bind to heparan sulfate, which is the most abundant glycosaminoglycan of the eGC (50–90% of the total glycosaminoglycan pool). Other glycosaminoglycans that contribute to eGC integrity and permeability control are chondroitin, dermatan, and keratan sulfates [17,18]. The final integral part of the eGC is hyaluronan. Hyaluronan—which is a longer disaccharide polymer compared to chondroitin and heparan sulfate—is synthesized directly on the cell surface and anchored with a CD_{44} cell receptor [19]. Preventing hyaluronan loss has been suggested as particularly important in preventing the vascular complications of diabetes mellitus [6,20]. Dogné et al. [20] suggest that one therapeutic approach could be the inhibition of hyaluronidase 1, an enzyme responsible for cleaving hyaluronan from CD_{44} cell receptors. However, given that proteoglycans and glycosaminoglycans are not specific to the endothelial cells, but are ubiquitous in the human organism, the inhibition of cleavage enzymes might negatively affect functions in other cells. Singh et al. [21] found that immunoglobulin G (IgG) *N*-glycosylation patterns in type 2 diabetes were associated with a faster decline of kidney function, thus reflecting a pro-inflammatory state of the IgG. In that regard, studying the glycosylation patterns of eGC components might provide a more precise way of combating vascular symptoms.

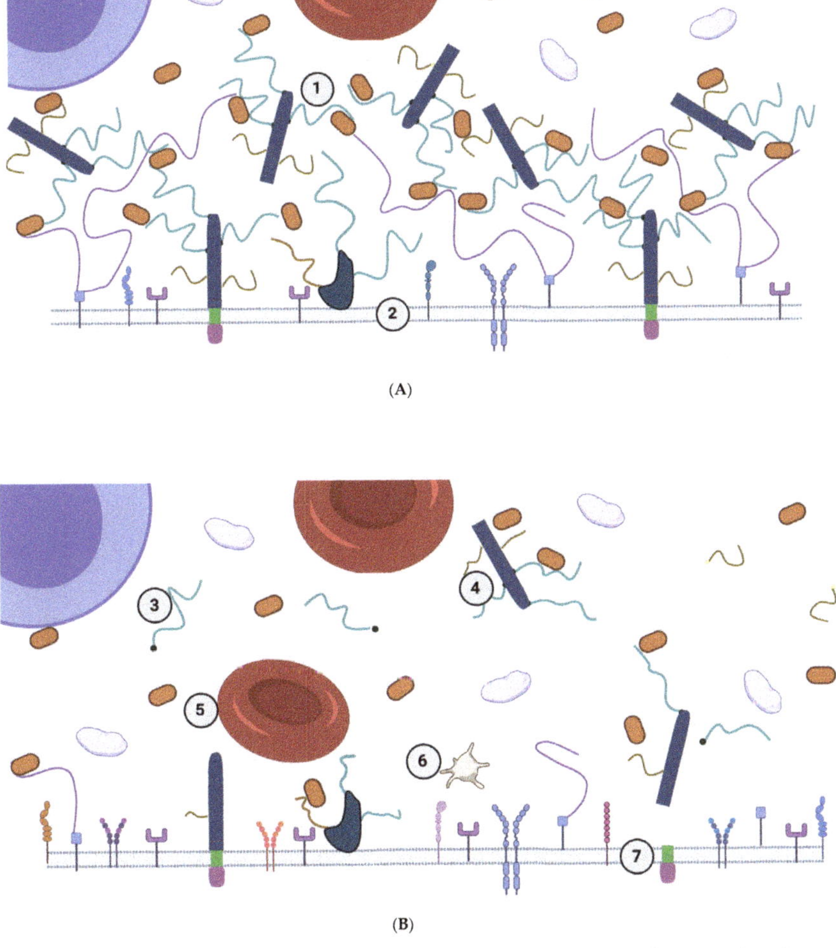

Figure 2. (**A**) The extended state of the endothelial glycocalyx. Due to their negative charge, the glycocalyx components are in dynamic equilibrium with positively charged plasma proteins, which are depicted as orange ellipses (1). Some of the components are not membrane-bound but contribute

to the thickness of the whole structure. The thickness and the net negative charge prevent the blood components from adhering to the cell receptors. The cell membrane (2) does not contain maɪ adherence receptors. (**B**) The collapsed state of the endothelial glycocalyx. Different disruptors can cause eGC to diminish or collapse, either directly or by triggering proteolytic enzymes that cleave different structures, which, therefore, destroy the equilibrium with plasma proteins. Those kinds of proteolytic enzymes are usually upregulated in inflammation. The eGC structures detach from the cell surface and become bloodborne. They may be measured in the blood. The high plasma concentration of, for example, heparan sulfate (3) or syndecan-1 (4) indicates eGC injury. In inflammation, blood components, such as red blood cells (5) or platelets (6), have easier access to the cell surface, and the expression of the adherence receptors on the cell membrane is upregulated (7). eGC–endothelial glycocalyx.

To date, the main functions of the eGC layer in homeostatic conditions are protection of the endothelial cell membrane, mechanotransduction, the regulation of vasodilatation, and the prevention of blood clot and plaque formation (Figure 3) [22]. Another important function of this layer is the promotion of blood flow homogeneity, which may be a useful strategy for improving tissue perfusion in many [23]. Most eGC components, especially the glycosaminoglycans, have anionic-ending molecules such as uronic or salicylic acid, thereby giving the eGC a net negative charge in homeostatic conditions. The net negative charge allows the eGC to interact with blood proteins through ionic interactions or create repulsion against platelets and leukocytes (Figures 1 and 2A) [22,24]. A healthy eGC may act as a sodium buffer by controlling the influx of sodium into the endothelial cells [25]. This was proposed because positively charged sodium is most likely to be bound to negatively charged organic material, which would provide for its osmotic inactivity. The 'perfect' sodium storage place would be the eGC, given its ubiquitous position and net negative charge in homeostatic conditions [25]. Related to this, eGC components, such as chondroitin sulfate, may have a therapeutic application in the targeted delivery of medical substances. This suggests an important role of eGC in the docking of oxidants, which may be the base for developing novel enzymatic antioxidant treatments [26]

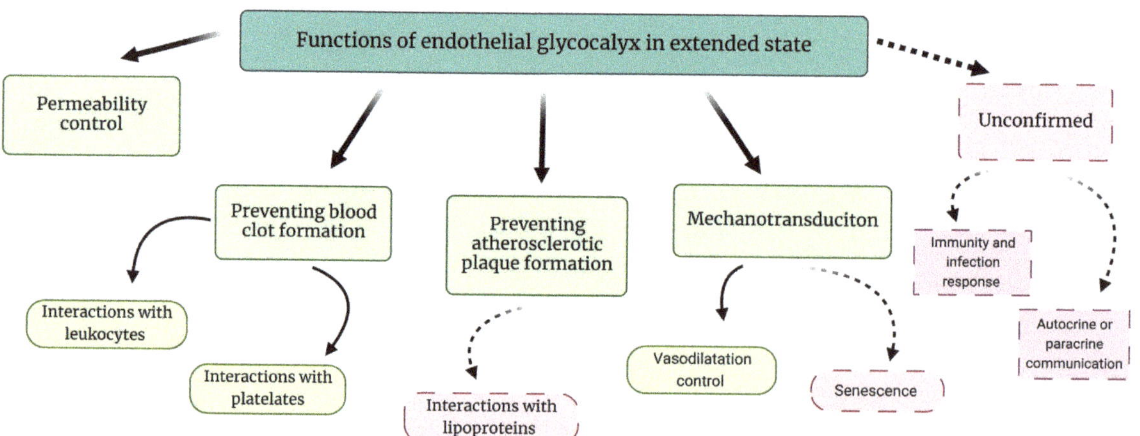

Figure 3. Functions of the extended state of the endothelial glycocalyx. The functions in green boxes have been studied in more depth. The functions in red and dashed boxes are currently not well understood. Solid lines are established connections, dashed lines are potential connections. In an eGC diminished/collapsed state, the functions of the endothelial glycocalyx are practically reversed. The collapsed state allows blood clot formation and plaque formation, while mechanotransduction function is reduced. Created with BioRender.com.

In a healthy human body, the eGC is present in an extended state [27]. This means that the glycocalyx components exist in equilibrium with blood proteins, thus forming a thick and dense structure (see Figure 2A) [28]. The thickness and density of the eGC provide significant protection for the endothelial cells by preventing the platelets and leukocytes from adhering to the cell receptors on the cell surface. This is possible because the eGC is much wider, and the proteoglycans extend much further into the lumen of the blood vessel than the active sites of adherence receptors [29]. In addition, in homeostatic conditions, the rate of synthesis of adherence receptors is lower than in acute or chronic diseases, which further reduces the chance of blood component adherence [30] (Figure 2B).

The homeostatic eGC is often described as a vascular gatekeeper. Indeed, the firm and thick, but flexible and adjustable, structure can be imagined as a protective wall surrounding the castle, providing great protection against intruders, with the castle being the endothelial cell.

3. Extended eGC Activates Mechanotransduction and Endothelial Nitric Oxide Release

The so-called wind-in-the-trees conceptual model is sometimes used to describe one important function of the eGC: the mechanotransduction of blood-flow-induced fluid shear stress (Figure 4) [22]. Mechanotransduction is a process by which the eGC senses forces caused by fluid passage through the blood vessels and translates that information inside the endothelial cell to initiate intracellular signaling cascades. Upon receiving the signal, the endothelial cell acts in accordance with the given information by implementing, for example, structural maintenance, vasodilatation, senescence, or an inflammatory response [31]. This conceptual model describes the eGC components as the trees in a forest and the blood flow as the wind. The branches are the glycosaminoglycans, the tree trunks are the proteoglycan backbones, the cell receptors are the grass, and the endothelial cell membrane is the ground [22].

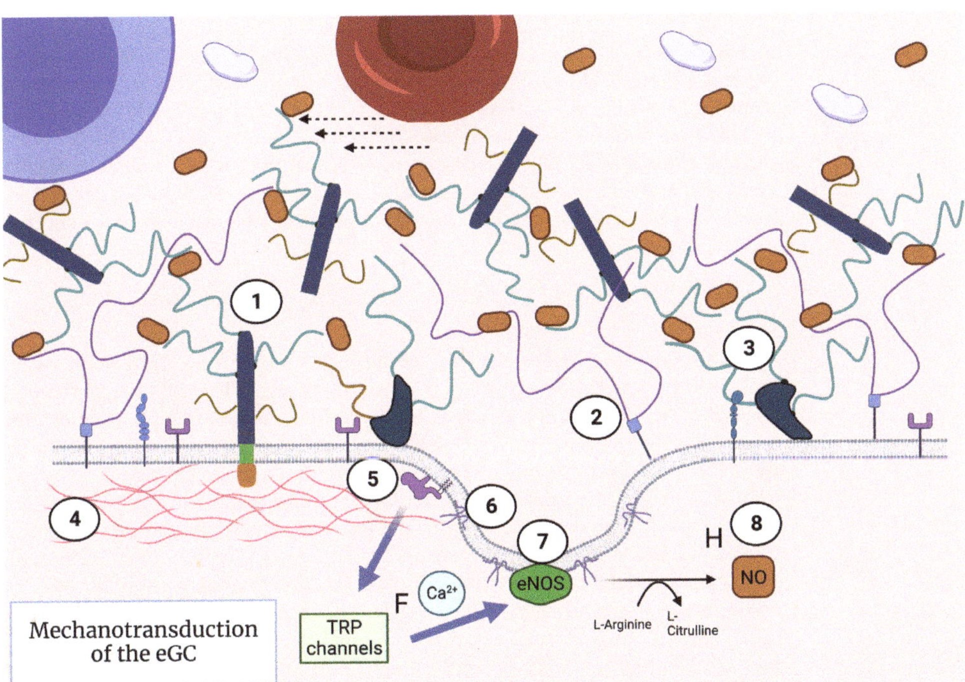

Figure 4. The mechanism of mechanotransduction of a healthy, extended eGC. The syndecans (1), hyaluronans (2), andother glycoconjugates (3) sense the changes in fluid shear force. This mechanical signal is transduced inside the endothelial cell and causes the movement of cytoskeletal structures (4).

Those cytoskeletal structures activate the G-protein (5), which activates the transient receptor potential (TRP) channels and causes Ca^{2+} influx (6). Ca^{2+} ions are needed to activate caveolin proteins on the cell membrane (7), which attach to the endothelial nitric oxide synthase (eNOS) system and activate it. Activated eNOS generates nitric oxide (NO) (8). NO can then diffuse into the lumen of the blood vessel or the smooth muscles and cause vasodilatation. Created with BioRender.com.

When the wind (fluid shear stress) blows through the forest, the treetops (glycosaminoglycans) sense it and move back and forth accordingly. The sensed force is then transduced onto the three trunks (proteoglycan core) and into the roots. The roots, in this case, would be the cytoskeletal structures and intracellular eGC domains. When the eGC is preserved (in the extended state), the winds' impact (blood flow shear stress) is scattered between the branches of the treetops, whereas the ground senses little to none of the wind's impact. However, the important information about the state of the blood flow is transduced via the trees into the roots. If the trees were not present or were sparse, the wind would directly interact with the ground and impact the ground, while the information could not be properly transmitted into the deeper structures.

Another theoretical review of the glycocalyx describes the mechanotransduction properties of the eGC by using a bumper car conceptual model. This model explains the processes inside the endothelial cell upon the arrival of signals from the blood. The bumper car analogy came to be after an experiment which showed that dense peripheral actin bands (DAPABs) were disrupted by uniform shear stress. In addition, the movement of vinculin, a cytoskeletal protein, closer to the cell membrane was detected. Both observations were enhanced when higher shear stress was implemented [32]. These observations suggest that cytoskeletal reorganization indeed occurs as a consequence of mechanical signal transduction of the eGC. The same effect of cytoskeletal reorganization inside the endothelial cell was not observed when the eGC was destroyed by proteolytic enzymes. Mechanotransduction is an important mechanism used to trigger different pathways in the endothelial cell. For example, eGC mechanotransduction initiates the process of nitric oxide (NO) production through the activation of endothelial nitric oxide synthase (eNOS) through calcium/calmodulin activation and consequential Ca^{2+} influx [31] (Figure 4). Endothelial-derived NO exerts powerful localized vasodilatory effects on the vascular smooth muscle cells [33]. Additionally, NO contributes to the inhibition of platelet aggregation and leukocyte adhesion. As NO is produced when eGC is preserved, the same functions can be contributed to a healthy eGC, and a proper NO production could be an indicator of eGC health.

4. The eGC in a State of Low-Grade Inflammation—Diminished eGC

4.1. eGC Response to Inflammation

The state of inflammation, led by the activation of the TNF-α pathway, triggers the collapse of the eGC (described in the chapter below). The collapse of the eGC is much needed in acute conditions such as injury or infection, as this promotes endothelial cell interaction with leukocytes/platelets and blood clot formation. However, certain eGC disruptors (described in detail later) can induce low-grade chronic inflammation, which may lead to the development of vascular consequences of chronic diseases if disruptors continuously trigger the shedding or endocytosis of eGC. Shedding and endocytosis are two main responses of eGC components to various disruptions from the environment. Both responses cause the collapse of the eGC and disrupt the dynamic equilibrium. For example, acting on an environmental cue, a family of enzymes collectively known as sheddases or secretases can be released and proteolytically cleave the proteoglycans' ectodomain, leaving it virtually intact (Figure 2B, element 4). Little is known about the exact structure of the eGC sheddases. Novel findings by Yang et al. [34] revealed that a sheddase called disintegrin and metalloproteinase 15 (ADAM15) cleaved the CD_{44} surface anchor for hyaluronan. These kinds of shedded ectodomains travel via the bloodstream, where they may function as autocrine or paracrine effectors. The shedding mechanism both generates soluble ectodomains and rapidly reduces the amount of cell surface heparan sulfate [35]. Heparan sulfate reduction is much needed in conditions following acute injury, where platelet aggregation is needed for wound healing [36]. On the other hand, different ligands from the blood can bind to surface glycosaminoglycans and be internalized into the endothelial cell and serve as the cell's nutrient source. Viruses can sometimes hijack this mechanism of ligand transport. For example, early-stage research has shown that the spike protein of SARS-CoV-2 interacts not only with angiotensin-converting enzyme 2 (ACE2), but with heparan sulfate as well, via the process of endocytosis or internalization [37]. These findings suggest that eGC knowledge might be important in formulating virus treatments, especially for the vascular consequences of the disease such as hypercoagulability and acute coronary symptoms.

4.2. Mechanisms of eGC Collapse

TNF-α and TNF signaling are hallmarks of inflammation and have been related to the cardiovascular pathophysiology of atherosclerosis, sepsis, diabetes, and obesity, among others [38]. TNF-α is a master regulator of proinflammatory mediators, and the activation of the TNF-α metabolic pathway has a major impact on the eGC structure [38]. Studies have demonstrated that acute exposure to TNF-α or thrombin, another inflammatory mediator, causes rapid shedding of the glycocalyx structure [16,39]. This process results in increased vascular permeability, tissue oedema, coronary leakage, and mast cell degranulation [40]. TNF-α activation has a negative impact on the eGC integrity through several mechanisms, e.g., by the induction of reactive oxygen and nitrogen species (ROS/RNS) [12] or by activating the NF-κB metabolic pathway [41]. Certain ROS/RNS can then cleave the ectodomains of eGC constituents via the activation of matrix metalloproteinases (MMPs) and inactivation of endogenous protease inhibitors [15]. Enhanced NF-κB signaling leads to the syndecan-4 domain synthesis [42]. In these inflammatory conditions, the synthesized syndecan-4 domains are rapidly shed from the endothelial cell surface. This shedding causes a disruption in the eGC integrity, but it is also proposed that, when present as a soluble molecule in the blood, syndecan-4 may facilitate tissue fibrosis [42]. SIRT-1-deficient endothelial cells have been shown to exhibit increased NF-κB pathways, thus shedding syndecan-4 ectodomains [43,44].

5. eGC in Chronic Diseases

Given that eGC destruction, due to environmental factors is inherent to several chronic pathologies, it is difficult to ascertain the contribution of eGC integrity to vascular disease progression/severity. As seen with metabolic syndrome, these kinds of diseases can be unified when constructing prevention tactics to preserve the eGC. The pathophysiology of important chronic conditions and the importance of eGC is highlighted in the following section.

5.1. Hypertension

As mentioned previously, one of the proposed eGC functions is to buffer plasma sodium and control its influx into the endothelial cells [25]. These findings are particularly important when studying hypertension. In one cross-sectional study, newly diagnosed hypertensive patients ($n = 320$) had decreased eGC thickness compared to healthy controls ($n = 160$). Reduced eGC thickness was related to signs of impaired vascular function, including increased central systolic blood pressure, as well as increased pulse wave velocity. These findings suggest that eGC thickness is reduced in untreated hypertension [5]. However, the question of the primary disruptor of eGC destabilization in hypertension is not yet definitively answered. Is it merely a consequence of mechanical destruction by increased blood pressure, or is it due to high salt intake, which may cause eGC collapse that has arterial stiffness and increased blood pressure as a consequence? The current evidence seems to suggest the latter. One study showed that a mere 2% increase in plasma sodium beyond 140 mM may stiffen the endothelial cells by approximately 20% [45]. In a related study, five days of sodium overload led to a ~50% eGC destabilization and 68% reduction of heparan sulfate [46]. It seems that, in the state of daily sodium excess, the eGC diminishes (Figure 2) and loses the sodium buffer capacity.

The first study to investigate the effect of salt (sodium chloride) on microvascular densities using in vivo sublingual imaging was performed by Rorije et al. [47]. The researchers aimed to determine if high salt intake would reduce sublingual microvascular density and, therefore, reduce eGC thickness in normotensive individuals ($n = 18$, all-male, mean age 29 ± 5 years). No blood pressure or sublingual microvascular differences were found when comparing high and low salt intake groups. However, an increase in salt consumption significantly correlated with a lower recruitment rate of sublingual capillaries after the administration of nitroglycerin, thus indicating lower structural microvascular density. These findings also suggest that a high salt load by itself might be one of the first disruptors of the eGC, which may lead to the cardiovascular damage observed in hypertensive patients. This study opens an array of questions and discussions worth exploring further in the attempt to prevent cardiovascular disease occurrence.

5.2. Atherosclerosis

Although research in humans is scarce, the role of the eGC in atherosclerosis in animal models is well recognized. A healthy, extended eGC prevents the occurrence of atherosclerotic plaque by ionically repelling macromolecules and mechanically preventing their adherence to the endothelial cell, thus lowering the chance for their migration and oxidation [4]. In animals, vascular sites with compromised eGC and lower mechanotransduction function seem to be more vulnerable to inflammatory consequences and the formation of atherosclerotic plaque [48,49]. The results from animal studies have led to an attractive hypothesis, which suggests that clinically assessing the thickness of the eGC might allow for the early detection of atherosclerosis. However, studies conducted in humans display contradicting results. For example, a clear connection between eGC thickness and vascular risk or progression/severity of vascular diseases has not been demonstrated. In one multiethnic cross-sectional study ($n = 6169$, 42.4% male, mean age 43.6 ± 13), the eGC thickness was assessed by using side-stream dark-field imagining, which is considered to be the most suited method according to some sources [50]. Reduced eGC thickness was associated with the female sex and diabetes after correcting for possible confounders such as age, diastolic blood pressure, and body mass index. Reduced eGC thickness was not associated with prevalent cardiovascular disease [51]. This study questions the viability of the used measurement technique as a proper measure of eGC thickness and the involvement of the eGC in atherosclerosis pathophysiology.

However, the cardiovascular biomarkers used in the study by Valerio et al. [51] (LDL cholesterol, HDL cholesterol, and triglycerides) might not show a full picture of the cardiovascular disease risk and, therefore, might not be reliable in comparing the consequences of reduced eGC thickness [52]. In a certain part of the population, namely, people with insulin resistance, the measures of LDL cholesterol can be in discordance with apolipoprotein B (apo B) and LDL particle concentration [53]. The number of apoB-containing lipoprotein particles is sometimes more predictive of high cardiovascular disease risk than the cholesterol content (LDL-C). In addition, the smaller size of the lipoproteins could be more indicative of future cardiovascular disease than the number of LDL-C packed in these molecules, due to their greater reactivity [54]. When assessing the connection between the eGC thickness and cardiovascular disease risk, other cardiovascular risk assessment parameters might be a good addition to the lipid panel.

5.3. Abdominal Obesity

The discordance in atherosclerosis risk assessment described in the chapter above is particularly important in people with diagnosed metabolic syndrome and type 2 diabetes [53]. Those conditions are usually accompanied by increased visceral and ectopic fat, which are metabolically different from subcutaneous fat and contribute to proinflammatory, proatherogenic, and procoagulant states [55]. Visceral fat in the abdomen and ectopic fat surrounding the liver, heart, kidneys, or pancreas accumulate as a consequence of a chronically positive energy balance. When a person reaches their so-called "fat threshold", the subcutaneous fat loses its ability to expand through adipocyte hyperplasia. In that state, the body responds by increasing visceral fat deposits, thus increasing the production of proinflammatory cytokines such as TNF-α and interleukin 6 [55]. In addition, ectopic fat deposits can increase the hepatic production of glucose, which is a process linked to glucose intolerance. The increased TNF-α activity leads to an eGC degradation by activating ROS/RNS and NF-κB metabolic pathway [55]. Furthermore, increased glucose production can lead to AGE formation and further eGC destruction (see 'The effects of elevated blood glucose on eGC').

5.4. Hyperglycemia, Type 2 Diabetes, and Metabolic Syndrome

Chronically increased blood glucose (i.e., hyperglycemia), a hallmark of diabetes, metabolic syndrome, and obesity, causes the increased production of advanced glycation end-products (AGE) [56]. AGEs are glycated proteins, lipids, and nucleic acids commonly present in homeostatic conditions but can be rapidly generated in pathological conditions, such as insulin resistance [56]. Generated AGEs upregulate the iNOS system in chronic uncontrolled hyperglycemia. When upregulated, the iNOS system causes eNOS dysfunction and contributes to the inflammatory conditions of the body [57]. Both eNOS and iNOS systems used L-arginine and molecular oxygen to produce L-citrulline and NO. However, when NO is produced in inflammatory conditions, it quickly reacts with superoxide ($O_2{}^-$) and generates peroxynitrite ($ONOO^-$), which causes nitrosative stress and damages proteins, lipids, and DNA [58]. When activated, iNOS further enhances the generation of oxidative stress, which is a powerful process in eliminating microbial infections and tumor cells but may also contribute to chronic disease development [59,60]. Reactions between $O_2{}^-$ and NO lower NO bioavailability, which can cause arterial stiffness and promote atherosclerotic events [61]. AGEs form a cross-link between the basement membrane molecules of the extracellular matrix [62], thus implying their close relationship with the eGC; however, the interaction between AGEs and eGC, as well as the implications for chronic disease development, are largely unknown.

AGEs activate various intercellular signals through receptor- and non-receptor-mediated mechanisms. For example, receptors for advanced glycation end-products (RAGEs) may, in part, explain the relationship between the eGC and AGEs. Accumulated AGEs activate RAGE receptors that have been shown to be present on the endothelial cells and can be shed from the surface in a manner similar to eGC components, thus forming soluble RAGEs (sRAGEs) [63]. sRAGEs upregulate the NF-κB pathway and facilitate the inflammatory cascade through the release of ROS/RNS. ROS/RNS contribute to the shedding of eGC components and enhance endothelial dysfunction. It has been shown that AGE-bound RAGEs increase endothelial permeability to macromolecules and block NO signaling activity [64]. This is perhaps explained by the destruction of the eGC (i.e., the removal of the endothelial gatekeeper), thus promoting the accessibility of blood component migration to the endothelial cell.

A growing body of evidence finds that AGEs can derive from food sources and tobacco, and they suggest dietary AGE restriction [65,66]. In addition to AGE pro-oxidative effects, hyperglycaemia might also be responsible for altering the sulfation patterns of glycosaminoglycan chains and can prevent hyaluronan binding to the glycocalyx [40]. We have shown that 14 days of sucrose supplementation (3×75 g of sucrose per day) impairs vascular function in young healthy male subjects that was indicated by blood flow during passive leg movement, which is a method suited for determining the impacts on mechanotransduction in vivo [67]. The high sucrose ingestion affected the vasodilatory properties of the vessels, reduced eNOS activation, and upregulated PECAM-1. Given that a healthy eGC is important for proper mechanotransduction and that PECAM-1 is upregulated in oxidative stress that might have been caused by increased sucrose intake, these findings suggest a disruption of the eGC. Future research is warranted to investigate the influence of dietary sugar on the eGC, particularly regarding eGC shedding and integrity.

The effect of accumulated abdominal fat is further augmented by chronically elevated glucose levels, which can cause insulin resistance, which is a hallmark of metabolic syndrome, diabetes, and prediabetes [68]. In addition, high postprandial glucose levels generate AGEs, which further activate proinflammatory signaling pathways [66] (see 'The effects of elevated blood glucose on eGC'). Pertynska-Marczewska and Merhi [69] researched the role of the AGE–RAGE axis in the prevention of atherosclerosis in women in menopause. They found that circulating sRAGE levels could be correlated with increased abdominal fat, insulin resistance, diabetes, and metabolic syndrome. They suggested that a therapeutic inhibition of the RAGE signaling pathway might be beneficial for decreasing cardiovascular disease risk in women in menopause [69]. Indeed, as discussed previously, Valerio et al. [51] reported a correlation between diabetes and low eGC thickness, thus implying "a small glycocalyx size in people with diabetes" (i.e., the diminished/collapsed state of the eGC). This is in accordance with other research that showed that both acute and chronic hyperglycaemia significantly reduced eGC size, particularly in patients with microalbuminuria [70]. Both acute and chronic effects of hyperglycaemia on vasculature have been recorded and are evident from the fact that the comorbidities of type 2 diabetes are closely related to the degradation of the vascular system and eGC health [70].

5.5. Chronic Kidney Disease

One of the adverse consequences of diabetes is diabetic nephropathy, which is, next to hypertension, a major cause of end-stage renal disease [71]. The role of the eGC in the pathophysiology of diabetic nephropathy is well recognized, and efforts have been currently made to produce specific therapies that target the regeneration of the glycocalyx of the fenestrated glomerular endothelial cells and the prevention of albuminuria [72]. Albuminuria is a pathologic state of increased urine albumin due to improper glomerular filtration partially caused by the destruction of the glomerular endothelial cell's glycocalyx [72]. It has been shown that people with albuminuria caused by diabetic nephropathy have drastically increased levels of heparinase and hyaluronidase, which cause shedding of the glycocalyx [73]. One of such efforts is the inhibition of monocyte chemotactic protein-1 (MCP-1), which is a protein that activates the migration of inflammatory cells such as monocytes and macrophages to the kidney. Those infiltrated glomerular macrophages can secrete cathepsin L, which is proposed to be responsible for heparinase activation [74]. Boels et al. [75] showed that MCP-1 inhibition significantly reduced albuminuria in diabetic nephropathy and restored glomerular eGC dimensions.

5.6. Chronic Inflammation

Lipopolysaccharides or endotoxins are components of the outer membrane of Gram-negative bacteria and are often used to trigger inflammation in experimental studies. In humans, lipopolysaccharides usually originate from the skin, local infections, and mucosal membranes. In some instances, lipopolysaccharides may cause endotoxemia, which is marked by the activation of TNF-α and iNOS inflammatory pathways and a consequential increase in oxidative stress and inflammation [76]. Inagawa et al. [77] noticed that the eGC of the lungs was severely diminished under experimental endotoxemia conditions. These findings suggest a causal relationship between the disruption of the eGC and microvascular endothelial dysfunction, which is a characteristic of sepsis-induced acute respiratory distress syndrome. In one pioneer research, Li et al. [78] showed that 100 ng of maresin conjugates in tissue regeneration 1 (MCTR1) increased the survival rates of mice from lipopolysaccharide-induced sepsis. These researchers also found a reduction in serum heparan sulfate and syndecan-1 levels in mice treated with MCTR1 compared to the control group, which indicates lower rates of eGC shedding. MCTR1 is produced in macrophages by the 14-lipoxygenation of docosahexaenoic acid (DHA) [79]. This fact may show importance in researching the connection between the eGC and the dietary intake of DHA (discussed later). eGC destruction might be connected to chronic inflammatory bowel diseases (IBDs) such as Crohn's disease and ulcerative colitis. IBDs are marked by chronic low-grade inflammation, which is believed to originate from the gut endotoxins [80]. People with IBD have higher rates of intestinal permeability and serum levels of lipopolysaccharides, which may negatively influence the eGC by activating the TNF-α inflammatory pathway, which destroys the eGC integrity [81].

Sodium glucose co transporter 2 (SGLT2) inhibitors are administered in the treatment of both type 2 diabetes and chronic kidney disease, and the positive effect of SLGT2 inhibitors on vascular function may be related to eGC recovery. Decreased oxidative stress seem to be one of the mediators of the effects of SGLT-2 inhibitors, and since ROS are known disruptors of the endothelial glycocalyx, SGLT2 inhibitors may improve mechanotransduction, restore nitric oxide production, and improve vasodilation. Future research should aim to confirm this hypothesis.

6. Perspectives of Nutritional Therapy for eGC Health

The presented evidence suggests a close connection between the pathophysiology of various chronic diseases and points towards eGC destruction as a driver of endothelial dysfunction and consequent vascular injury. Any chronic damage to the eGC can result in vascular permeability, oedema, platelet aggregation, and aprothrombotic environment, which are consequences that are well recognized in the end stages of chronic conditions such as diabetes, hypertension, metabolic syndrome, and obesity [40]. Defining the eGC disruptors and regenerative compounds might be the future of eGC prevention or recovery therapy in both healthy and diseased individuals. A few potential nutritional and behavior-related therapies are discussed in the chapter below.

6.1. Preventing Vitamin D Deficiency

The first described functions of vitamin D were related to immunity, viral disease, and autoimmune disease prevention [82]. This ancient function is being rediscovered, with findings suggesting an association between vitamin D deficiency and COVID-19 symptoms, particularly thrombosis and coagulopathy, which are the same symptoms related to heparan sulfate loss [83,84]. Given that heparan sulfate is one of the SARS-CoV-2 receptors, a logical postulate would be that improper heparan sulfate synthesis in COVID-19 is responsible for some of the observed symptoms. In recent times, vitamin D has also been connected to the preservation of the endothelial function through monocyte adhesion prevention and inflammation reduction [85]. This suggests that vitamin D could be an essential element for preserving or regenerating the eGC. Future research studies are warranted to investigate the role of vitamin D on eGC health and patient groups that will administer vitamin D as therapy for, e.g., hyperparathyroidism or osteoporosis, which might provide the first line of evidence.

6.2. Vitamin D and eGC Connection Hypothesis

For vitamin D to be activated, it needs to be converted into the hormonal 1,25-OH vitamin D. The enzymes needed for that conversion are the vitamin D receptor (VDR) and 1-a-hydroxylase, both of which can be found in cardiovascular tissues. Particularly, the VDR is expressed on the endothelial cells and is upregulated under stress [86]. When activated, the VDR affects calcium influx across the endothelial cell membrane, which is needed for eNOS activation and proper nitric oxide release. As mechanotransduction signals arising from the eGC activate this process, a close interaction between vitamin D and eGC has been suggested [87].

Vitamin D has a role in immunity; particularly, it has been shown that vitamin D regulates apoptosis and autophagy. One of its protective mechanisms is thought to be the inhibition of superoxide anion generation, NF-κB, and TNF-α [88]. In one randomized controlled trial, Omidian et al. [89] found a significant reduction in TNF-α levels when supplementing diabetic patients with 4000 IU/day of vitamin D for three months. This evidence is another argument in favor of the protective role of vitamin D on eGC, as TNF-α degrades the glycocalyx.

6.3. Supplementing with Omega-3 Fatty Acids and Probiotics

Combined supplementation with probiotics and omega-3 (Ω-3) fatty acids might be important in eGC regeneration. Probiotic strains such as Lactobacilli and Bifidobacteria lower lipopolysaccharide-dependent chronic low-grade inflammation by inhibiting the binding of lipopolysaccharide to the CD_{14} receptor, thereby reducing the overall activation of NF-$\kappa\beta$. Ω-3 fatty acids have been shown to increase Bifidobacteria—via unclarified mechanisms—which then suppress lipopolysaccharide and decrease lipopolysaccharide-producing bacteria, such as Enterobacteria [90]. Supplementation with Ω-3 has repeatedly been shown to decrease endothelial dysfunction and increase vasodilation and vessel elasticity, as well as decrease inflammatory pathways [91–93].

Taken together with the proposed vitamin D connection, the presented evidence suggests that interactions between lipid metabolism and eGC might be particularly relevant to research further. When conducting such studies, the type of dietary fat is an important factor to consider. Some studies suggest that fat-rich and energy-rich diets are the main source of increased endotoxemia, whereas unsaturated fatty acids have been associated with lower postprandial circulating levels of lipopolysaccharides [90,94,95]. Caution must be taken with the potential cofounders. Dietary sugars and salt might also be eGC disruptors, so their influence should be considered when forming further hypotheses and experiments.

6.4. Providing the eGC Building Blocks

Currently, there are two nutraceuticals on the market that have been developed with the purpose of eGC regeneration. Various other therapies such as metformin, rosuvastatin, hydrocortisone, sulodexide, and heparin have been proposed as having eGC regenerative properties [96]. The main premises of the developing therapies are to provide the eGC with the building blocks for quicker regeneration or to remove eGC disruptors. An important animal study showed that a 10-week treatment that targeted the eGC by using high molecular weight hyaluronan and other eGC components improved eGC properties and ameliorated age-related arterial dysfunction in old mice. The findings suggest that the eGC may be a potential therapeutic target for treating age-related arterial dysfunction [97].

Healthy dietary patterns, particularly the Mediterranean diet (MedDiet), Nordic diet, Traditional Asian diet, and Dietary Approaches to Stop Hypertension (DASH), have been shown to be beneficial in reducing the risk of arterial dysfunction and other diet-related chronic diseases [98–100]. Compared to a typical Western-style diet, these kinds of dietary patterns are characterized by lower trans fat and lower excess sodium and sugar consumption (lower meat and processed food intake), higher fiber intake (from whole grains and legumes), higher fruit and vegetables content, and higher Ω-3 content (from fish and nuts) [101]. Studies suggest that the MedDiet is suitable as a type 2 diabetes therapy, as it was associated with improved glycemic control when compared to a control dietary pattern [102]. In addition, some long-term randomized controlled trials and meta-analyses showed a greater chance of remission from metabolic syndrome following the MedDiet and a significant reduction in stroke incidence [102,103]. One possible explanation for the observed health benefits could be that healthy dietary patterns contain the building blocks for preventing the pathological destruction and/or regenerating the eGC, thus decreasing cardiovascular disease risk. Further large-scale studies are needed to confirm the connection to healthy dietary patterns, which might also be useful as a source of the eGC building blocks and, therefore, may decrease cardiovascular disease risk. Further large-scale studies are needed to confirm the connection [104]. Other potential beneficial dietary components are discussed in the chapters that follow.

6.5. Dietary Sulfur

Sulfur-containing compounds might help reduce eGC damage by exhibiting antioxidative and anti-inflammatory properties and are, therefore, good candidates to be implemented in preventative nutritional therapy [105]. High levels of sulfur can be found in meat and fish, as sulfur is a part of the sulfur-containing amino acids (methionine and cysteine). However, Doleman et al. [106] state that an impressive 89.5% of dietary sulfur in a typical diet derives from other sources due to the differences in the distribution of types of food. Intake from alliaceous and cruciferous vegetables contributed to almost half of the total sulfur intake. Both alliaceous vegetables (for example onion, garlic, or leek) and cruciferous vegetables (such as broccoli, kale, asparagus, or mangold) are abundant in various healthy dietary patterns. In alliaceous vegetables, sulfur is a part of organosulfur compounds and is known for its effectiveness in eliminating viral and bacterial infections [107]. Interestingly, one double-blinded placebo-controlled randomized study found that aged garlic extract may protect and slightly improve the microcirculation in patients with a Framingham Risk Score ≥ 10 (increased risk for cardiovascular disease) [108]. To date, this is the largest study researching the effects of garlic extract on microcirculation.

Sulfur is also a structural part of isothiocyanates, which are, in a broad range, found in cruciferous vegetables, which are staples of healthy dietary patterns [105]. Sulforaphane, the most researched isothiocyanate, exhibits anti-inflammatory and antioxidative properties that have been confirmed in various in vivo and epidemiological studies and may reduce levels of fasting blood glucose and glycated hemoglobin [109], as well as AGE concentration [110]. When researching the effects of sulforaphane in mice models of skin cancer, Alyoussef and Taha [111] found that sulforaphane blocked sulfatase-2 activity which, when activated, significantly elevated heparan sulfate proteoglycan concentration in plasma. The main function of sulforaphane is thought to be the activation of the antioxidative nuclear factor E2-related factor 2 (Nrf2) metabolically pathway, which ameliorates excess oxidative stress upon activating numerous cytoprotective proteins [111]. One of those proteins is metallothionein, a cysteine protein that binds copper and zinc as cofactors. When activated, metallothionein can extinguish ROS/RNS due to its high thiol content [112]. In animal models of type 1 diabetes, sulforaphane and zinc have proven to be more cardioprotective when combined [112]. Human studies are needed to further substantiate this evidence. The eGC is a ubiquitous structure with a relatively high amount of sulfur-containing components such as heparan and chondroitin sulfates, which are in an almost constant state of metabolic turnover [17]. Aside from its antioxidative properties, dietary sulfur might be an important sulfur donor and could provide the building blocks for eGC regeneration.

6.6. Dietary Nitrates

Dietary nitrates are high in green leafy vegetables and some root vegetables, such as beetroot. In research, dietary nitrates are usually provided as beetroot juice or sodium nitrate and have been shown to reduce inflammation and thrombosis; however, the findings are not conclusive [113]. The explanation behind the beetroot juice intervention is based on the hypothesis that providing an exogenous source of nitric oxide might improve endothelial function. Most of the acute and short-term research on hypertensive patients in this area shows substantial improvements in resting blood pressure and muscle microvascular function, as well as reduced arterial stiffness [114,115]. On the contrary, longer (7 days) but similarly designed studies did not find improvements in cardiac function or endothelial integrity in healthy non-smoking adults [116]. Furthermore, two-week supplementation of beetroot juice was insufficient to improve blood pressure or endothelial function in type 2 diabetics [117].

One of the possible explanations for the discordance in evidence might be the fact that a healthy and extended eGC is responsible for proper nitric oxide production [31]. Providing the end product (nitric oxide) will not regenerate the eGC but can be beneficial in providing nitric oxide in, for example, people who are newly diagnosed with a chronic vascular disease, based on the observation that eGC is destroyed in these conditions. In that case, beetroot juice provides a secondary source of nitric oxide and acutely helps in the preservation of vascular function until the eGC is fully restored and regains the ability to produce endogenous nitric oxide. This hypothesis would explain the acute beneficial effect and long-term stagnation of the results seen in some studies and in healthy subjects.

6.7. Lifestyle Changes

Dietary and lifestyle modifications that promote weight loss in overweight and obese individuals have been shown to be beneficial in decreasing vascular fat storage and inflammatory molecule production [118]. In recent years, professional-led weight loss even helped in completely reversing type 2 diabetes by impacting secondary insulin resistance to reduced hyperinsulinemia [119]. In that sense, weight loss could be a powerful tool in preventing eGC damage, as it combats many eGC disruptors simultaneously. Collectively, weight loss can eliminate inflammatory cytokines generated by visceral fat, decrease blood glucose levels, and prevent the formation of AGEs [54]. The combination of a healthy dietary pattern, caloric restriction, and physical activity has been used in clinical settings and shows promising results for treating non-alcoholic fatty liver diseases, which are frequently present in patients diagnosed with diabetes, insulin resistance, and obesity [120].

Intermittent fasting, prolonged fasting, time-restricted eating, and similar dietary strategies are based on restricting daily time defined for eating and prolonging fasting time [121]. The beneficial effect of intermittent fasting on vascular health parameters, microcirculation, and vasodilatation has been detected, even in the absence of weight loss in both healthy individuals and men with prediabetes [122,123]. The observed effects were related to lower blood pressure, increased insulin sensitivity, decreased oxidative stress, and higher levels of nitric oxide release. However, the mechanisms of the positive effects are not yet fully discovered [124].

The eGC might hold the key in explaining the observed positive effects of time-restricted dietary regimes; however, repeated uniform intermittent fasting studies with a reliable measure for eGC thickness are needed to confirm this hypothesis. If confirmed, intermittent fasting (or other interventions based on time restriction of eating) might be useful in nutritional therapy for vascular consequences of chronic diseases. Recent findings in this area suggest that setting the eating window earlier in the day may be optimal due to the natural circadian rhythms of humans which control hormone release, thus influencing stress levels and insulin resistance [125,126].

7. Conclusions

Deep and precise knowledge regarding the eGC seems to be the missing link in finding novel treatments for lifestyle-related diseases such as atherosclerosis, type 2 diabetes, hypertension, obesity, and metabolic syndrome. Changes in the eGC integrity might be an early sign of cardiovascular disease development. The detection of these changes could particularly be important for raising adherence to healthy dietary patterns and lifestyle interventions that have proven to be cardio-protective. Studying vitamin D, omega-3 fatty acids, probiotics, dietary sulfurs, and dietary nitrates could be promising in exploring the connection between diet components and eGC regeneration. Lifestyle interventions such as weight loss and time management of eating might be important for eGC regeneration. Increasing research in this area could provide more precise guidelines in chronic cardiovascular disease prevention.

Author Contributions: Both authors conceived the outline of this review. P.F. drafted the first version, and L.G. revised the manuscript. All authors have read and agreed to the published version of the manuscript.

Funding: This research received no external funding.

Institutional Review Board Statement: Not applicable.

Informed Consent Statement: Not applicable.

Data Availability Statement: No new data were created.

Conflicts of Interest: The authors declare no conflict of interest.

References

1. Alphonsus, C.S.; Rodseth, R.N. The endothelial glycocalyx: A review of the vascular barrier. *Anaesthesia* **2014**, *69*, 777–784. [CrossRef] [PubMed]
2. Jacob, M.; Rehm, M.; Loetsch, M.; Paul, J.O.; Bruegger, D.; Welsch, U.; Conzen, P.; Becker, B.F. The endothelial glycocalyx prefers albumin for evoking shear stress-induced, nitric oxide-mediated coronary dilatation. *J. Vasc. Res.* **2007**, *44*, 435–443. [CrossRef] [PubMed]
3. DellaValle, B.; Hasseldam, H.; Johansen, F.F.; Iversen, H.K.; Rungby, J.; Hempel, C. Multiple soluble components of the glycocalyx are increased in patient plasma after ischemic stroke. *Stroke* **2019**, *50*, 2948–2951. [CrossRef]
4. Mitra, R.; O'Neil, G.L.; Harding, I.C.; Cheng, M.J.; Mensah, S.A.; Ebong, E.E. Glycocalyx in Atherosclerosis-Relevant Endothelium Function and as a Therapeutic Target. *Curr. Atheroscler. Rep.* **2017**, *1*, 63. [CrossRef]
5. Ikonomidis, I.; Voumvourakis, A.; Makavos, G.; Triantafyllidi, H.; Pavlidis, G.; Katogiannis, K.; Benas, D.; Vlastos, D.; Trivilou, P.; Varoudi, M.; et al. Association of impaired endothelial glycocalyx with arterial stiffness, coronary microcirculatory dysfunction, and abnormal myocardial deformation in untreated hypertensives. *J. Clin. Hypertens.* **2018**, *20*, 672–679. [CrossRef] [PubMed]
6. Nieuwdorp, M.; van Haeften, T.W.; Gouverneur, M.C.L.G.; Mooij, H.L.; van Lieshout, M.H.P.; Levi, M.; Meijers, J.C.M.; Holleman, F.; Hoekstra, J.B.L.; Vink, H.; et al. Loss of endothelial glycocalyx during acute hyperglycemia coincides with endothelial dysfunction and coagulation activation in vivo. *Diabetes* **2006**, *55*, 480–486. [CrossRef]
7. Yilmaz, O.; Afsar, B.; Ortiz, A.; Kanbay, M. The role of endothelial glycocalyx in health and disease. *Clin. Kidney J.* **2019**, *12*, 611–619. [CrossRef]
8. Annecke, T.; Chappell, D.; Chen, C.; Jacob, M.; Welsch, U.; Sommerhoff, C.P.; Rehm, M.; Conzen, P.F.; Becker, B.F. Sevoflurane preserves the endothelial glycocalyx against ischaemia-reperfusion injury. *Br. J. Anaesth.* **2010**, *104*, 414–421. [CrossRef]
9. Iba, T.; Levy, J.H. Derangement of the endothelial glycocalyx in sepsis. *J. Thromb. Haemost.* **2019**, *17*, 283–294. [CrossRef]
10. Astapenko, D.; Benes, J.; Pouska, J.; Lehmann, C.; Islam, S.; Cerny, V. Endothelial glycocalyx in acute care surgery—What anaesthesiologists need to know for clinical practice. *BMC Anesthesiol.* **2019**, *19*, 238. [CrossRef]
11. Maldonado, F.; Morales, D.; Gutiérrez, R.; Barahona, M.; Cerda, O.; Cáceres, M. Effect of sevoflurane and propofol on tourniquet-induced endothelial damage: A pilot randomized controlled trial for knee-ligament surgery. *BMC Anesthesiol.* **2020**, *20*, 121. [CrossRef] [PubMed]
12. Kolářová, H.; Ambrůzová, B.; Švihálková Šindlerová, L.; Klinke, A.; Kubala, L. Modulation of endothelial glycocalyx structure under inflammatory conditions. *Mediat. Inflamm.* **2014**, *2014*, 694312. [CrossRef]
13. Drake-Holland, A.J.; Noble, M.I.M. Update on the Important New Drug Target in Cardiovascular Medicine—The Vascular Glycocalyx. *Cardiovasc. Hematol. Disord. Drug Targets* **2012**, *12*, 76–81. [CrossRef]
14. Groner, W.; Winkelman, J.W.; Harris, A.G.; Ince, C.; Bouma, G.J.; Messmer, K.; Nadeau, R.G. Orthogonal polarization spectral imaging: A new method for study of the microcirculation. *Nat. Med.* **1999**, *5*, 1209–1213. [PubMed]
15. Vlahu, C.A.; Lemkes, B.A.; Struijk, D.G.; Koopman, M.G.; Krediet, R.T.; Vink, H. Damage of the endothelial glycocalyx in dialysis patients. *J. Am. Soc. Nephrol.* **2012**, *23*, 1900–1908. [CrossRef]
16. Chappell, D.; Hofmann-Kiefer, K.; Jacob, M.; Rehm, M.; Briegel, J.; Welsch, U.; Conzen, P.; Becker, B.F. TNF-α induced shedding of the endothelial glycocalyx is prevented by hydrocortisone and antithrombin. *Basic Res. Cardiol.* **2009**, *104*, 78–89. [CrossRef]
17. Oohira, A.; Wight, T.N.; Bornstein, P. Sulfated proteoglycans synthesized by vascular endothelial cells in culture. *J. Biol. Chem.* **1983**, *258*, 2014–2021. [CrossRef]
18. Ushiyama, A.; Kataoka, H.; Iijima, T. Glycocalyx and its involvement in clinical pathophysiologies. *J. Intensive Care* **2016**, *4*, 59. [CrossRef] [PubMed]
19. Dogné, S.; Flamion, B.; Caron, N. Endothelial glycocalyx as a shield against diabetic vascular complications: Involvement of hyaluronan and hyaluronidases. *Arterioscler. Thromb. Vasc. Biol.* **2018**, *38*, 1427–1439. [CrossRef]
20. Dogné, S.; Flamion, B. Endothelial Glycocalyx Impairment in Disease: Focus on Hyaluronan Shedding. *Am. J. Pathol. Am. Soc. Investig. Pathol.* **2020**, *190*, 768–780. [CrossRef]
21. Singh, S.S.; Heijmans, R.; Meulen, C.K.E.; Lieverse, A.G.; Gornik, O.; Sijbrands, E.J.G.; Lauc, G.; van Hoek, M. Association of the IgG N-glycome with the course of kidney function in type 2 diabetes. *BMJ Open Diabetes Res. Care* **2020**, *8*, e001026. [CrossRef]
22. Tarbell, J.M.; Pahakis, M.Y. Mechanotransduction and the glycocalyx. *J. Intern. Med.* **2006**, *259*, 339–350. [CrossRef]
23. McClatchey, P.M.; Schafer, M.; Hunter, K.S.; Reusch, J.E. The endothelial glycocalyx promotes homogenous blood flow distribution within the microvasculature. *Am. J. Physiol. Heart Circ. Physiol.* **2016**, *311*, H168–H176. [CrossRef] [PubMed]
24. Choi, S.J.; Lillicrap, D. A sticky proposition: The endothelial glycocalyx and von Willebrand factor. *J. Thromb. Haemost.* **2020**, *18*, 781–785. [CrossRef] [PubMed]
25. Oberleithner, H. Vascular endothelium: A vulnerable transit zone for merciless sodium. *Nephrol. Dial. Transplant.* **2014**, *29*, 440–446. [CrossRef]

26. Chappell, D.; Jacob, M.; Paul, O.; Rehm, M.; Welsch, U.; Stoeckelhuber, M.; Conzen, P.; Becker, B.F. The glycocalyx of the human umbilical vein endothelial cell: An impressive structure ex vivo but not in culture. *Circ. Res.* **2009**, *104*, 1313–1317. [CrossRef] [PubMed]

27. Becker, B.F.; Jacob, M.; Leipert, S.; Salmon, A.H.J.; Chappell, D. Degradation of the endothelial glycocalyx in clinical settings: Searching for the sheddases. *Br. J. Clin. Pharmacol.* **2015**, *80*, 389–402. [CrossRef] [PubMed]

28. Pries, A.R.; Secomb, T.W.; Gaehtgens, P. The endothelial surface layer. *Pflug. Arch. Eur. J. Physiol.* **2000**, *440*, 653–666. [CrossRef]

29. Van Haaren, P.M.A.; VanBavel, E.; Vink, H.; Spaan, J.A.E. Charge modification of the endothelial surface layer modulates the permeability barrier of isolated rat mesenteric small arteries. *Am. J. Physiol. Heart Circ. Physiol.* **2005**, *289*, 2503–2507. [CrossRef]

30. Constantinescu, A.A.; Vink, H.; Spaan, J.A.E. Endothelial cell glycocalyx modulates immobilization of leukocytes at the endothelial surface. *Arterioscler. Thromb. Vasc. Biol.* **2003**, *23*, 1541–1547. [CrossRef]

31. Dragovich, M.A.; Chester, D.; Fu, B.M.; Wu, C.; Xu, Y.; Goligorsky, M.S.; Zhang, X.F. Mechanotransduction of the endothelial glycocalyx mediates nitric oxide production through activation of TRP channels. *Am. J. Physiol. Cell Physiol.* **2016**, *311*, C846–C853. [CrossRef] [PubMed]

32. Thi, M.M.; Tarbell, J.M.; Weinbaum, S.; Spray, D.C. The role of the glycocalyx in reorganization of the actin cytoskeleton under fluid shear stress: A "bumper-car" model. *Proc. Natl. Acad. Sci. USA* **2004**, *101*, 16483–16488. [CrossRef] [PubMed]

33. Cohen, R.A.; Adachi, T. Nitric-Oxide-Induced Vasodilatation: Regulation by Physiologic S-Glutathiolation and Pathologic Oxidation of the Sarcoplasmic Endoplasmic Reticulum Calcium ATPase. *Trends Cardiovasc. Med.* **2006**, *16*, 109–114. [CrossRef]

34. Yang, X.; Meegan, J.E.; Jannaway, M.; Coleman, D.C.; Yuan, S.Y. A disintegrin and metalloproteinase 15-mediated glycocalyx shedding contributes to vascular leakage during inflammation. *Cardiovasc. Res.* **2018**, *114*, 1752–1763. [CrossRef]

35. Teng, Y.H.F.; Aquino, R.S.; Park, P.W. Molecular functions of syndecan-1 in disease. *Matrix Biol.* **2012**, *31*, 3–16. [CrossRef] [PubMed]

36. Melrose, J. Glycosaminoglycans in Wound Healing. *Bone Tissue Regen. Insights* **2016**, *7*, BTRI.S38670. [CrossRef]

37. Zhang, Q.; Chen, C.Z.; Swaroop, M.; Xu, M.; Wang, L.; Lee, J.; Wang, A.Q.; Pradhan, M.; Hagen, N.; Chen, L.; et al. Heparan sulfate assists SARS-CoV-2 in cell entry and can be targeted by approved drugs in vitro. *Cell Discov.* **2020**, *6*, 80. [CrossRef]

38. Parameswaran, N.; Patial, S. Tumor necrosis factor-α signaling in macrophages. *Crit. Rev. Eukaryot. Gene Expr.* **2010**, *20*, 87–103. [CrossRef]

39. Jannaway, M.; Yang, X.; Meegan, J.E.; Coleman, D.C.; Yuan, S.Y. Thrombin-cleaved syndecan-3/-4 ectodomain fragments mediate endothelial barrier dysfunction. *PLoS ONE* **2019**, *14*, e0214737. [CrossRef]

40. Pillinger, N.L.; Kam, P.C.A. Endothelial glycocalyx: Basic science and clinical implications. *Anaesth. Intensive Care* **2017**, *45*, 295–307. [CrossRef]

41. Hayden, M.S.; Ghosh, S. Regulation of NF-κB by TNF Family Cytokines. *Semin. Immunol.* **2014**, *26*, 253–266. [CrossRef]

42. Strand, M.E.; Herum, K.M.; Rana, Z.A.; Skrbic, B.; Askevold, E.T.; Dahl, C.P.; Vistnes, M.; Hasic, A.; Kvaløy, H.; Sjaastad, I.; et al. Innate immune signaling induces expression and shedding of the heparan sulfate proteoglycan syndecan-4 in cardiac fibroblasts and myocytes, affecting inflammation in the pressure-overloaded heart. *FEBS J.* **2013**, *280*, 2228–2247. [CrossRef]

43. Lipphardt, M.; Dihazi, H.; Müller, G.A.; Goligorsky, M.S. Fibrogenic secretome of sirtuin 1-deficient endothelial cells: Wnt, notch and glycocalyx rheostat. *Front. Physiol.* **2018**, *9*, 1–7. [CrossRef] [PubMed]

44. Sieve, I.; Münster-Kühnel, A.K.; Hilfiker-Kleiner, D. Regulation and function of endothelial glycocalyx layer in vascular diseases. *Vasc. Pharmacol.* **2018**, *100*, 26–33. [CrossRef]

45. Oberleithner, H.; Riethmüller, C.; Schillers, H.; MacGregor, G.A.; de Wardener, H.E.; Hausberg, M. Plasma sodium stiffens vascular endothelium and reduces nitric oxide release. *Proc. Natl. Acad. Sci. USA* **2007**, *104*, 16281. [CrossRef] [PubMed]

46. Oberleithner, H.; Peters, W.; Kusche-Vihrog, K.; Korte, S.; Schillers, H.; Kliche, K.; Oberleithner, K. Salt overload damages the glycocalyx sodium barrier of vascular endothelium. *Pflug. Arch. Eur. J. Physiol.* **2011**, *462*, 519–528. [CrossRef]

47. Rorije, N.M.G.; Rademaker, E.; Schrooten, E.M.; Wouda, R.D.; van der Heide, J.J.H.; van den Born, B.J.H.; Vogt, L. High-salt intake affects sublingual microcirculation and is linked to body weight change in healthy volunteers: A randomized cross-over trial. *J. Hypertens.* **2019**, *37*, 1254–1261. [CrossRef] [PubMed]

48. Van den Berg, B.M.; Spaan, J.A.E.; Rolf, T.M.; Vink, H. Atherogenic region and diet diminish glycocalyx dimension and increase intima-to-media ratios at murine carotid artery bifurcation. *Am. J. Physiol. Heart Circ. Physiol.* **2006**, *290*, 915–920. [CrossRef] [PubMed]

49. Chiu, J.J.; Usami, S.; Chien, S. Vascular endothelial responses to altered shear stress: Pathologic implications for atherosclerosis. *Ann. Med.* **2009**, *41*, 19–28. [CrossRef]

50. Hahn, R.G.; Patel, V.; Dull, R.O. Human glycocalyx shedding: Systematic review and critical appraisal. *Acta Anaesthesiol. Scand.* **2021**, *65*, 590–606. [CrossRef]

51. Valerio, L.; Peters, R.J.; Zwinderman, A.H.; Pinto-Sietsma, S.J. Sublingual endothelial glycocalyx and atherosclerosis. A cross-sectional study. *PLoS ONE* **2019**, *14*, e0213097. [CrossRef]

52. Giles, J.T.; Wasko, M.C.M.; Chung, C.P.; Szklo, M.; Blumenthal, R.S.; Kao, A.; Bokhari, S.; Zartoshti, A.; Stein, C.M.; Bathon, J.M. Exploring the Lipid Paradox Theory in Rheumatoid Arthritis: Associations of Low Circulating Low-Density Lipoprotein Concentration with Subclinical Coronary Atherosclerosis. *Arthritis Rheumatol.* **2019**, *71*, 1426–1436. [CrossRef] [PubMed]

53. Varvel, S.A.; Dayspring, T.D.; Edmonds, Y.; Thiselton, D.L.; Ghaedi, L.; Voros, S.; McConnell, J.P.; Sasinowski, M.; Dall, T.; Warnick, G.R. Discordance between apolipoprotein B and low-density lipoprotein particle number is associated with insulin resistance in clinical practice. *J. Clin. Lipidol.* **2015**, *9*, 247–255. [CrossRef]

54. Lechner, K.; Mckenzie, A.L.; Kränkel, N.; von Schacky, C.; Worm, N.; Nixdorff, U.; Lechner, B.; Scherr, J.; Weingärtner, O.; Krauss, R.M. High-Risk Atherosclerosis and Metabolic Phenotype: The Roles of Ectopic Adiposity, Atherogenic Dyslipidemia, and Inflammation. *Metab. Syndr. Relat. Disord.* **2020**, *18*, 176–185. [CrossRef]

55. Neeland, I.J.; Ross, R.; Després, J.P.; Matsuzawa, Y.; Yamashita, S.; Shai, I.; Seidell, J.; Magni, P.; Santos, R.D.; Arsenault, B.; et al. Visceral and ectopic fat, atherosclerosis, and cardiometabolic disease: A position statement. *Lancet Diabetes Endocrinol.* **2019**, *7*, 715–725. [CrossRef] [PubMed]

56. Piperi, C.; Adamopoulos, C.; Dalagiorgou, G.; Diamanti-Kandarakis, E.; Papavassiliou, A.G. Crosstalk between advanced glycation and endoplasmic reticulum stress: Emerging therapeutic targeting for metabolic diseases. *J. Clin. Endocrinol. Metab.* **2012**, *97*, 2231–2242. [CrossRef] [PubMed]

57. Jamwal, S.; Sharma, S. Vascular endothelium dysfunction: A conservative target in metabolic disorders. *Inflamm. Res.* **2018**, *67*, 391–405. [CrossRef]

58. Förstermann, U.; Sessa, W.C. Nitric oxide synthases: Regulation and function. *Eur. Heart J.* **2012**, *33*, 829–837. [CrossRef]

59. Fehsel, K.; Jalowy, A.; Qi, S.; Burkart, V.; Hartmann, B.; Kolb, H. Islet cell DNA is a target of inflammatory attack by nitric oxide. *Diabetes* **1993**, *42*, 496–500. [CrossRef] [PubMed]

60. Udi, S.; Hinden, L.; Ahmad, M.; Drori, A.; Iyer, M.R.; Cinar, R.; Herman-Edelstein, M.; Tam, J. Dual inhibition of cannabinoid CB1 receptor and inducible NOS attenuates obesity-induced chronic kidney disease. *Br. J. Pharmacol.* **2020**, *177*, 110–127. [CrossRef]

61. Łuczak, A.; Madej, M.; Kasprzyk, A.; Doroszko, A. Role of the eNOS Uncoupling and the Nitric Oxide Metabolic Pathway in the Pathogenesis of Autoimmune Rheumatic Diseases. *Oxidative Med. Cell. Longev.* **2020**, *2020*, 1417981. [CrossRef] [PubMed]

62. Goldin, A.; Beckman, J.A.; Schmidt, A.M.; Creager, M.A. Advanced glycation end products: Sparking the development of diabetic vascular injury. *Circulation* **2006**, *114*, 597–605. [CrossRef] [PubMed]

63. Gryszczyńska, B.; Budzyń, M.; Begier-Krasińska, B.; Osińska, A.; Boruczkowski, M.; Kaczmarek, M.; Bukowska, A.; Iskra, M.; Kasprzak, M.P. Association between advanced glycation end products, soluble RAGE receptor, and endothelium dysfunction, evaluated by circulating endothelial cells and endothelial progenitor cells in patients with mild and resistant hypertension. *Int. J. Mol. Sci.* **2019**, *20*, 3942. [CrossRef]

64. Mudau, M.; Genis, A.; Lochner, A.; Strijdom, H. Endothelial dysfunction: The early predictor of atherosclerosis. *Cardiovasc. J. Afr.* **2012**, *23*, 222–231. [CrossRef]

65. Cai, W.; He, J.C.; Zhu, L.; Peppa, M.; Lu, C.; Uribarri, J.; Vlassara, H. High levels of dietary advanced glycation end products transform low-density lipoprotein into a potent redox-sensitive mitogen-activated protein kinase stimulant in diabetic patients. *Circulation* **2004**, *110*, 285–291. [CrossRef] [PubMed]

66. Vlassara, H.; Uribarri, J. Advanced glycation end products (AGE) and diabetes: Cause, effect, or both? *Curr. Diabetes Rep.* **2014**, *14*, 1–17. [CrossRef] [PubMed]

67. Hellsten, Y.; Rufener, N.; Nielsen, J.J.; Høier, B.; Krustrup, P.; Bangsbo, J. Passive leg movement enhances interstitial VEGF protein, endothelial cell proliferation, and eNOS mRNA content in human skeletal muscle. *Am. J. Physiol. Regul. Integr. Comp. Physiol.* **2008**, *294*, 975–982. [CrossRef] [PubMed]

68. Roberts, C.K.; Hevener, A.L.; Barnard, R.J. Metabolic syndrome and insulin resistance: Underlying causes and modification by exercise training. *Compr. Physiol.* **2013**, *3*, 1–58. [PubMed]

69. Pertynska-Marczewska, M.; Merhi, Z. Relationship of advanced glycation end products with cardiovascular disease in menopausal women. *Reprod. Sci.* **2015**, *22*, 774–782. [CrossRef]

70. Nieuwdorp, M.M.; Kroon, H.L.; Atasever, J.; Spaan, B.; Ince, J.A.E.; Holleman, C.; Diamant, F.; Heine, M.; Hoekstra, R.J.; Joost, B.L.; et al. Endothelial Glycocalyx Damage Coincides with Microalbuminuria in Type 1 Diabetes. *Diabetes* **2005**, *55*, 1127–1132. [CrossRef]

71. Primary Causes of End-Stage Renal Disease. Available online: https://www.uspharmacist.com/article/primary-causes-of-endstage-renal-disease (accessed on 30 May 2023).

72. Stephen, R.; Jolly, S.E.; Nally, J.V.; Navaneethan, S.D. Albuminuria: When urine predicts kidney and cardiovascular disease. *Clevel. Clin. J. Med.* **2014**, *81*, 41–50. [CrossRef]

73. Van den Hoven, M.J.; Rops, A.L.; Bakker, M.A.; Aten, J.; Rutjes, N.; Roestenberg, P.; Goldschmeding, R.; Zcharia, E.; Vlodavsky, I.; van der Vlag, J.; et al. Increased expression of heparanase in overt diabetic nephropathy. *Kidney Int.* **2006**, *70*, 2100–2108. [CrossRef]

74. Garsen, M.; Rops, A.L.W.M.M.; Dijkman, H.; Willemsen, B.; van Kuppevelt, T.H.; Russel, F.G.; Rabelink, T.J.; Berden, J.H.M.; Reinheckel, T.; van der Vlag, J. Cathepsin L is crucial for the development of early experimental diabetic nephropathy. *Kidney Int.* **2016**, *90*, 1012–1022. [CrossRef] [PubMed]

75. Boels, M.G.S.; Koudijs, A.; Avramut, M.C.; Sol, W.M.P.J.; Wang, G.; van Oeveren-Rietdijk, A.M.; van Zonneveld, A.J.; de Boer, H.C.; van der Vlag, J.; van Kooten, C.; et al. Systemic Monocyte Chemotactic Protein-1 Inhibition Modifies Renal Macrophages and Restores Glomerular Endothelial Glycocalyx and Barrier Function in Diabetic Nephropathy. *Am. J. Pathol.* **2017**, *187*, 2430–2440. [CrossRef] [PubMed]

76. Wassenaar, T.M.; Zimmermann, K. Lipopolysaccharides in Food, Food Supplements, and Probiotics: Should We be Worried? *Eur. J. Microbiol. Immunol.* **2018**, *8*, 63. [CrossRef] [PubMed]

77. Inagawa, R.; Okada, H.; Takemura, G.; Suzuki, K.; Takada, C.; Yano, H.; Ando, Y.; Usui, T.; Hotta, Y.; Miyazaki, N.; et al. Ultrastructural Alteration of Pulmonary Capillary Endothelial Glycocalyx During Endotoxemia. *Chest* **2018**, *154*, 317–325. [CrossRef]

78. Li, H.; Hao, Y.; Yang, L.L.; Wang, X.Y.; Li, X.Y.; Bhandari, S.; Han, J.; Liu, Y.J.; Gong, Y.Q.; Scott, A.; et al. MCTR1 alleviates lipopolysaccharide-induced acute lung injury by protecting lung endothelial glycocalyx. *J. Cell. Physiol.* **2020**, *235*, 7283–7294. [CrossRef]

79. Pan, J.; Li, X.; Wang, X.; Yang, L.; Chen, H.; Su, N.; Wu, C.; Hao, Y.; Jin, S.; Li, H. MCTR1 Intervention Reverses Experimental Lung Fibrosis in Mice. *J. Inflamm. Res.* **2021**, *14*, 1873–1881. [CrossRef]

80. Gardiner, K.R.; Halliday, M.I.; Barclay, G.R.; Milne, L.; Brown, D.; Stephens, S.; Maxwell, R.J.; Rowlands, B.J. Significance of systemic endotoxaemia in inflammatory bowel disease. *Gut BMJ* **1995**, *36*, 897. [CrossRef]

81. Oliveira Magro, D.; Kotze, P.G.; Real Martinez, C.A.; Camargo, M.G.; Guadagnini, D.; Ramos Calixto, A.; Carolina, A.; Vasques, J.; de Lourdes, M.; Ayrizono, S.; et al. Changes in serum levels of lipopolysaccharides and CD26 in patients with Crohn's disease. *Intest. Res.* **2017**, *15*, 352–357. [CrossRef]

82. Koivisto, O.; Hanel, A.; Carlberg, C. Key vitamin D target genes with functions in the immune system. *Nutrients* **2020**, *12*, 1140. [CrossRef]

83. Clausen, T.M.; Sandoval, D.R.; Spliid, C.B.; Pihl, J.; Perrett, H.R.; Painter, C.D.; Narayanan, A.; Majowicz, S.A.; Kwong, E.M.; McVicar, R.N.; et al. SARS-CoV-2 Infection Depends on Cellular Heparan Sulfate and ACE2. *Cell* **2020**, *183*, 1043–1057.e15. [CrossRef] [PubMed]

84. Zhang, J.; McCullough, P.A.; Tecson, K.M. Vitamin D deficiency in association with endothelial dysfunction: Implications for patients with COVID-19. *Rev. Cardiovasc. Med.* **2020**, *21*, 339–344. [CrossRef] [PubMed]

85. Kanikarla-Marie, P.; Jain, S.K. 1,25(OH) 2 D 3 inhibits oxidative stress and monocyte adhesion by mediating the upregulation of GCLC and GSH in endothelial cells treated with acetoacetate (ketosis). *J. Steroid Biochem. Mol. Biol.* **2016**, *159*, 94–101. [CrossRef] [PubMed]

86. Al Mheid, I.; Quyyumi, A.A. Vitamin D and Cardiovascular Disease: Controversy Unresolved. *J. Am. Coll. Cardiol.* **2017**, *70*, 89–100. [CrossRef]

87. Kim, D.H.; Meza, C.A.; Clarke, H.; Kim, J.S.; Hickner, R.C. Vitamin D and endothelial function. *Nutrients* **2020**, *12*, 575. [CrossRef]

88. Stio, M.; Martinesi, M.; Bruni, S.; Treves, C.; Mathieu, C.; Verstuyf, A.; d'Albasio, G.; Bagnoli, S.; Bonanomi, A.G. The Vitamin D analogue TX 527 blocks NF-κB activation in peripheral blood mononuclear cells of patients with Crohn's disease. *J. Steroid Biochem. Mol. Biol.* **2007**, *103*, 51–60. [CrossRef] [PubMed]

89. Omidian, M.; Mahmoudi, M.; Javanbakht, M.H.; Eshraghian, M.R.; Abshirini, M.; Daneshzad, E.; Hasani, H.; Alvandi, E.; Djalali, M. Effects of vitamin D supplementation on circulatory YKL-40 and MCP-1 biomarkers associated with vascular diabetic complications: A randomized, placebo-controlled, double-blind clinical trial. *Diabetes Metab. Syndr. Clin. Res. Rev.* **2019**, *13*, 2873–2877. [CrossRef]

90. Hutchinson, A.N.; Tingö, L.; Brummer, R.J. The Potential Effects of Probiotics and ω-3 Fatty Acids on Chronic Low-Grade Inflammation. *Nutrients* **2020**, *12*, 2402. [CrossRef]

91. Goodfellow, J.; Bellamy, M.F.; Ramsey, M.W.; Jones, C.J.H.; Lewis, M.J. Dietary supplementation with marine omega-3 fatty acids improve systemic large artery endothelial function in subjects with hypercholesterolemia. *J. Am. Coll. Cardiol.* **2000**, *35*, 265–270. [CrossRef]

92. Miyoshi, T.; Noda, Y.; Ohno, Y.; Sugiyama, H.; Oe, H.; Nakamura, K.; Kohno, K.; Ito, H. Omega-3 fatty acids improve postprandial lipemia and associated endothelial dysfunction in healthy individuals—A randomized cross-over trial. *Biomed. Pharmacother.* **2014**, *68*, 1071–1077. [CrossRef]

93. Huang, F.; del-Río-Navarro, B.E.; Leija-Martinez, J.; Torres-Alcantara, S.; Ruiz-Bedolla, E.; Hernández-Cadena, L.; Barraza-Villarreal, A.; Romero-Nava, R.; Sanchéz-Muñoz, F.; Villafaña, S.; et al. Effect of omega-3 fatty acids supplementation combined with lifestyle intervention on adipokines and biomarkers of endothelial dysfunction in obese adolescents with hypertriglyceridemia. *J. Nutr. Biochem.* **2019**, *64*, 162–169. [CrossRef] [PubMed]

94. Pendyala, S.; Walker, J.M.; Holt, P.R. A High-Fat Diet Is Associated with Endotoxemia That Originates from the Gut. *Gastroenterology* **2012**, *142*, 1100–1101.e2. [CrossRef] [PubMed]

95. André, P.; Laugerette, F.; Féart, C. Metabolic Endotoxemia: A Potential Underlying Mechanism of the Relationship between Dietary Fat Intake and Risk for Cognitive Impairments in Humans? *Nutrients* **2019**, *11*, 1887. [CrossRef]

96. Kazuma, S.; Tokinaga, Y.; Kimizuka, M.; Azumaguchi, R.; Hamada, K.; Yamakage, M. Sevoflurane Promotes Regeneration of the Endothelial Glycocalyx by Upregulating Sialyltransferase. *J Surg Res.* **2019**, *241*, 40–47. [CrossRef] [PubMed]

97. Machin, D.R.; Trott, D.W.; Gogulamudi, V.R.; Islam, M.T.; Bloom, S.I.; Vink, H.; Lesniewski, L.A.; Donato, A.J. Glycocalyx-targeted therapy ameliorates age-related arterial dysfunction. *Geroscience* **2023**, *Online ahead of print*. [CrossRef]

98. Ndanuko, R.N.; Tapsell, L.C.; Charlton, K.E.; Neale, E.P.; Batterham, M.J. Dietary Patterns and Blood Pressure in Adults: A Systematic Review and Meta-Analysis of Randomized Controlled Trials. *Adv. Nutr.* **2016**, *7*, 76. [CrossRef]

99. Neuhouser, M.L. The Importance of Healthy Dietary Patterns in Chronic Disease Prevention. *Nutr. Res.* **2019**, *70*, 3. [CrossRef]

100. Galbete, C.; Kröger, J.; Jannasch, F.; Iqbal, K.; Schwingshackl, L.; Schwedhelm, C.; Weikert, C.; Boeing, H.; Schulze, M.B. Nordic diet, Mediterranean diet, and the risk of chronic diseases: The EPIC-Potsdam study. *BMC Med.* **2018**, *16*, 99. [CrossRef]
101. Rees, K.; Takeda, A.; Martin, N.; Ellis, L.; Wijesekara, D.; Vepa, A.; Das, A.; Hartley, L.; Stranges, S. Mediterranean-style diet for the primary and secondary prevention of cardiovascular disease. *Cochrane Database Syst. Rev.* **2019**, *3*, CD009825. [CrossRef]
102. Esposito, K.; Maiorino, M.I.; Bellastella, G.; Chiodini, P.; Panagiotakos, D.; Giugliano, D. A journey into a Mediterranean diet and type 2 diabetes: A systematic review with meta-analyses. *BMJ Open* **2015**, *5*, e008222. [CrossRef]
103. Saulle, R.; Lia, L.; de Giusti, M.; la Torre, G. A systematic overview of the scientific literature on the association between Mediterranean Diet and the Stroke prevention. *Clin. Ter.* **2019**, *170*, e396–e408. [PubMed]
104. Nazarian, B.; Fazeli Moghadam, E.; Asbaghi, O.; Zeinali Khosroshahi, M.; Choghakhori, R.; Abbasnezhad, A. Effect of L-arginine supplementation on C-reactive protein and other inflammatory biomarkers: A systematic review and meta-analysis of randomized controlled trials. *Complement. Ther. Med.* **2019**, *47*, 102226. [CrossRef]
105. Bai, Y.; Wang, X.; Zhao, S.; Ma, C.; Cui, J.; Zheng, Y. Sulforaphane Protects against Cardiovascular Disease via Nrf2 Activation. *Oxid. Med. Cell. Longev.* **2015**, *2015*, 407580. [CrossRef]
106. Doleman, J.F.; Grisar, K.; van Liedekerke, L.; Saha, S.; Roe, M.; Tapp, H.S.; Mithen, R.F. The contribution of alliaceous and cruciferous vegetables to dietary sulphur intake. *Food Chem.* **2017**, *234*, 38–45. [CrossRef]
107. Wilson, E.A.; Demmig-Adams, B. Antioxidant, anti-inflammatory, and antimicrobial properties of garlic and onions. *Nutr. Food Sci.* **2007**, *37*, 178–183. [CrossRef]
108. Wlosinska, M.; Nilsson, A.C.; Hlebowicz, J.; Malmsjö, M.; Fakhro, M.; Lindstedt, S. Aged garlic extract preserves cutaneous microcirculation in patients with increased risk for cardiovascular diseases: A double-blinded placebo-controlled study. *Int. Wound J.* **2019**, *16*, 1487–1493. [CrossRef]
109. Axelsson, A.S.; Tubbs, E.; Mecham, B.; Chacko, S.; Nenonen, H.A.; Tang, Y.; Fahey, J.W.; Derry, J.M.J.; Wollheim, C.B.; Wierup, N.; et al. Sulforaphane reduces hepatic glucose production and improves glucose control in patients with type 2 diabetes. *Sci. Transl. Med.* **2017**, *9*, 1–13. [CrossRef]
110. Pereira, A.; Fernandes, R.; Crisóstomo, J.; Seiça, R.M.; Sena, C.M. The Sulforaphane and pyridoxamine supplementation normalize endothelial dysfunction associated with type 2 diabetes. *Sci. Rep.* **2017**, *7*, 14375. [CrossRef] [PubMed]
111. Alyoussef, A.; Taha, M. Antitumor activity of sulforaphane in mice model of skin cancer via blocking sulfatase-2. *Exp. Dermatol.* **2019**, *28*, 28–34. [CrossRef] [PubMed]
112. Wang, J.; Wang, S.; Wang, W.; Chen, J.; Zhang, Z.; Zheng, Q.; Liu, Q.; Cai, L. Protection against diabetic cardiomyopathy is achieved using a combination of sulforaphane and zinc in type 1 diabetic OVE26 mice. *J. Cell. Mol. Med.* **2019**, *23*, 6319–6330. [CrossRef] [PubMed]
113. Raubenheimer, K.; Hickey, D.; Leveritt, M.; Fassett, R.; Munoz, J.O.D.Z.; Allen, J.D.; Briskey, D.; Parker, T.J.; Kerr, G.; Peake, J.M.; et al. Acute effects of nitrate-rich beetroot juice on blood pressure, hemostasis and vascular inflammation markers in healthy older adults: A randomized, placebo-controlled crossover study. *Nutrients* **2017**, *9*, 1270. [CrossRef] [PubMed]
114. Zafeiridis, A.; Triantafyllou, A.; Papadopoulos, S.; Koletsos, N.; Touplikioti, P.; Zafeiridis, A.S.; Gkaliagkousi, E.; Dipla, K.; Douma, S. Dietary nitrate improves muscle microvascular reactivity and lowers blood pressure at rest and during isometric exercise in untreated hypertensives. *Microcirculation* **2019**, *26*, e12525. [CrossRef]
115. Kapil, V.; Khambata, R.S.; Robertson, A.; Caulfield, M.J.; Ahluwalia, A. Dietary nitrate provides sustained blood pressure lowering in hypertensive patients: A randomized, phase 2, double-blind, placebo-controlled study. *Hypertension* **2015**, *65*, 320–327. [CrossRef]
116. Oggioni, C.; Jakovljevic, D.G.; Klonizakis, M.; Ashor, A.W.; Ruddock, A.; Ranchordas, M.; Williams, E.; Siervo, M. Dietary nitrate does not modify blood pressure and cardiac output at rest and during exercise in older adults: A randomised cross-over study. *Int. J. Food Sci. Nutr.* **2018**, *69*, 74–83. [CrossRef]
117. Gilchrist, M.; Winyard, P.G.; Aizawa, K.; Anning, C.; Shore, A.; Benjamin, N. Effect of dietary nitrate on blood pressure, endothelial function, and insulin sensitivity in type 2 diabetes. *Free Radic. Biol. Med.* **2013**, *60*, 89–97. [CrossRef] [PubMed]
118. Castro-Barquero, S.; Ruiz-León, A.M.; Sierra-Pérez, M.; Estruch, R.; Casas, R. Dietary strategies for metabolic syndrome: A comprehensive review. *Nutrients* **2020**, *12*, 2983. [CrossRef]
119. Howatson, A.; Wall, C.R.; Turner-Benny, P. The contribution of dietitians to the primary health care workforce. *J. Prim. Health Care* **2015**, *7*, 324–332. [CrossRef]
120. Romero-Gómez, M.; Zelber-Sagi, S.; Trenell, M. Treatment of NAFLD with diet, physical activity and exercise. *J. Hepatol.* **2017**, *67*, 829–846. [CrossRef]
121. Fanti, M.; Mishra, A.; Longo, V.D.; Brandhorst, S. Time-Restricted Eating, Intermittent Fasting, and Fasting-Mimicking Diets in Weight Loss. *Curr. Obes. Rep. Curr. Obes. Rep.* **2021**, *10*, 70–80. [CrossRef]
122. Esmaeilzadeh, F.; van de Borne, P. Does intermittent fasting improve microvascular endothelial function in healthy middle-aged subjects? *Biol. Med.* **2016**, *8*, 6. [CrossRef]
123. Sutton, E.F.; Beyl, R.; Early, K.S.; Cefalu, W.T.; Ravussin, E.; Peterson, C.M. Early Time-Restricted Feeding Improves Insulin Sensitivity, Blood Pressure, and Oxidative Stress Even without Weight Loss in Men with Prediabetes. *Cell Metab. Cell Press* **2018**, *27*, 1212–1221.e3. [CrossRef] [PubMed]
124. Malinowski, B.; Zalewska, K.; Węsierska, A.; Sokołowska, M.M.; Socha, M.; Liczner, G.; Pawlak-Osińska, K.; Wiciński, M. Intermittent fasting in cardiovascular disorders—An overview. *Nutrients* **2019**, *11*, 673. [CrossRef] [PubMed]

125. Qian, J.; Dalla Man, C.; Morris, C.J.; Cobelli, C.; Scheer, F.A.J.L. Differential effects of the circadian system and circadian misalignment on insulin sensitivity and insulin secretion in humans. *Diabetes Obes. Metab.* **2018**, *20*, 2481–2485. [CrossRef]
126. Poggiogalle, E.; Jamshed, H.; Peterson, C.M. Circadian regulation of glucose, lipid, and energy metabolism in humans. *Metab. Clin. Exp.* **2018**, *84*, 11–27. [CrossRef]

 *nutrients*

Review

The Role of Omega-3 Polyunsaturated Fatty Acids and Their Lipid Mediators on Skeletal Muscle Regeneration: A Narrative Review

Sebastian Jannas-Vela [1], Alejandra Espinosa [2], Alejandro A. Candia [1], Marcelo Flores-Opazo [1], Luis Peñailillo [3] and Rodrigo Valenzuela [4,*]

[1] Instituto de Ciencias de la Salud, Universidad de O'Higgins, Rancagua 2820000, Chile
[2] Escuela de Medicina, Campus San Felipe, Universidad de Valparaíso, San Felipe 2170000, Chile
[3] Exercise and Rehabilitation Sciences Institute, School of Physical Therapy, Faculty of Rehabilitation Sciences, Universidad Andres Bello, Las Condes, Santiago 7591538, Chile
[4] Department of Nutrition, Faculty of Medicine, University of Chile, Santiago 8380000, Chile
* Correspondence: rvalenzuelab@uchile.cl

Abstract: Skeletal muscle is the largest tissue in the human body, comprising approximately 40% of body mass. After damage or injury, a healthy skeletal muscle is often fully regenerated; however, with aging and chronic diseases, the regeneration process is usually incomplete, resulting in the formation of fibrotic tissue, infiltration of intermuscular adipose tissue, and loss of muscle mass and strength, leading to a reduction in functional performance and quality of life. Accumulating evidence has shown that omega-3 (n-3) polyunsaturated fatty acids (PUFAs) and their lipid mediators (i.e., oxylipins and endocannabinoids) have the potential to enhance muscle regeneration by positively modulating the local and systemic inflammatory response to muscle injury. This review explores the process of muscle regeneration and how it is affected by acute and chronic inflammatory conditions, focusing on the potential role of n-3 PUFAs and their derivatives as positive modulators of skeletal muscle healing and regeneration.

Keywords: omega-3; skeletal muscle; oxylipins; endocannabinoids; regeneration

Citation: Jannas-Vela, S.; Espinosa, A.; Candia, A.A.; Flores-Opazo, M.; Peñailillo, L.; Valenzuela, R. The Role of Omega-3 Polyunsaturated Fatty Acids and Their Lipid Mediators on Skeletal Muscle Regeneration: A Narrative Review. *Nutrients* **2023**, *15*, 871. https://doi.org/10.3390/nu15040871

Academic Editors: Sandro Massao Hirabara, Gabriel Nasri Marzuca Nassr and Maria Fernanda Cury Boaventura

Received: 11 January 2023
Revised: 1 February 2023
Accepted: 6 February 2023
Published: 8 February 2023

1. Introduction

Skeletal muscle is the largest tissue in the human body, comprising approximately 40% of body mass, and has an essential role in energy metabolism, locomotion, and stability [1]. These actions can be compromised by muscle damage or injury resulting from a variety of events, including lacerations, contusions, strains, or exercise [2]. Following damage, a healthy skeletal muscle is often fully regenerated; however, with aging and chronic diseases, such as obesity, type 2 diabetes, rheumatoid arthritis, chronic obstructive pulmonary disease, and muscular dystrophies, the regeneration process may be incomplete, resulting in the formation of fibrotic tissue, infiltration of intermuscular ectopic adipose tissue, and loss of muscle mass and strength [3–6], which might lead to a reduction in functional performance and quality of life.

Satellite cells (SCs) are considered the major players in the regeneration process after muscle injury [7]. They are located between the sarcolemma and the basal lamina, representing approximately 2–7% of skeletal muscle cells [3,8,9]. In homeostatic conditions, these cells reside in their niche in a quiescent state, whereas upon muscle damage, SCs become activated and proliferate to form myogenic precursor cells (i.e., myoblasts). An efficient regenerative capacity is supported by the maintenance of a promyogenic muscle niche by additional cells in the local milieu, including mesenchymal fibroadipogenic progenitors (FAPs), immune cells (i.e., eosinophils, macrophages, Treg), and endothelial cells [10,11]. In the presence of a favorable promyogenic cellular microenvironment, most

---

of the SCs differentiate and form new myocytes that fuse with damaged myofibers, leading to muscle tissue repair, while a fraction of the SC population, generated by asymmetric division, self-renew and return to quiescence. In parallel, a tightly regulated and time-dependent recruitment of immune cells occurs, releasing inflammatory factors (e.g., TNF-α, IL-6), which also have the capacity to promote SC activation and proliferation. Later in the regeneration process, immune cells together with FAP cells contribute to the removal of cell debris and necrotic tissue [12–14]. This response is followed by the clearance of proinflammatory cytokines and the recruitment of anti-inflammatory immune cells, promoting SC differentiation, tissue repair, and the return of tissue homeostasis [15–17].

In chronic inflammatory conditions, this process is usually dysregulated and impaired, preventing the healing of damaged tissue, leading to loss of muscle function and decreased quality of life [18,19]. In consequence, it is of major importance to develop strategies that can mitigate and adequately regulate the inflammatory response after muscle injury. Accumulating evidence has shown that the omega-3 (n-3) polyunsaturated fatty acids (PUFAs) and their lipid mediators (i.e., oxylipins and endocannabinoids) have the potential to enhance muscle regeneration by positively modulating the local and systemic inflammatory response to muscle injury [20–22]. This review will explore the process of muscle regeneration and how it is affected by acute and chronic inflammatory conditions, focusing on the potential role of n-3 PUFAs and their derivatives as positive modulators of skeletal muscle healing and regeneration.

2. An Inflammatory Process Initiates Skeletal Muscle Repair

The inflammatory response triggered by an acute muscle injury is required to command SCs proliferation and myogenesis, and coordinates scavenger activity and phagocytosis to properly eradicate cellular debris and progress to tissue regeneration (Figure 1). The initiation, development, and resolution of the inflammatory response involves interactions between circulating and resident immune cells and SCs within the muscle tissue [11], of which the most abundant are mast cells and macrophages [13]. These resident cells act as the primary responders to injury by secreting proinflammatory molecules, including tumor necrosis factor alpha (TNF-α), and macrophage inflammatory protein-2 (MIP-2) [23]. This initial burst promotes vasodilation, vascular permeability, and rapid recruitment of neutrophils, which ingest and remove cellular debris from the damaged tissue and release a myriad of chemokines and cytokines that attract the recruitment of monocytes [24], and activation of FAPs via the interleukin 4 signaling pathway [14]. The infiltrated monocytes divide into two main categories: (1) Ly6C+ monocytes, which peak during the first few days of injury (days 1–3); and (2) the Ly6C− monocytes, which are recruited later (days 3–7) for the optimum regeneration of muscle tissue. The Ly6C+ and Ly6C− monocytes rapidly differentiate into the proinflammatory M1 and anti-inflammatory M2 macrophages, respectively. The M1 macrophage phenotype expresses proinflammatory cytokines (IL-1β, IL-6, IL-8, and TNF-α), responsible for the activation and proliferation of SCs, while the M2 macrophage phenotype is responsible for the release of anti-inflammatory molecules such as IL-10 and transforming growth factor-beta (TGF-β), promoting the differentiation and fusion of myotubes into damaged myofibers, leading to proper regeneration [11]. On the other hand, the activated resident FAPs will proliferate within a narrow time window, generally no longer than 5 days postinjury, and secrete a plethora of promyogenic cytokines and growth factors, including Follistatin, IL-6, WNT1 inducible signaling pathway protein 1 (WISP1), and IGF-1 [25]. Later, FAP numbers rapidly return to basal levels by the induction of apoptosis, a mechanism mediated by the release of TNF- α by M1 macrophages [26].

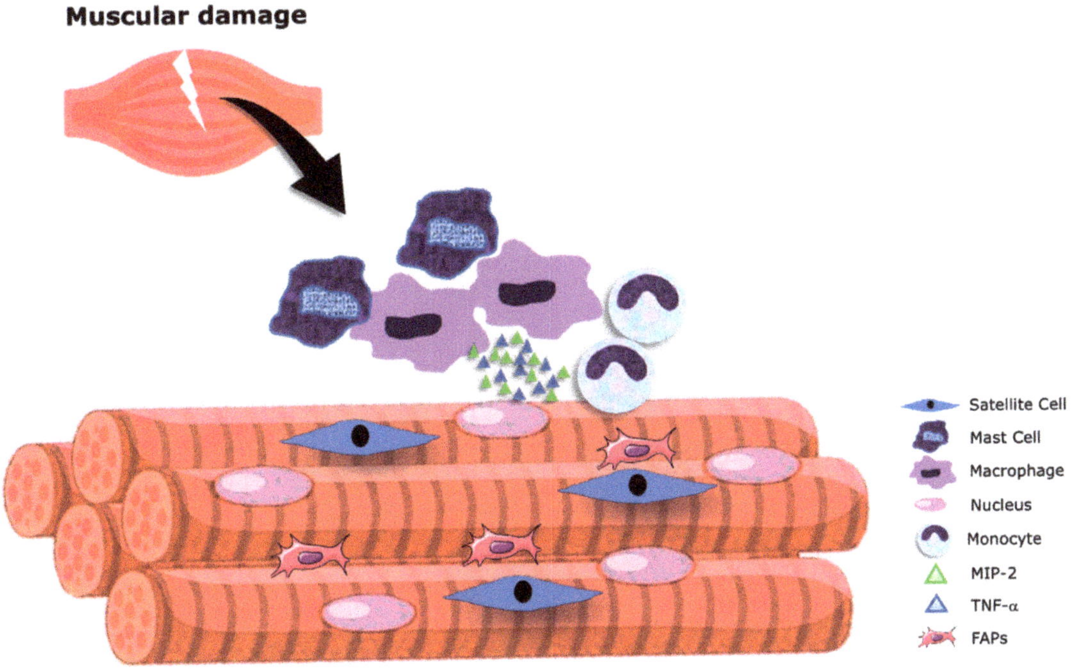

Figure 1. Inflammatory response to muscle damage. After an acute muscular injury, resident cells (mast cells and macrophages) secrete proinflammatory molecules such as tumor necrosis factor alpha (TNF-α) and macrophage inflammatory protein-2 (MIP-2), promoting neutrophil and monocyte recruitment to the injury region. FAPs: fibroadipogenic precursors. The Figure was partly generated using Servier Medical Art, provided by Servier, licensed under a Creative Commons Attribution 3.0 unported license.

The inflammatory response to damage also leads to the production of reactive oxygen species (ROS) and reactive nitrogen species (RNS) (collectively known as RONS), from circulating and resident immune cells. The RONS are molecules with one or more unpaired electrons in atomic or molecular orbitals derived from oxygen or nitrogen [27]. The main RONS formed in cells include superoxide (O2-), and nitric oxide (NO). Both O2- and NO can rapidly combine to produce peroxynitrite, and its reaction is three times faster than the dismutation of superoxide to hydrogen peroxide via superoxide dismutase (SOD) [28]. In healthy muscle, RONS activate signaling pathways essential for proper muscle regeneration [27]; however, when RONS are exacerbated (i.e., chronic inflammatory conditions), it can lead to impaired regeneration through the inhibition of myogenesis, cell death, and loss of muscle function [29]. Collectively, these processes highlight the importance of an efficient and coordinated inflammatory response after muscle injury, as slight shifts in the inflammatory and oxidative stress responses could decrease muscle regenerative capacity as observed in obese and aged populations.

A crucial yet overlooked metabolic process regulating the inflammatory response after muscle damage is the synthesis of bioactive lipid metabolites derived from n-3 and n-6 PUFAs, the oxylipins and endocannabinoids. These lipid mediators are suggested to regulate the muscle regeneration process via autocrine and paracrine inflammatory signaling of immune cells.

3. Role of PUFAs on Inflammation Resolution

Fatty acids (FAs) have diverse functions of physiological importance as they are major components of cellular membranes, precursors for the synthesis of bioactive lipids, and major sources of energy [30]. Saturated FAs contain no double bonds in their hydrocarbon chain, have a rigid structure, and are the most abundant FAs in the human diet, with palmitic and stearic acid as the most popular. Conversely, unsaturated FAs have one or more double bonds in their hydrocarbon chain; those with one double bond are classified as monounsaturated fatty acids (MUFAs). The most common MUFAs are omega-7 (n-7) and omega-9 (n-9), with palmitoleic acid and oleic acid being the most consumed in the diet, respectively. Both MUFAs have been reported to reduce the risk of heart disease and inflammation [31,32].

FAs with two or more double bonds are classified as polyunsaturated fatty acids (PUFAs). PUFAs are mainly composed by n-3 and n-6 fatty acids. N-3 PUFAs have their first double bond on the third hydrocarbon chain when counted from the methyl terminus, whereas the n-6 PUFAs have their first double bond on the sixth hydrocarbon chain. The shortest members of each family are α-linolenic acid (C18:3n-3, ALA) and linoleic acid (C18:2n-6, LA), which are 18 hydrocarbons in acyl-chain length [33–35]. Through a series of reactions catalyzed by the same enzymes in the liver, ALA and LA can be further metabolized by desaturation and elongation to eicosapentaenoic acid (C20:5n-3, EPA) and docosahexaenoic acid (C22:6n-3, DHA), and to arachidonic acid (C20:4n-6, AA), respectively [35,36]. As n-6 PUFAs are consumed in the diet 15 times more than n-3 PUFAs, this process results in a higher synthesis of AA at the expense of EPA and DHA [30]. In consequence, the presence of n-3 PUFAs in cell membranes and tissues is typically low and depends on the amount of EPA and DHA ingested in the diet. Thus, when consumed in high quantities, EPA and DHA are incorporated into cell membrane phospholipids [37], resulting in a decreased synthesis of n-6 PUFAs derivatives and an increased synthesis of n-3 PUFA-lipid mediators [38]. As the n-6 PUFAs are associated with proinflammatory actions [39] and the n-3 lipid derivatives are known for their anti-inflammatory and pro-resolving actions [40–42], their concentrations in cell membranes could determine the regeneration process after muscle injury.

There is emerging evidence that bioactive lipids derived from n-3 and n-6 PUFAs play a key role in the initiation and resolution of the inflammatory response [12,43]. After a muscle injury, n-3 and n-6 PUFAs are rapidly released from immune cell membrane phospholipids via phospholipase enzymes and metabolized via enzymatic reactions to the lipid mediators oxylipins and endocannabinoids [44]. During the early stages of injury, the classical n-6 PUFA-derived lipid mediators are synthesized and released, promoting acute inflammation by regulating local blood flow, vascular permeability, cytokine production, and leukocyte chemotaxis [12]. Later, a shift in the profile of these mediators results in the generation of oxylipins and endocannabinoids mainly derived from n-3 PUFAs, whose functions are to actively resolve and terminate inflammation, leading to tissue regeneration and return to homeostasis [45,46].

The pathways involved in the formation of oxylipins and endocannabinoids are a complex network of time-specific enzymatic reactions, and to date have not been fully elucidated (Figure 2). These metabolites act as intercellular messengers and mediators of the muscle regeneration process by regulating the inflammatory response to injury [12]. The oxylipins represent the most common and widest family of bioactive lipids synthesized from the long-chain n-6 and n-3 precursors, including LA, AA, ALA, EPA, and DHA [45]. These PUFAs are released upon endogenous and exogenous stimuli from membrane phospholipids via phospholipase A2, although they can be released by phospholipase C, and metabolized into their bioactive products via cyclooxygenase (COX), lipooxygenase (LOX), and cytochrome P450 (CYP450) enzymes [15]. Once formed, oxylipins can mediate their biological effects via interactions with G protein-coupled receptors or intracellular effectors, including peroxisome proliferator-activated receptor gamma (PPARγ) [15]. The oxylipins derived from AA (n-6) are the most common and include the two-series prostaglandins (PGs), thromboxanes (TXs), and the four-series leukotrienes (LTs); meanwhile, EPA (n-3) is the precursor of the three-series PGs and TXs and five-series LTs. The PGs, TXs, and LTs are all proinflammatory in nature; however, those that are EPA-derived are less potent compared with those synthesized from AA [44]. Oxylipins with anti-inflammatory and proresolving activities are mostly derived from EPA and DHA, including the resolvins (Rv) from EPA (E series) and DHA (D series) and the DHA-derived maresins (Ma) and protectins (Pr), collectively termed "specialized pro-resolving mediators" (SPMs). AA-derived lipoxins also contribute to this group [43].

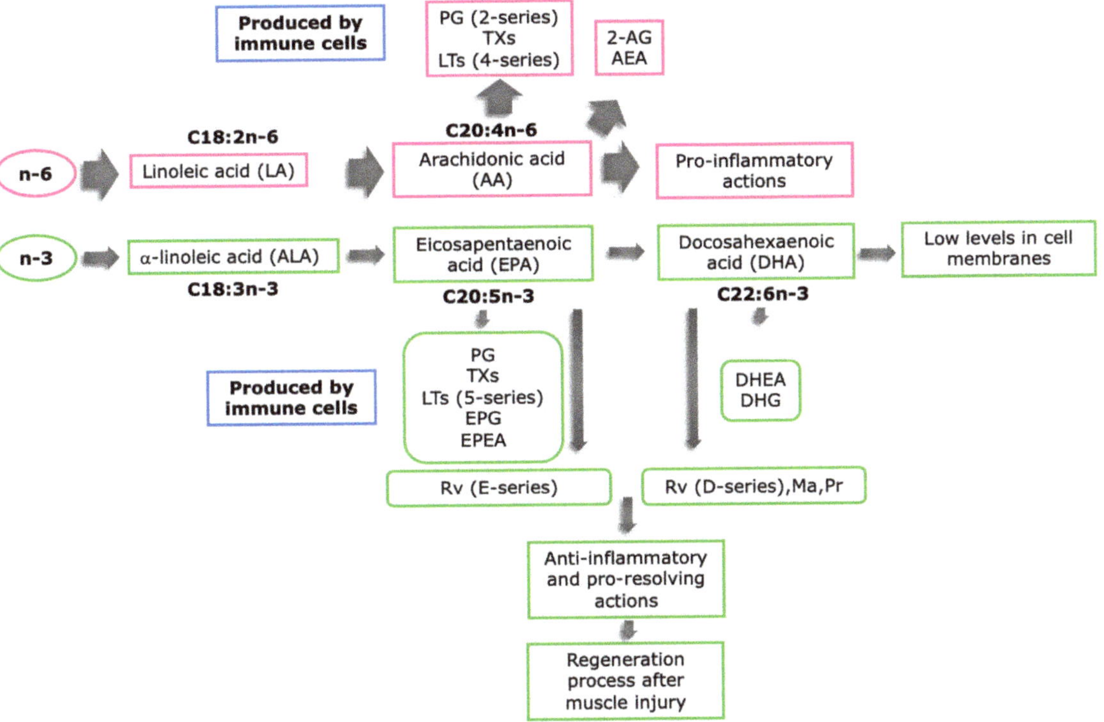

Figure 2. Omega-6 (n-6) and omega-3 (n-3) polyunsaturated fatty acid (PUFA) lipid mediators. The n-6 PUFA, linoleic acid (LA), is elongated and desaturated in the liver to arachidonic acid (AA). AA is processed by immune cells and converted into the two-series prostaglandins (PGs), thromboxanes (Txs), the four-series leukotrienes (LTs), and to the endocannabinoids, 2-acylglycerols (2-AG) and n-arachidonoylethanolamine (AEA). This metabolic pathway is associated with proinflammatory actions. On the other hand, the n-3 PUFA, α-linoleic acid (ALA), is elongated and desaturated in the liver to eicosapentaenoic acid (EPA) and docosahexaenoic acid (DHA). Immune cells convert EPA to the three-series PG, TXs, five-series LTs, to the endocannabinoids, eicosapentanoylglycerol (EPG), cicosapentaenoylethanolamide (EPEA), and to the e-series derived resolvins (Rv). DHA is converted to docosahexaenoyl ethanolamide (DHEA) and docosaheanoyl-glycerol (DHG) and to the D-series Rv, maresins (Ma) and protectins (Pr). The n-3-derived lipid mediators have anti-inflammatory and proresolving actions, which may accelerate the muscle regeneration process.

The family of endocannabinoids consists of the precursors 2-acylglycerols and ethanolamides, of which the most abundant and best characterized are the n-6-derived 2-arachidonoylglycerol (2-AG) and n-arachidonoylethanolamine (AEA, anandamide), respectively [47]. The 2-AG are agonists with low-to-moderate affinity for cannabinoid receptors 1 and 2 (CB1-2), while AEA is a partial agonist with a higher affinity to bind to CB1 relative to CB2. Both CB1 and CB2 are expressed in the central nervous system (CNS) and peripheral tissues [48]. CB1 are predominantly expressed in the CNS and are responsible for mediating neurobehavioral activities such as the regulation of appetite and executive functions [49]. Meanwhile, CB2 are mainly expressed in peripheral tissues, including cells from the immune system, and therefore are in part responsible for mediating the inflammatory response to injury or pathogens [50]. Notably, exogenous anandamide during lactation increases body fat content and CB1 receptor levels in adipose tissue [51].

Recent evidence has shown that the n-3 PUFAs can also be converted into 2-acylglycerols and ethanolamides, including eicosapentanoylglycerol (EPG), eicosapentaenoyl ethanolamide (EPEA), docosahexaenoyl ethanolamide (DHEA), and docosahexanoyl-glycerol (DHG) [52]. Moreover, they can be further metabolized by COX, LOX, and CYP450 enzymes into bioactive endocannabinoid epoxides [53,54], agonists with greater affinity for CB receptors than their former metabolites [46]. As the endocannabinoids derived from n-3 and n-6 PUFAs have been shown to have anti-inflammatory activity, they could play an important role in the muscle regeneration process; however, evidence to date is scarce.

Overall, oxylipins and endocannabinoids are synthesized from membrane-bound n-3 and n-6 PUFAs of immune cells and peripheral tissues. These metabolites share some actions, including the immune cells' inflammatory response to stress or injury, and thus play an important role in the muscle regeneration process [20,21,46]. Indeed, in vitro and rodent studies have recently shown that n-3 PUFAs and their derived metabolites positively modulate the inflammatory response to muscle injury [21,55–59]. Furthermore, daily ingestion of >1 g/d EPA and DHA increases n-3-derived oxylipin [38] and endocannabinoid [60,61] levels in humans. Interestingly, when administered in periods longer than eight weeks, EPA and DHA can increase muscle protein synthesis [62] and significantly reduce symptoms after eccentric exercise-induced damage [63], which may indicate an enhanced muscle regeneration. In turn, it has been described that oxylipin levels depend on the skeletal muscle type; for example, in the soleus, a muscle with predominant composition (~80%) of slow-twitch fibers [64], there is more oxylipin content than in the gastrocnemius muscle, which contains a mixed-fiber composition [65]. The latter suggests that the accumulation of oxylipins could be regulated according to muscle fiber type. This fact could be relevant in the phenotypic muscle fiber shift observed during ageing, that features a distinctive fast-to-slow fiber type transition and subsequent muscle weakness [66]. On this basis, we speculate that the improvements in muscle regeneration observed after n-3 PUFA supplementation could be in part mediated by increased production of n-3 PUFA-derived oxylipins and endocannabinoids, which may reflect a redistribution of muscle fiber composition.

4. Potential Beneficial Effects of n3-PUFA-Derived Metabolites in Muscle Regeneration: Evidence from In Vitro Muscular Cell Lines, Adult Skeletal Fibers, and Animal Models

Following myofiber damage, a rise in the production of proinflammatory n-3 and n-6 PUFA-derived oxylipins and endocannabinoids is necessary for the proper regulation of muscle regeneration [21]. However, in chronic proinflammatory conditions, commonly observed with aging and metabolic diseases, the muscle regeneration process is impaired and linked to fibrosis and infiltration of lipids [19]. Moreover, in aging, there is a deficiency of intramuscular pro-resolving lipid mediator biosynthesis, and RvD1 treatment does not rescue age-related defects in myofiber regeneration [67]. These alterations may be associated with an overproduction of n-6 PUFA-derived metabolites and a decreased synthesis of the bioactive pro-resolving anti-inflammatory lipids oxylipins and endocannabinoids derived from the n-3 PUFAs. In this context, a rise in the concentrations of n-3 PUFA-derived metabolites, through administration of EPA, DHA, and/or n-3 PUFA-derived metabolites, could be a potential strategy to enhance muscle regeneration [20,21,46].

The incorporation of n3-PUFA into cell membranes affects the skeletal muscle proliferation and differentiation processes via changes in membrane fluidity, membrane microdomains involved in cellular signaling, and via the regulation of inflammation through the production of n-3 PUFA-derived metabolites [37,68]. Interestingly, to date, there is little evidence regarding the beneficial role of n3-PUFAs and their derivatives in the myogenic process, mostly because experiments have been performed in uninjured or inflammation-free conditions [69]. In these conditions, both EPA and DHA downregulate the proliferation and differentiation of C2C12 skeletal muscle cells when compared to a fatty-acid free control condition [55,56,70–72]. These effects appear to be dose- and time- dependent with higher doses (>50 µM) and longer incubation times (>48 h), yielding increased inhibition of SC proliferation and differentiation, which has been suggested to be mediated by the accumulation of lipid droplets [70]. To the best of our knowledge, only one study has examined the role of the n-3 PUFA-derived metabolites on myogenesis [21]. Murine C2C12 cells incubated with a supraphysiological dose of RvD1 (1 µM) resulted in increased myotube diameter [21]. Similarly, mature C2C12 myotubes incubated with very high concentrations of EPA (400–600 µM) and DHA (300–700 µM) decreased protein degradation through partial inhibition of the proinflammatory NF-κB pathway [73]. Thus, it appears that very high doses of n-3 PUFAs and/or their derived metabolites promote myofiber hypertrophy by decreasing inflammation and protein degradation.

In vitro treatment with proinflammatory molecules (e.g., palmitate, TNF-α, lipopolysaccharide [LPS]) mimics some of the metabolic abnormalities associated with chronic inflammation, including decreased protein synthesis, increased protein degradation, and muscle atrophy [57,74–76]. Under these conditions, the effects of n-3 PUFA and their derivatives on myogenesis are promising. First, DHA treatment protects against muscle palmitate-induced myofiber atrophy [76–78]. Similarly, the addition of 25 µM of EPA or DHA suppresses the decline in myotube diameter and myofibrillar protein content induced by LPS [79], and cotreatment with 700 µM PA and 50 µM of EPA or DHA blunts the expression of IL-6 and TNF-α induced by PA alone in C2C12 myotubes [80]. Secondly, exposure to EPA-derived oxylipin RvE1 decreased LPS-induced IL-6 and MCP-1 expression in C2C12 myotubes [58]. Moreover, RvE1 attenuated secreted IL-6 protein levels and prevented LPS-induced myotube atrophy. Likewise, the DHA-derived oxylipin RvD1 protected against TNF-α-mediated myotube atrophy [21]. Mechanistically, these effects appear to be mediated in part by the restoration of the Akt/mTOR/FoxO3 pathway involved in the muscle differentiation process [76,77], as well as the inhibition of the proinflammatory transcription factors activation protein-1 and NF-κB [80–82]. Moreover, in dystrophic myoblasts derived from dystrophin-deficient mdx mice exposed to proinflammatory macrophage-conditioned medium, treatment with RvD2 enhanced myoblast differentiation compared to control and prednisone-treated groups [83]. This effect was associated with a 2-fold increase in myogenin-expressing differentiated myoblasts along with a concomitant decrease in the proportion of undifferentiated Pax7+ cells. In addition, RvD2 increased the fusion index and the expression of myosin heavy chain (MyHC), a marker of terminal differentiation, while knockdown of the cannabinoid receptor Gpr18 blocked RvD2 promyogenic effects via downregulation of the Akt-mTOR pathway [83].

Altogether, these results show the promising effects of n-3 PUFAs and their derived metabolites, specifically the oxylipins, to sustain myogenesis during inflammatory conditions, at least in part via activation of the Akt-mTOR pathway and inhibition of proinflammatory signals. Whether n-3 derived endocannabinoids share similar actions in muscle cells remains to be elucidated; however, there is a growing body of evidence showing that these metabolites have greater anti-inflammatory properties compared with EPA, DHA, or the n-6-derived endocannabinoids. Future studies should determine whether n-3 and n-6-derived endocannabinoids positively regulate the myogenic process.

The positive effects of n-3 PUFAs and their derived metabolites on myogenesis and regeneration in muscle cells have also been observed in rodent muscle tissue. For example, Machado et al. [59] observed that 14-day-old mdx mice—a model of Duchenne muscle dystrophy—treated with 300 mg/kg of EPA for 16 days resulted in a decrease in plasma creatine kinase levels and TNF-α muscle protein content concomitantly with a decline in myonecrotic fibers. Further studies from this group corroborated these findings and provided new evidence showing that EPA and/or DHA in mdx mice attenuated the loss of muscle function [84], and increased muscle regenerative capacity by augmenting the levels of MyoD [85]. In parallel, this treatment increased M1-to-M2 macrophage phenotype transition [86,87], decreased inflammation via changes in serum levels of proinflammatory (IFN-γ) and anti-inflammatory (IL-10) cytokines [86], and diminished muscle oxidative stress through downregulation of inducible nitric oxide synthase protein [86] and 4-HNE-protein adducts [84]. Not surprisingly, similar findings have also been observed in other rodent models of muscle damage such as high-fat diet [88] and cardiotoxin-induced gastrocnemius muscle injury [89]. Nevertheless, to our knowledge, no study has identified whether the positive changes in myogenesis observed after n-3 PUFA administration are mediated by their derived metabolites, the oxylipins and endocannabinoids; however, recent evidence has shown a promising role of the oxylipin derived from n-3 PUFA, the RVs, as a potential molecule to enhance muscle regeneration.

Mdx mice treated with 5 µg/kg/d of RvD2 or prednisone, the gold-standard treatment of Duchenne muscle dystrophy, showed reduced neutrophil accumulation and levels of the proinflammatory M1-macrophages (40–50%), and concomitantly increased presence of proregenerative M2- macrophages [83]. Surprisingly, while prednisone administration did not affect the pool of myogenic cells, RvD2 administration induced a ~2-fold increase in the total number of myogenic cells and 2–3-fold increase in the number of Myog+ differentiated myoblasts. Furthermore, assessment of global physical function using the hang test revealed that RvD2 treated mice showed greater global physical function on days 7 and 21 compared with prednisone-treated mdx mice [83]. In a mouse model of barium chloride ($BaCl_2$)-induced muscle injury, an intraperitoneal injection of RvD1 resulted in diminished accumulation of CD68+ macrophages; reduced mRNA expression of the proinflammatory molecules Il-6, IL-1β MCP-1, and TNF-α; expedited clearance of polymorphonuclear cells; and enhanced macrophage phagocytosis [21]. These changes translated in enhanced myofiber regeneration and improved recovery (+15%) of muscle strength. Similarly, cardiotoxin-induced muscle (i.e., tibialis anterior, TA) injury, intramuscular administration of 200 pg RvD2 increased M2 and decreased M1 macrophages after 2 and 3 days of muscle injury [20]. As a result, RvD2 administration improved muscle (i.e., TA) force recovery by 50% and muscle mass by 17% after 8 and 14 days of cardiotoxin injury. Finally, in aged mice, daily intraperitoneal injection of RvD1 after intramuscular injection of $BaCl_2$ blunted inflammatory cytokine expression and accumulation of fibrotic tissue in TA muscle [67]. This strategy improved muscle-specific force recovery. However, myofiber regeneration was not enhanced when assessed by centrally located nuclei and expression of embryonic myosin heavy chain.

Collectively, these results suggest that in different rodent models of muscle damage, n-3 PUFAs and the D-series resolvins may enhance skeletal muscle regeneration via decreased inflammation and reduced oxidative stress. Whether other oxylipins and endocannabinoids derived from the n-3 PUFAs, including the RvE series, MaRs, PrTs, EPG, EPEA, and DHEA, play a role in the muscle regeneration process remains to be elucidated, as well as whether the beneficial effects of n-3 PUFAs on myogenesis are mediated by an increased production of oxylipins and endocannabinoids.

5. Clinical Interventions Supporting the Consumption of n-3 PUFAs to Aid Muscle Recovery

A common approach to investigating the muscle regeneration capacity is by inducing damage through repeated efforts of maximal to near-maximal eccentric lengthening contractions—a force applied to the muscle that exceeds the torque produced by the muscle itself [90]. The high mechanical stress induced by these contractions leads to focal microlesions of the sarcomeres as well as the extracellular matrix and connective tissue of the muscle fibers [91], which manifests itself by a range of clinical symptoms such as delayed-onset muscle soreness (DOMS), muscle stiffness, swelling, decreased proprioceptive function, and loss in maximal force-generating capacity [92]. Moreover, eccentric exercise-induced muscle damage (EIMD) leads to systemic and local inflammatory responses that initially were considered detrimental, albeit similar to the response to pathogens or local injuries. It is now well established that the inflammatory stages are crucial for optimal recovery, as they ensure the removal of tissue debris and promote muscle regeneration via the regulation of different immune cells and activation of SCs. In this context, n-3 PUFAs and their lipid mediators may play a significant role in this process.

The effect of n-3 PUFA supplementation after eccentric damaging exercise has been assessed in some studies. Interestingly, the majority have been performed in healthy young to middle-aged males and have used a wide range of supplementation doses (<1 g up to 6 g/day) and supplementation times (from days to 8 weeks). Despite this, most of the studies have shown that n-3 PUFAs induce slightly faster recovery of muscle function and muscle soreness after EIMD [63,93–101]. For instance, Kyriakidou et al. [102] showed that 4 weeks of n-3 PUFA supplementation successfully attenuated minor aspects of EIMD, although it did not improve performance. Recent systematic reviews and meta-analyses have corroborated the efficacy of n-3 PUFAs in reducing DOMS and markers of muscle damage [103,104]; however, only a few studies have found significantly lower maximal muscle strength loss or recovery (i.e., maximal voluntary contraction; MVC), indicated as the best indirect marker of muscle damage, probably because of a poor control of the participants' diet characteristics affecting the n-6/n-3 ratio and the use of a broad range of supplementation doses and times. Therefore, more studies assessing MVC after damage are warranted.

Systemically, EIMD is paralleled by an inflammatory response involving many mediators, such as interleukin −1 receptor antagonist (IL-1ra), interleukin (IL)-6, IL-10, and acute phase proteins [105,106]. Kyriakidou et al. [102] supplemented healthy young individuals with 3.9 g/d of fish oil containing 3 g of n-3 PUFA (2.145 g of EPA and 0.858 g DHA) per day for a period of 4 weeks ($n = 7$) and induced damaging exercise (downhill running; 60 min at 65% $\dot{V}O2max$ with a −10% gradient), and found a reduced increase in IL-6 and a small protective effect of supplementation with n-3 supplementation in muscle function markers. To the best of our knowledge, no study has yet associated the faster recovery rates after n-3 PUFA supplementation with changes in the levels of oxylipins and endocannabinoids. Future research is warranted to determine the effect of longer time supplementations and larger sample sizes.

The positive effects of n-3 PUFAs have also been observed in patients with Duchenne muscular dystrophy (DMD). These patients lack dystrophin, an important structural skeletal muscle protein. This leads to progressive muscle weakness, chronic muscle degeneration, infiltration of fat, and fibrosis, resulting in the loss of muscle mass and aberrant muscle regeneration, decreasing muscle function, and causing premature death. In a placebo-controlled, double-blinded, randomized study, 28 patients with DMD were supplemented with 2.9 g/d of n-3 PUFAs ($n = 14$) or sunflower oil (placebo, $n = 14$) for 6 months. the results showed that there was a tendency to decrease the loss of lean mass in patients supplemented with n-3 PUFAs compared with the placebo group [107]. In another study by the same group and same set of patients, leukocyte levels of IL-1β and IL-6 were decreased after n-3 PUFA supplementation [108]. In support of these findings, in a 24-week supplementation with a multi-ingredient nutraceutical, including high concentrations of DHA (1.2 g/d) and EPA (0.36 g/d), there was a reduced 6 min walk distance and increased isokinetic knee extension in facioscapulohumeral and limb girdle muscle dystrophy patients when compared to a placebo group [109]. Serum CK levels decreased in all treated groups, with a significant difference in DMD subjects. Conversely, an n3-PUFA-rich diet performed worse than a MUFA-rich diet in MDX mice, suggesting that n3 PUFA may exacerbate stress in dystrophic skeletal muscle [110] potentially by increased fluidity of muscle membranes. However, in general, the evidence provided suggests that EPA and DHA slow the progression of muscle loss and decrease muscle damage potentially via enhanced muscle regeneration through the attenuation of proinflammatory mediators (Figure 3). Whether these changes are mediated by n-3 PUFA oxylipins or endocannabinoids remains to be elucidated.

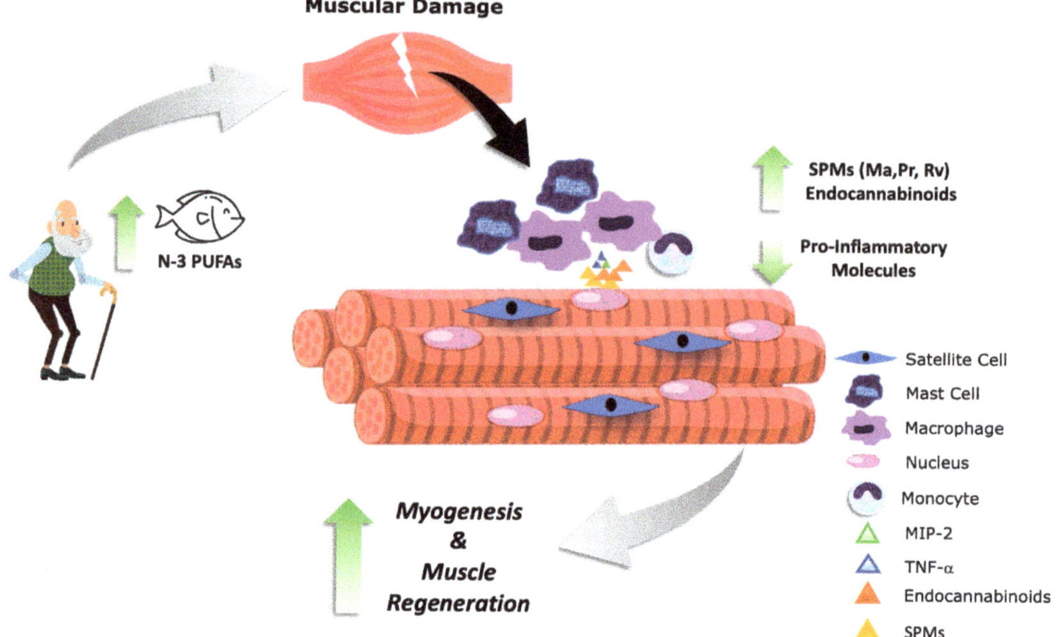

Figure 3. Potential benefits of the omega-3 (n-3) polyunsaturated fatty acid (PUFA) lipid mediators after muscular damage. N-3 PUFA consumption leads to increased synthesis of endocannabinoids and specialized proresolving mediators (SPMs) such as resolvins (Rv), maresins (Ma), and protectins (Pr), leading to a decrease in the production of proinflammatory molecules. These effects may accelerate inflammation resolution, improving myogenesis and muscle regeneration. TNF-α: tumor necrosis factor alpha; MIP-2: macrophage inflammatory protein-2. The Figure was partly generated using Servier Medical Art, provided by Servier, licensed under a Creative Commons Attribution 3.0 unported license.

6. Conclusions

We have summarized the existing data that support the potential role of n-3 PUFAs and their lipid mediators (i.e., oxylipins and endocannabinoids) on skeletal muscle healing and regeneration. There is a gap in knowledge regarding which n-3 PUFAs (EPA, DHA, or a combination of both) and specific lipid mediators are involved in this process. Future human studies should be focused in establishing the relationship between changes in membrane composition, endocannabinoid and oxylipin levels, and the changes in regeneration capacity after muscle-damaging protocols using a long-term n-3 PUFA supplementation period (>3 months) of at least 1 g/d (EPA and/or DHA) in healthy and/or diseased populations.

Author Contributions: Conceptualization, S.J.-V. and R.V.; writing—original draft preparation, S.J.-V., A.E. and R.V.; writing—review and editing, S.J.-V., A.E., A.A.C., M.F.-O., L.P. and R.V.; funding acquisition, S.J.-V. and R.V. All authors have read and agreed to the published version of the manuscript.

Funding: This research was funded by Fondo Nacional de Desarrollo Científico y Tecnológico (FONDECYT), grant number 11220333 (SJ-V), grant number 11190971 (MF-O), grant number 1211962 (LP) and grant number 1221098 (RV).

Institutional Review Board Statement: Not applicable.

Informed Consent Statement: Not applicable.

Data Availability Statement: No new data were created or analyzed in this study. Data sharing is not applicable to this article.

Conflicts of Interest: The authors declare no conflict of interest.

References

1. Janssen, I.; Heymsfield, S.B.; Wang, Z.M.; Ross, R. Skeletal Muscle Mass and Distribution in 468 Men and Women Aged 18–88 Yr. *J. Appl. Physiol.* **2000**, *89*, 81–88. [CrossRef] [PubMed]
2. Järvinen, T.A.H.; Järvinen, T.L.N.; Kääriäinen, M.; Kalimo, H.; Järvinen, M. Muscle Injuries: Biology and Treatment. *Am. J. Sports Med.* **2005**, *33*, 745–764. [CrossRef] [PubMed]
3. Yin, H.; Price, F.; Rudnicki, M.A. Satellite Cells and the Muscle Stem Cell Niche. *Physiol. Rev.* **2013**, *93*, 23–67. [CrossRef] [PubMed]
4. Sinha, I.; Sakthivel, D.; Varon, D.E. Systemic Regulators of Skeletal Muscle Regeneration in Obesity. *Front. Endocrinol.* **2017**, *8*, 29. [CrossRef] [PubMed]
5. Domingues-Faria, C.; Vasson, M.-P.; Goncalves-Mendes, N.; Boirie, Y.; Walrand, S. Skeletal Muscle Regeneration and Impact of Aging and Nutrition. *Ageing Res. Rev.* **2016**, *26*, 22–36. [CrossRef]
6. Almada, A.E.; Wagers, A.J. Molecular Circuitry of Stem Cell Fate in Skeletal Muscle Regeneration, Ageing and Disease. *Nat. Rev. Mol. Cell Biol.* **2016**, *17*, 267–279. [CrossRef]
7. Tedesco, F.S.; Dellavalle, A.; Diaz-Manera, J.; Messina, G.; Cossu, G. Repairing Skeletal Muscle: Regenerative Potential of Skeletal Muscle Stem Cells. *J. Clin. Investig.* **2010**, *120*, 11–19. [CrossRef]
8. Allbrook, D.B.; Han, M.F.; Hellmuth, A.E. Population of Muscle Satellite Cells in Relation to Age and Mitotic Activity. *Pathology* **1971**, *3*, 223–243. [CrossRef]
9. Mauro, A. Satellite Cell of Skeletal Muscle Fibers. *J. Biophys. Biochem. Cytol.* **1961**, *9*, 493–495. [CrossRef]
10. Theret, M.; Rossi, F.M.V.; Contreras, O. Evolving Roles of Muscle-Resident Fibro-Adipogenic Progenitors in Health, Regeneration, Neuromuscular Disorders, and Aging. *Front. Physiol.* **2021**, *12*, 673404. [CrossRef]
11. Wosczyna, M.N.; Rando, T.A. A Muscle Stem Cell Support Group: Coordinated Cellular Responses in Muscle Regeneration. *Dev. Cell* **2018**, *46*, 135–143. [CrossRef]
12. Markworth, J.F.; Maddipati, K.R.; Cameron-Smith, D. Emerging Roles of Pro-Resolving Lipid Mediators in Immunological and Adaptive Responses to Exercise-Induced Muscle Injury. *Exerc. Immunol. Rev.* **2016**, *22*, 110–134.
13. Bentzinger, C.F.; Wang, Y.X.; Dumont, N.A.; Rudnicki, M.A. Cellular Dynamics in the Muscle Satellite Cell Niche. *EMBO Rep.* **2013**, *14*, 1062–1072. [CrossRef]
14. Heredia, J.E.; Mukundan, L.; Chen, F.M.; Mueller, A.A.; Deo, R.C.; Locksley, R.M.; Rando, T.A.; Chawla, A. Type 2 Innate Signals Stimulate Fibro/Adipogenic Progenitors to Facilitate Muscle Regeneration. *Cell* **2013**, *153*, 376–388. [CrossRef]
15. Gabbs, M.; Leng, S.; Devassy, J.G.; Monirujjaman, M.; Aukema, H.M. Advances in Our Understanding of Oxylipins Derived from Dietary PUFAs. *Adv. Nutr.* **2015**, *6*, 513–540. [CrossRef]
16. Serhan, C.N.; Dalli, J.; Colas, R.A.; Winkler, J.W.; Chiang, N. Protectins and Maresins: New pro-Resolving Families of Mediators in Acute Inflammation and Resolution Bioactive Metabolome. *Biochim. Biophys. Acta* **2015**, *1851*, 397–413. [CrossRef]
17. Tidball, J.G.; Villalta, S.A. Regulatory Interactions between Muscle and the Immune System during Muscle Regeneration. *Am. J. Physiol. Regul. Integr. Comp. Physiol.* **2010**, *298*, R1173–R1187. [CrossRef]
18. McKenna, C.F.; Fry, C.S. Altered Satellite Cell Dynamics Accompany Skeletal Muscle Atrophy during Chronic Illness, Disuse, and Aging. *Curr. Opin. Clin. Nutr. Metab. Care* **2017**, *20*, 447–452. [CrossRef]
19. Tidball, J.G. Regulation of Muscle Growth and Regeneration by the Immune System. *Nat. Rev. Immunol.* **2017**, *17*, 165–178. [CrossRef]
20. Giannakis, N.; Sansbury, B.E.; Patsalos, A.; Hays, T.T.; Riley, C.O.; Han, X.; Spite, M.; Nagy, L. Dynamic Changes to Lipid Mediators Support Transitions among Macrophage Subtypes during Muscle Regeneration. *Nat. Immunol.* **2019**, *20*, 626–636. [CrossRef]
21. Markworth, J.F.; Brown, L.A.; Lim, E.; Floyd, C.; Larouche, J.; Castor-Macias, J.A.; Sugg, K.B.; Sarver, D.C.; Macpherson, P.C.; Davis, C.; et al. Resolvin D1 Supports Skeletal Myofiber Regeneration via Actions on Myeloid and Muscle Stem Cells. *JCI Insight* **2020**, *5*, e137713. [CrossRef] [PubMed]
22. Phillips, T.; Childs, A.C.; Dreon, D.M.; Phinney, S.; Leeuwenburgh, C. A Dietary Supplement Attenuates IL-6 and CRP after Eccentric Exercise in Untrained Males. *Med. Sci. Sports Exerc.* **2003**, *35*, 2032–2037. [CrossRef] [PubMed]
23. Wang, Y.; Thorlacius, H. Mast Cell-Derived Tumour Necrosis Factor-Alpha Mediates Macrophage Inflammatory Protein-2-Induced Recruitment of Neutrophils in Mice. *Br. J. Pharmacol.* **2005**, *145*, 1062–1068. [CrossRef] [PubMed]
24. Tecchio, C.; Micheletti, A.; Cassatella, M.A. Neutrophil-Derived Cytokines: Facts beyond Expression. *Front. Immunol.* **2014**, *5*, 508. [CrossRef]
25. Molina, T.; Fabre, P.; Dumont, N.A. Fibro-Adipogenic Progenitors in Skeletal Muscle Homeostasis, Regeneration and Diseases. *Open Biol.* **2021**, *11*, 210110. [CrossRef]
26. Lemos, D.R.; Babaeijandaghi, F.; Low, M.; Chang, C.-K.; Lee, S.T.; Fiore, D.; Zhang, R.-H.; Natarajan, A.; Nedospasov, S.A.; Rossi, F.M. V Nilotinib Reduces Muscle Fibrosis in Chronic Muscle Injury by Promoting TNF-Mediated Apoptosis of Fibro/Adipogenic Progenitors. *Nat. Med.* **2015**, *21*, 786–794. [CrossRef]

27. Cheng, A.J.; Yamada, T.; Rassier, D.E.; Andersson, D.C.; Westerblad, H.; Lanner, J.T. Reactive Oxygen/Nitrogen Species and Contractile Function in Skeletal Muscle during Fatigue and Recovery. *J. Physiol.* **2016**, *594*, 5149–5160. [CrossRef]
28. Powers, S.K.; Morton, A.B.; Ahn, B.; Smuder, A.J. Redox Control of Skeletal Muscle Atrophy. *Free Radic. Biol. Med.* **2016**, *98*, 208–217. [CrossRef]
29. Vasilaki, A.; Jackson, M.J. Role of Reactive Oxygen Species in the Defective Regeneration Seen in Aging Muscle. *Free Radic. Biol. Med.* **2013**, *65*, 317–323. [CrossRef]
30. Calder, P.C. Fatty Acids and Inflammation: The Cutting Edge between Food and Pharma. *Eur. J. Pharmacol.* **2011**, *668*, 50–58. [CrossRef]
31. Tang, J.; Yang, B.; Yan, Y.; Tong, W.; Zhou, R.; Zhang, J.; Mi, J.; Li, D. Palmitoleic Acid Protects against Hypertension by Inhibiting NF-KB-Mediated Inflammation. *Mol. Nutr. Food Res.* **2021**, *65*, e2001025. [CrossRef]
32. Gorzynik-Debicka, M.; Przychodzen, P.; Cappello, F.; Kuban-Jankowska, A.; Marino Gammazza, A.; Knap, N.; Wozniak, M.; Gorska-Ponikowska, M. Potential Health Benefits of Olive Oil and Plant Polyphenols. *Int. J. Mol. Sci.* **2018**, *19*, 686. [CrossRef]
33. González-Mañán, D.; Tapia, G.; Gormaz, J.G.; D'Espessailles, A.; Espinosa, A.; Masson, L.; Varela, P.; Valenzuela, A.; Valenzuela, R. Bioconversion of α-Linolenic Acid to n-3 LCPUFA and Expression of PPAR-Alpha, Acyl Coenzyme A Oxidase 1 and Carnitine Acyl Transferase I Are Incremented after Feeding Rats with α-Linolenic Acid-Rich Oils. *Food Funct.* **2012**, *3*, 765–772. [CrossRef]
34. Rincón-Cervera, M.Á.; Valenzuela, R.; Hernandez-Rodas, M.C.; Barrera, C.; Espinosa, A.; Marambio, M.; Valenzuela, A. Vegetable Oils Rich in Alpha Linolenic Acid Increment Hepatic N-3 LCPUFA, Modulating the Fatty Acid Metabolism and Antioxidant Response in Rats. *Prostaglandins. Leukot. Essent. Fatty Acids* **2016**, *111*, 25–35. [CrossRef]
35. Videla, L.A.; Hernandez-Rodas, M.C.; Metherel, A.H.; Valenzuela, R. Influence of the Nutritional Status and Oxidative Stress in the Desaturation and Elongation of N-3 and n-6 Polyunsaturated Fatty Acids: Impact on Non-Alcoholic Fatty Liver Disease. *Prostaglandins. Leukot. Essent. Fatty Acids* **2022**, *181*, 102441. [CrossRef]
36. Calder, P.C.; Grimble, R.F. Polyunsaturated Fatty Acids, Inflammation and Immunity. *Eur. J. Clin. Nutr.* **2002**, *56*, S14–S19. [CrossRef]
37. Gerling, C.J.; Mukai, K.; Chabowski, A.; Heigenhauser, G.J.F.; Holloway, G.P.; Spriet, L.L.; Jannas-Vela, S. Incorporation of Omega-3 Fatty Acids into Human Skeletal Muscle Sarcolemmal and Mitochondrial Membranes Following 12 Weeks of Fish Oil Supplementation. *Front. Physiol.* **2019**, *10*, 348. [CrossRef]
38. Calder, P.C. Eicosapentaenoic and Docosahexaenoic Acid Derived Specialised Pro-Resolving Mediators: Concentrations in Humans and the Effects of Age, Sex, Disease and Increased Omega-3 Fatty Acid Intake. *Biochimie* **2020**, *178*, 105–123. [CrossRef]
39. Egan, R.W.; Kuehl, F.A. Prostaglandins, Arachidonic Acid, and Inflammation. *Science* **1980**, *210*, 978–984. [CrossRef]
40. Kuda, O. Bioactive Metabolites of Docosahexaenoic Acid. *Biochimie* **2017**, *136*, 12–20. [CrossRef]
41. Serhan, C.N. Novel Omega—3-Derived Local Mediators in Anti-Inflammation and Resolution. *Pharmacol. Ther.* **2005**, *105*, 7–21. [CrossRef] [PubMed]
42. Echeverría, F.; Valenzuela, R.; Espinosa, A.; Bustamante, A.; Álvarez, D.; Gonzalez-Mañan, D.; Ortiz, M.; Soto-Alarcon, S.A.; Videla, L.A. Reduction of High-Fat Diet-Induced Liver Proinflammatory State by Eicosapentaenoic Acid plus Hydroxytyrosol Supplementation: Involvement of Resolvins RvE1/2 and RvD1/2. *J. Nutr. Biochem.* **2019**, *63*, 35–43. [CrossRef] [PubMed]
43. Zúñiga-Hernández, J.; Sambra, V.; Echeverría, F.; Videla, L.A.; Valenzuela, R. N-3 PUFAs and Their Specialized pro-Resolving Lipid Mediators on Airway Inflammatory Response: Beneficial Effects in the Prevention and Treatment of Respiratory Diseases. *Food Funct.* **2022**, *13*, 4260–4272. [CrossRef] [PubMed]
44. Das, U.N. Bioactive Lipids in Age-Related Disorders. *Adv. Exp. Med. Biol.* **2020**, *1260*, 33–83. [CrossRef] [PubMed]
45. Serhan, C.N. Pro-Resolving Lipid Mediators Are Leads for Resolution Physiology. *Nature* **2014**, *510*, 92–101. [CrossRef]
46. McDougle, D.R.; Watson, J.E.; Abdeen, A.A.; Adili, R.; Caputo, M.P.; Krapf, J.E.; Johnson, R.W.; Kilian, K.A.; Holinstat, M.; Das, A. Anti-Inflammatory ω-3 Endocannabinoid Epoxides. *Proc. Natl. Acad. Sci. USA* **2017**, *114*, E6034–E6043. [CrossRef]
47. Simopoulos, A.P. Omega-6 and Omega-3 Fatty Acids: Endocannabinoids, Genetics and Obesity. *OCL* **2020**, *27*, 7. [CrossRef]
48. Mackie, K. Cannabinoid Receptors: Where They Are and What They Do. *J. Neuroendocrinol.* **2008**, *20*, 10–14. [CrossRef]
49. Carnevale, L.N.; Das, A. Novel Anti-Inflammatory and Vasodilatory ω-3 Endocannabinoid Epoxide Regioisomers. *Adv. Exp. Med. Biol.* **2019**, *1161*, 219–232. [CrossRef]
50. Shahbazi, F.; Grandi, V.; Banerjee, A.; Trant, J.F. Cannabinoids and Cannabinoid Receptors: The Story so Far. *iScience* **2020**, *23*, 101301. [CrossRef]
51. Aguirre, C.A.; Castillo, V.A.; Llanos, M.N. The Endocannabinoid Anandamide during Lactation Increases Body Fat Content and CB1 Receptor Levels in Mice Adipose Tissue. *Nutr. Diabetes* **2015**, *5*, e167. [CrossRef] [PubMed]
52. Watson, J.E.; Kim, J.S.; Das, A. Emerging Class of Omega-3 Fatty Acid Endocannabinoids & Their Derivatives. *Prostaglandins Other Lipid Mediat.* **2019**, *143*, 106337. [CrossRef]
53. Zelasko, S.; Arnold, W.R.; Das, A. Endocannabinoid Metabolism by Cytochrome P450 Monooxygenases. *Prostaglandins Other Lipid Mediat.* **2015**, *116–117*, 112–123. [CrossRef] [PubMed]
54. Rouzer, C.A.; Marnett, L.J. Endocannabinoid Oxygenation by Cyclooxygenases, Lipoxygenases, and Cytochromes P450: Cross-Talk between the Eicosanoid and Endocannabinoid Signaling Pathways. *Chem. Rev.* **2011**, *111*, 5899–5921. [CrossRef] [PubMed]
55. Risha, M.A.; Siengdee, P.; Dannenberger, D.; Wimmers, K.; Ponsuksili, S. PUFA Treatment Affects C2C12 Myocyte Differentiation, Myogenesis Related Genes and Energy Metabolism. *Genes* **2021**, *12*, 192. [CrossRef] [PubMed]

56. Hsueh, T.-Y.; Baum, J.I.; Huang, Y. Effect of Eicosapentaenoic Acid and Docosahexaenoic Acid on Myogenesis and Mitochondrial Biosynthesis during Murine Skeletal Muscle Cell Differentiation. *Front. Nutr.* **2018**, *5*, 15. [CrossRef]
57. Haghani, K.; Pashaei, S.; Vakili, S.; Taheripak, G.; Bakhtiyari, S. TNF-α Knockdown Alleviates Palmitate-Induced Insulin Resistance in C2C12 Skeletal Muscle Cells. *Biochem. Biophys. Res. Commun.* **2015**, *460*, 977–982. [CrossRef]
58. Baker, L.A.; Martin, N.R.W.; Kimber, M.C.; Pritchard, G.J.; Lindley, M.R.; Lewis, M.P. Resolvin E1 (R(v) E(1)) Attenuates LPS Induced Inflammation and Subsequent Atrophy in C2C12 Myotubes. *J. Cell. Biochem.* **2018**, *119*, 6094–6103. [CrossRef]
59. Machado, R.V.; Mauricio, A.F.; Taniguti, A.P.T.; Ferretti, R.; Neto, H.S.; Marques, M.J. Eicosapentaenoic Acid Decreases TNF-α and Protects Dystrophic Muscles of Mdx Mice from Degeneration. *J. Neuroimmunol.* **2011**, *232*, 145–150. [CrossRef]
60. Yang, B.; Lin, L.; Bazinet, R.P.; Chien, Y.-C.; Chang, J.P.-C.; Satyanarayanan, S.K.; Su, H.; Su, K.-P. Clinical Efficacy and Biological Regulations of ω-3 PUFA-Derived Endocannabinoids in Major Depressive Disorder. *Psychother. Psychosom.* **2019**, *88*, 215–224. [CrossRef]
61. Ramsden, C.E.; Zamora, D.; Makriyannis, A.; Wood, J.T.; Mann, J.D.; Faurot, K.R.; MacIntosh, B.A.; Majchrzak-Hong, S.F.; Gross, J.R.; Courville, A.B.; et al. Diet-Induced Changes in n-3- and n-6-Derived Endocannabinoids and Reductions in Headache Pain and Psychological Distress. *J. Pain* **2015**, *16*, 707–716. [CrossRef]
62. Smith, G.I.; Atherton, P.; Reeds, D.N.; Mohammed, B.S.; Rankin, D.; Rennie, M.J.; Mittendorfer, B. Omega-3 Polyunsaturated Fatty Acids Augment the Muscle Protein Anabolic Response to Hyperinsulinaemia-Hyperaminoacidaemia in Healthy Young and Middle-Aged Men and Women. *Clin. Sci.* **2011**, *121*, 267–278. [CrossRef]
63. Tsuchiya, Y.; Yanagimoto, K.; Nakazato, K.; Hayamizu, K.; Ochi, E. Eicosapentaenoic and Docosahexaenoic Acids-Rich Fish Oil Supplementation Attenuates Strength Loss and Limited Joint Range of Motion after Eccentric Contractions: A Randomized, Double-Blind, Placebo-Controlled, Parallel-Group Trial. *Eur. J. Appl. Physiol.* **2016**, *116*, 1179–1188. [CrossRef]
64. Gollnick, P.D.; Sjödin, B.; Karlsson, J.; Jansson, E.; Saltin, B. Human Soleus Muscle: A Comparison of Fiber Composition and Enzyme Activities with Other Leg Muscles. *Pflugers Arch.* **1974**, *348*, 247–255. [CrossRef]
65. Penner, A.L.; Waytt, V.; Winter, T.; Leng, S.; Duhamel, T.A.; Aukema, H.M. Oxylipin Profiles and Levels Vary by Skeletal Muscle Type, Dietary Fat and Sex in Young Rats. *Appl. Physiol. Nutr. Metab. Physiol. Appl. Nutr. Metab.* **2021**, *46*, 1378–1388. [CrossRef]
66. Yamazaki, H.; Nishimura, M.; Uehara, M.; Kuribara-Souta, A.; Yamamoto, M.; Yoshikawa, N.; Morohashi, K.-I.; Tanaka, H. Eicosapentaenoic Acid Changes Muscle Transcriptome and Intervenes in Aging-Related Fiber Type Transition in Male Mice. *Am. J. Physiol. Endocrinol. Metab.* **2021**, *320*, E346–E358. [CrossRef]
67. Markworth, J.F.; Brown, L.A.; Lim, E.; Castor-Macias, J.A.; Larouche, J.; Macpherson, P.C.D.; Davis, C.; Aguilar, C.A.; Maddipati, K.R.; Brooks, S.V. Metabolipidomic Profiling Reveals an Age-Related Deficiency of Skeletal Muscle pro-Resolving Mediators That Contributes to Maladaptive Tissue Remodeling. *Aging Cell* **2021**, *20*, e13393. [CrossRef]
68. Jeromson, S.; Mackenzie, I.; Doherty, M.K.; Whitfield, P.D.; Bell, G.; Dick, J.; Shaw, A.; Rao, F.; Ashcroft, S.; Philp, A.; et al. Lipid Remodelling and an Altered Membrane Proteome May Drive the Effects of EPA and DHA Treatment on Skeletal Muscle Glucose Uptake and Protein Accretion. *Am. J. Physiol.-Endocrinol. Metab.* **2018**, *314*, E605–E619. [CrossRef]
69. Isesele, P.O.; Mazurak, V.C. Regulation of Skeletal Muscle Satellite Cell Differentiation by Omega-3 Polyunsaturated Fatty Acids: A Critical Review. *Front. Physiol.* **2021**, *12*, 682091. [CrossRef]
70. Ghnaimawi, S.; Shelby, S.; Baum, J.; Huang, Y. Effects of Eicosapentaenoic Acid and Docosahexaenoic Acid on C2C12 Cell Adipogenesis and Inhibition of Myotube Formation. *Anim. Cells Syst.* **2019**, *23*, 355–364. [CrossRef]
71. Zhang, J.; Xu, X.; Liu, Y.; Zhang, L.; Odle, J.; Lin, X.; Zhu, H.; Wang, X.; Liu, Y. EPA and DHA Inhibit Myogenesis and Downregulate the Expression of Muscle-Related Genes in C2C12 Myoblasts. *Genes* **2019**, *10*, 64. [CrossRef] [PubMed]
72. Lacham-Kaplan, O.; Camera, D.M.; Hawley, J.A. Divergent Regulation of Myotube Formation and Gene Expression by E2 and EPA during In-Vitro Differentiation of C2C12 Myoblasts. *Int. J. Mol. Sci.* **2020**, *21*, 745. [CrossRef] [PubMed]
73. Wang, Y.; Lin, Q.; Zheng, P.; Zhang, J.; Huang, F. DHA Inhibits Protein Degradation More Efficiently than EPA by Regulating the PPARγ/NFκB Pathway in C2C12 Myotubes. *Biomed Res. Int.* **2013**, *2013*, 318981. [CrossRef] [PubMed]
74. Espinosa, A.; Campos, C.; Díaz-Vegas, A.; Galgani, J.E.; Juretic, N.; Osorio-Fuentealba, C.; Bucarey, J.L.; Tapia, G.; Valenzuela, R.; Contreras-Ferrat, A.; et al. Insulin-Dependent H2O2 Production Is Higher in Muscle Fibers of Mice Fed with a High-Fat Diet. *Int. J. Mol. Sci.* **2013**, *14*, 15740–15754. [CrossRef] [PubMed]
75. Roseno, S.L.; Davis, P.R.; Bollinger, L.M.; Powell, J.J.S.; Witczak, C.A.; Brault, J.J. Short-Term, High-Fat Diet Accelerates Disuse Atrophy and Protein Degradation in a Muscle-Specific Manner in Mice. *Nutr. Metab.* **2015**, *12*, 39. [CrossRef]
76. Woodworth-Hobbs, M.E.; Hudson, M.B.; Rahnert, J.A.; Zheng, B.; Franch, H.A.; Price, S.R. Docosahexaenoic Acid Prevents Palmitate-Induced Activation of Proteolytic Systems in C2C12 Myotubes. *J. Nutr. Biochem.* **2014**, *25*, 868–874. [CrossRef]
77. Bryner, R.W.; Woodworth-Hobbs, M.E.; Williamson, D.L.; Alway, S.E. Docosahexaenoic Acid Protects Muscle Cells from Palmitate-Induced Atrophy. *ISRN Obes.* **2012**, *2012*, 647348. [CrossRef]
78. Saini, A.; Sharples, A.P.; Al-Shanti, N.; Stewart, C.E. Omega-3 Fatty Acid EPA Improves Regenerative Capacity of Mouse Skeletal Muscle Cells Exposed to Saturated Fat and Inflammation. *Biogerontology* **2017**, *18*, 109–129. [CrossRef]
79. Yamaguchi, A.; Nishida, Y.; Maeshige, N.; Moriguchi, M.; Uemura, M.; Ma, X.; Miyoshi, M.; Kondo, H.; Fujino, H. Preventive Effect of Docosahexaenoic Acid (DHA) and Eicosapentaenoic Acid (EPA) against Endotoxin-Induced Muscle Atrophy. *Clin. Nutr. ESPEN* **2021**, *45*, 503–506. [CrossRef]

80. Chen, S.-C.; Chen, P.-Y.; Wu, Y.-L.; Chen, C.-W.; Chen, H.-W.; Lii, C.-K.; Sun, H.-L.; Liu, K.-L. Long-Chain Polyunsaturated Fatty Acids Amend Palmitate-Induced Inflammation and Insulin Resistance in Mouse C2C12 Myotubes. *Food Funct.* **2016**, *7*, 270–278. [CrossRef]

81. Wei, H.-K.; Deng, Z.; Jiang, S.-Z.; Song, T.-X.; Zhou, Y.-F.; Peng, J.; Tao, Y.-X. Eicosapentaenoic Acid Abolishes Inhibition of Insulin-Induced MTOR Phosphorylation by LPS via PTP1B Downregulation in Skeletal Muscle. *Mol. Cell. Endocrinol.* **2017**, *439*, 116–125. [CrossRef]

82. Jung, T.W.; Kim, H.-C.; Abd El-Aty, A.M.; Jeong, J.H. Protectin DX Ameliorates Palmitate- or High-Fat Diet-Induced Insulin Resistance and Inflammation through an AMPK-PPARα-Dependent Pathway in Mice. *Sci. Rep.* **2017**, *7*, 1397. [CrossRef]

83. Dort, J.; Orfi, Z.; Fabre, P.; Molina, T.; Conte, T.C.; Greffard, K.; Pellerito, O.; Bilodeau, J.-F.; Dumont, N.A. Resolvin-D2 Targets Myogenic Cells and Improves Muscle Regeneration in Duchenne Muscular Dystrophy. *Nat. Commun.* **2021**, *12*, 6264. [CrossRef]

84. Fogagnolo Mauricio, A.; Minatel, E.; Santo Neto, H.; Marques, M.J. Effects of Fish Oil Containing Eicosapentaenoic Acid and Docosahexaenoic Acid on Dystrophic Mdx Mice. *Clin. Nutr.* **2013**, *32*, 636–642. [CrossRef]

85. Apolinário, L.M.; De Carvalho, S.C.; Santo Neto, H.; Marques, M.J. Long-Term Therapy with Omega-3 Ameliorates Myonecrosis and Benefits Skeletal Muscle Regeneration in Mdx Mice. *Anat. Rec.* **2015**, *298*, 1589–1596. [CrossRef]

86. Carvalho, S.C.D.; Apolinário, L.M.; Matheus, S.M.M.; Santo Neto, H.; Marques, M.J. EPA Protects against Muscle Damage in the Mdx Mouse Model of Duchenne Muscular Dystrophy by Promoting a Shift from the M1 to M2 Macrophage Phenotype. *J. Neuroimmunol.* **2013**, *264*, 41–47. [CrossRef]

87. de Carvalho, S.C.; Hindi, S.M.; Kumar, A.; Marques, M.J. Effects of Omega-3 on Matrix Metalloproteinase-9, Myoblast Transplantation and Satellite Cell Activation in Dystrophin-Deficient Muscle Fibers. *Cell Tissue Res.* **2017**, *369*, 591–602. [CrossRef]

88. Pinel, A.; Rigaudière, J.P.; Jouve, C.; Montaurier, C.; Jousse, C.; LHomme, M.; Morio, B.; Capel, F. Transgenerational Supplementation with Eicosapentaenoic Acid Reduced the Metabolic Consequences on the Whole Body and Skeletal Muscle in Mice Receiving an Obesogenic Diet. *Eur. J. Nutr.* **2021**, *60*, 3143–3157. [CrossRef]

89. Wang, Z.-G.; Zhu, Z.-Q.; He, Z.-Y.; Cheng, P.; Liang, S.; Chen, A.-M.; Yang, Q. Endogenous Conversion of N-6 to n-3 Polyunsaturated Fatty Acids Facilitates the Repair of Cardiotoxin-Induced Skeletal Muscle Injury in Fat-1 Mice. *Aging* **2021**, *13*, 8454–8466. [CrossRef]

90. Chalchat, E.; Gaston, A.-F.; Charlot, K.; Peñailillo, L.; Valdés, O.; Tardo-Dino, P.-E.; Nosaka, K.; Martin, V.; Garcia-Vicencio, S.; Siracusa, J. Appropriateness of Indirect Markers of Muscle Damage Following Lower Limbs Eccentric-Biased Exercises: A Systematic Review with Meta-Analysis. *PLoS ONE* **2022**, *17*, e0271233. [CrossRef]

91. Lieber, R.L.; Fridén, J. Mechanisms of Muscle Injury after Eccentric Contraction. *J. Sci. Med. Sport* **1999**, *2*, 253–265. [CrossRef] [PubMed]

92. Clarkson, P.M. Exercise-Induced Muscle Damage–Animal and Human Models. *Med. Sci. Sports Exerc.* **1992**, *24*, 510–511. [CrossRef] [PubMed]

93. Ochi, E.; Tsuchiya, Y.; Yanagimoto, K. Effect of Eicosapentaenoic Acids-Rich Fish Oil Supplementation on Motor Nerve Function after Eccentric Contractions. *J. Int. Soc. Sports Nutr.* **2017**, *14*, 23. [CrossRef] [PubMed]

94. Lenn, J.; Uhl, T.; Mattacola, C.; Boissonneault, G.; Yates, J.; Ibrahim, W.; Bruckner, G. The Effects of Fish Oil and Isoflavones on Delayed Onset Muscle Soreness. *Med. Sci. Sports Exerc.* **2002**, *34*, 1605–1613. [CrossRef]

95. Gray, P.; Chappell, A.; Jenkinson, A.M.E.; Thies, F.; Gray, S.R. Fish Oil Supplementation Reduces Markers of Oxidative Stress but Not Muscle Soreness after Eccentric Exercise. *Int. J. Sport Nutr. Exerc. Metab.* **2014**, *24*, 206–214. [CrossRef]

96. Rajabi, A.; Lotfi, N.; Abdolmaleki, A.; Rashid-Amiri, S. The Effects of Omega-3 Intake on Delayed Onset Muscle Sorness in Non-Athlet Men. *Pedagog. Psychol. Med. -Biol. Probl. Phys. Train. Sport.* **2013**, *17*, 91–95.

97. Tartibian, B.; Maleki, B.H.; Abbasi, A. The Effects of Ingestion of Omega-3 Fatty Acids on Perceived Pain and External Symptoms of Delayed Onset Muscle Soreness in Untrained Men. *Clin. J. Sport Med. Off. J. Can. Acad. Sport Med.* **2009**, *19*, 115–119. [CrossRef]

98. Tsuchiya, Y.; Yanagimoto, K.; Ueda, H.; Ochi, E. Supplementation of Eicosapentaenoic Acid-Rich Fish Oil Attenuates Muscle Stiffness after Eccentric Contractions of Human Elbow Flexors. *J. Int. Soc. Sports Nutr.* **2019**, *16*, 19. [CrossRef]

99. Philpott, J.D.; Donnelly, C.; Walshe, I.H.; MacKinley, E.E.; Dick, J.; Galloway, S.D.R.; Tipton, K.D.; Witard, O.C. Adding Fish Oil to Whey Protein, Leucine, and Carbohydrate over a Six-Week Supplementation Period Attenuates Muscle Soreness Following Eccentric Exercise in Competitive Soccer Players. *Int. J. Sport Nutr. Exerc. Metab.* **2018**, *28*, 26–36. [CrossRef]

100. McKinley-Barnard, S.K.; Andre, T.L.; Gann, J.J.; Hwang, P.S.; Willoughby, D.S. Effectiveness of Fish Oil Supplementation in Attenuating Exercise-Induced Muscle Damage in Women during Midfollicular and Midluteal Menstrual Phases. *J. Strength Cond. Res.* **2018**, *32*, 1601–1612. [CrossRef]

101. Lembke, P.; Capodice, J.; Hebert, K.; Swenson, T. Influence of Omega-3 (N3) Index on Performance and Wellbeing in Young Adults after Heavy Eccentric Exercise. *J. Sports Sci. Med.* **2014**, *13*, 151–156.

102. Kyriakidou, Y.; Wood, C.; Ferrier, C.; Dolci, A.; Elliott, B. The Effect of Omega-3 Polyunsaturated Fatty Acid Supplementation on Exercise-Induced Muscle Damage. *J. Int. Soc. Sports Nutr.* **2021**, *18*, 9. [CrossRef]

103. Xin, G.; Eshaghi, H. Effect of Omega-3 Fatty Acids Supplementation on Indirect Blood Markers of Exercise-Induced Muscle Damage: Systematic Review and Meta-Analysis of Randomized Controlled Trials. *Food Sci. Nutr.* **2021**, *9*, 6429–6442. [CrossRef]

104. Lv, Z.T.; Zhang, J.M.; Zhu, W.T. Omega-3 Polyunsaturated Fatty Acid Supplementation for Reducing Muscle Soreness after Eccentric Exercise: A Systematic Review and Meta-Analysis of Randomized Controlled Trials. *Biomed Res. Int.* **2020**, *2020*, 8062017. [CrossRef]

105. Bloomer, R.J.; Larson, D.E.; Fisher-Wellman, K.H.; Galpin, A.J.; Schilling, B.K. Effect of Eicosapentaenoic and Docosahexaenoic Acid on Resting and Exercise-Induced Inflammatory and Oxidative Stress Biomarkers: A Randomized, Placebo Controlled, Cross-over Study. *Lipids Health Dis.* **2009**, *8*, 36. [CrossRef]

106. Cornish, S.M.; Johnson, S.T. Systemic Cytokine Response to Three Bouts of Eccentric Exercise. *Results Immunol.* **2014**, *4*, 23–29. [CrossRef]

107. Rodríguez-Cruz, M.; Atilano-Miguel, S.; Barbosa-Cortés, L.; Bernabé-García, M.; Almeida-Becerril, T.; Cárdenas-Conejo, A.; Del Rocío Cruz-Guzmán, O.; Maldonado-Hernández, J. Evidence of Muscle Loss Delay and Improvement of Hyperinsulinemia and Insulin Resistance in Duchenne Muscular Dystrophy Supplemented with Omega-3 Fatty Acids: A Randomized Study. *Clin. Nutr.* **2019**, *38*, 2087–2097. [CrossRef]

108. Rodríguez-Cruz, M.; Cruz-Guzmán, O.D.R.; Almeida-Becerril, T.; Solís-Serna, A.D.; Atilano-Miguel, S.; Sánchez-González, J.R.; Barbosa-Cortés, L.; Ruíz-Cruz, E.D.; Huicochea, J.C.; Cárdenas-Conejo, A.; et al. Potential Therapeutic Impact of Omega-3 Long Chain-Polyunsaturated Fatty Acids on Inflammation Markers in Duchenne Muscular Dystrophy: A Double-Blind, Controlled Randomized Trial. *Clin. Nutr.* **2018**, *37*, 1840–1851. [CrossRef]

109. Sitzia, C.; Meregalli, M.; Belicchi, M.; Farini, A.; Arosio, M.; Bestetti, D.; Villa, C.; Valenti, L.; Brambilla, P.; Torrente, Y. Preliminary Evidences of Safety and Efficacy of Flavonoids- and Omega 3-Based Compound for Muscular Dystrophies Treatment: A Randomized Double-Blind Placebo Controlled Pilot Clinical Trial. *Front. Neurol.* **2019**, *10*, 755. [CrossRef]

110. Henderson, G.C.; Evans, N.P.; Grange, R.W.; Tuazon, M.A. Compared with That of MUFA, a High Dietary Intake of n-3 PUFA Does Not Reduce the Degree of Pathology in Mdx Mice. *Br. J. Nutr.* **2014**, *111*, 1791–1800. [CrossRef]

Systematic Review

Effectiveness of Whey Protein Supplementation during Resistance Exercise Training on Skeletal Muscle Mass and Strength in Older People with Sarcopenia: A Systematic Review and Meta-Analysis

Iván Cuyul-Vásquez [1,2], José Pezo-Navarrete [1], Cristina Vargas-Arriagada [1], Cynthia Ortega-Díaz [1], Walter Sepúlveda-Loyola [3], Sandro Massao Hirabara [4] and Gabriel Nasri Marzuca-Nassr [5,6,*]

[1] Departamento de Procesos Terapéuticos, Facultad de Ciencias de la Salud, Universidad Católica de Temuco, Temuco 4813302, Chile; icuyul@uct.cl (I.C.-V.); jpezo2017@alu.uct.cl (J.P.-N.); cristina.vargas2018@alu.uct.cl (C.V.-A.); cynthia.ortega2018@alu.uct.cl (C.O.-D.)
[2] Facultad de Ciencias de la Salud, Universidad Autónoma de Chile, Temuco 4810101, Chile
[3] Faculty of Health and Social Sciences, Universidad de Las Americas, Santiago 8370040, Chile; wsepulveda@udla.cl
[4] Interdisciplinary Post-graduate Program in Health Sciences, Cruzeiro do Sul University, São Paulo 01506-000, Brazil; sandro.hirabara@cruzeirodosul.edu.br
[5] Departamento de Ciencias de la Rehabilitación, Facultad de Medicina, Universidad de La Frontera, Claro Solar 115, Temuco 4811230, Chile
[6] Interuniversity Center for Healthy Aging, Talca 3460000, Chile
* Correspondence: gabriel.marzuca@ufrontera.cl

Citation: Cuyul-Vásquez, I.; Pezo-Navarrete, J.; Vargas-Arriagada, C.; Ortega-Díaz, C.; Sepúlveda-Loyola, W.; Hirabara, S.M.; Marzuca-Nassr, G.N. Effectiveness of Whey Protein Supplementation during Resistance Exercise Training on Skeletal Muscle Mass and Strength in Older People with Sarcopenia: A Systematic Review and Meta-Analysis. *Nutrients* 2023, 15, 3424. https://doi.org/10.3390/nu15153424

Academic Editors: David C. Nieman and David Rowlands

Received: 27 May 2023
Revised: 11 July 2023
Accepted: 13 July 2023
Published: 2 August 2023

Abstract: Objective: To determine the effectiveness of whey protein (WP) supplementation during resistance exercise training (RET) vs. RET with or without placebo supplementation on skeletal muscle mass, strength, and physical performance in older people with Sarcopenia. Methods: Electronic searches in the PubMed, Embase, Scopus, Web of Science, LILACS, SPORTDiscus, Epistemonikos, and CINAHL databases were performed until 20 January 2023. Randomized clinical trials conducted on sarcopenic adults aged 60 or older were included. The studies had to compare the effectiveness of the addition of supplements based on concentrated, isolated, or hydrolyzed whey protein during RET and compare it with RET with or without placebo supplementation on skeletal muscle mass and strength changes. The study selection process, data extraction, and risk of bias assessment were carried out by two independent reviewers. Results: Seven randomized clinical trials (591 participants) were included, and five of them provided data for quantitative synthesis. The overall pooled standardized mean difference (SMD) estimate showed a small effect size in favor of RET plus WP for skeletal muscle mass according to appendicular muscle index, with statistically significant differences compared with RET with or without the placebo group (SMD = 0.24; 95% CI, 0.05 to 0.42; $p = 0.01$; $I^2 = 0\%$, $p = 0.42$). The overall pooled mean difference (MD) estimate showed a significant difference of +2.31 kg (MD = 2.31 kg; 95% CI, 0.01 to 4.6; $p = 0.05$; $I^2 = 81\%$, $p < 0.001$) in handgrip strength in the RET plus WP group compared with the RET group with or without placebo. The narrative synthesis revealed discordance between the results of the studies on physical performance. Conclusions: WP supplementation during RET is more effective in increasing handgrip strength and skeletal muscle mass in older people with Sarcopenia compared with RET with or without placebo supplementation. However, the effect sizes were small, and the MD did not exceed the minimally important clinical difference. The quality of the evidence was low to very low according, to the GRADE approach. Further research is needed in this field.

Keywords: Sarcopenia; elderly; whey protein; resistance exercise; strength training

1. Background

Sarcopenia is a condition characterized by decreased skeletal muscle mass, muscle strength, and physical performance [1]. Sarcopenia has been positively correlated with elevated rates of disability, hospitalization, falls, fractures, and mortality risk in older adults [2,3]. Compared with older adults without Sarcopenia, the deterioration of physical performance is 3 times higher, and the mortality rate is 3.6 times higher in older adults with Sarcopenia [3]. Individuals with Sarcopenia present an increased risk of hospitalization for different adverse events [4]. For these reasons, preventing and treating Sarcopenia is essential to reducing healthcare costs [4]. In this sense, different interventions have been recommended, including exercise training and nutritional supplementation [5]. Progressive resistance exercise training (RET) of moderate to high intensity is one of the interventions with the highest degree of recommendation for the prevention and treatment of Sarcopenia [6–8].

RET has been shown to increase skeletal muscle mass, muscle strength, and physical performance in healthy older people or those with an increased risk of Sarcopenia [9–12]. Furthermore, RET has been shown to be superior in improving muscle strength of the upper and lower extremities, handgrip strength, depressive symptoms, physical performance, walking speed, and distance compared with other exercise modalities in healthy, sarcopenic, and hospitalized older people [9–13]. In addition, RET is an excellent, cost-effective modality for reducing frailty and the risk of falls in healthy older people [14,15]. Similar results have been observed in older people with Sarcopenia on muscle strength and physical performance; however, the effects of RET on skeletal muscle mass are heterogeneous [16–18]. According to a recent meta-analysis, the effect of RET on skeletal muscle mass is still controversial since RET showed a small effect size (SMD = 0.28) on lower limb skeletal muscle mass but not overall or on upper limb skeletal muscle mass compared with education or maintaining the daily lifestyle in people with Sarcopenia [16].

Another important recommendation to treat and/or prevent Sarcopenia is protein supplementation [5]. As in healthy older people [19,20], it has been suggested that oral Whey Protein (WP) could maximize the effects of exercise and positively influence skeletal muscle anabolism in people with Sarcopenia [21]. In older people, protein supplementation has been shown to increase overall lean mass and handgrip strength only when combined with RET [22]. For example, WP plus leucine intake has been shown to induce postprandial increases in plasma amino acid levels and stimulate muscle protein synthesis in people with Sarcopenia to a greater extent than any other protein source [23]. Regarding the use of WP, a daily dose of 20–40 g combined with RET has been shown to increase biceps strength and lower limb lean mass in post-menopausal women [24].

Although there is plausibility for the potential benefit of the combination of WP with RET in older people with Sarcopenia, the magnitude of the effects on skeletal muscle mass, muscle strength, and physical performance is still unknown. Therefore, the aim of this systematic review was to determine the effectiveness of WP supplementation during RET vs. RET with or without placebo supplementation on skeletal muscle mass, muscle strength, and physical performance in older people with Sarcopenia.

2. Methods

2.1. Protocol and Registration

The report of this research was carried out according to the Preferred Reporting Items for Systematic Reviews and Meta-analysis (PRISMA) and the recommendations of the Cochrane Manual of Systematic Reviews of Interventions [25,26]. The protocol of this systematic review was published in PROSPERO with the registration number CRD42023391714.

2.2. Eligibility Criteria

Studies that met the following inclusion criteria were eligible: (1) population: adults 60 years of age or older, diagnosed with Sarcopenia (low skeletal muscle mass, muscle strength, and/or physical performance, according to the criteria of the international consensus of EWGSOP [1] or AWGSOP [27]), with or without concomitant diseases; (2) intervention: addition of supplements based on concentrated, isolated, or hydrolyzed WP during RET; RET was considered when the program training used machines, elastic bands, or free weights with moderate to high intensity (equal to or greater than 60% of 1 repetition maximum (1RM), for a minimum of 6 weeks); (3) comparison: moderate and high-intensity RET with or without placebo supplementation; (4) primary outcomes: skeletal muscle mass was measured using dual computed tomography scan, nuclear magnetic resonance, dual energy X-ray absorptiometry (DEXA), bioimpedance, or anthropometry; upper- and lower-limb muscle and grip strength were measured using dynamometry, 1RM, or load cell; physical performance was considered a secondary outcome: measurements conducted with the Short Physical Performance Battery (SPPB), Timed Up and Go (TUG), and walking speed were considered physical performance; and (5) types of studies: controlled clinical trials or randomized clinical trials published in English, Spanish, or Portuguese.

Studies were excluded if: (1) They were conducted in people with presarcopenia or dynapenia; (2) They were carried out with mixed samples of people with and without Sarcopenia; or (3) They were published only in conference proceedings.

2.3. Information Sources

Electronic searches were performed in Pubmed, Scopus, Web of Science (WOS), Embase, LILACS, SPORTDiscus, CINAHL, and Epistemonikos from the beginning of each database until January 2023. In addition, manual searches were performed on the references of the articles included in the electronic searches.

2.4. Search Strategies

Two independent reviewers (JP-N and CO-D) performed the electronic search in the databases. The search strategy was composed of the following MESH and free terms: "Sarcopenia", "Sarcopenic", "Sarcopaenia", "Sarcopen", "Strength training", "Strength exercise", "Weightlifting", "Resistance training", "Resistance exercise", "Whey protein", "Protein supplement", "Whey supplement", "Whey intake", "Protein intake", "Controlled Clinical Trial", "Randomized controlled trial", and "Clinical trial". The search strategies for each database can be reviewed in Table S1 (Supplementary Materials).

2.5. Study Selection

The study selection process was carried out through the Rayyan collaborative web application [28]. Duplicates were eliminated before starting the article selection process. Subsequently, two independent reviewers (JP-N and CO-D) reviewed the titles and abstracts of the studies. Studies that did not meet the eligibility criteria were discarded. Potentially eligible studies were reviewed for full text. The agreement rate between reviewers for the study selection process was calculated using the Kappa statistic. Discrepancies between the reviewers' assessments were discussed with a third reviewer (CV-A).

2.6. Data Collection Process

Data extraction was performed independently by two reviewers (JP-N and I-CV). The following information was extracted: Characteristics of the population (sample size, age, health status, and level of physical activity); intervention (type of intervention, supplements, dosage); and results (skeletal muscle mass, muscle strength, walking speed, physical performance, and dynamic balance).

2.7. Risk of Bias

Two independent reviewers (JP-N and WS-L) assessed studies using the Cochrane Risk of Bias 2 (RoB 2) tool [29]. ROB 2 has six domains: bias arising from the randomization process; bias due to deviations from intended interventions; bias due to missing outcome data; bias in measurement of the outcome; bias in selection of the reported result; and overall bias. Each domain could be considered "low risk", "some concerns", or "high risk" [29]. Discrepancies between reviewers' assessments were discussed with a third reviewer (IC-V).

2.8. Statistical Methods

A quantitative synthesis was performed if there were at least three studies with comparable data. There were insufficient data to perform a meta-analysis of upper and lower limb strength (1RM), walking speed, SPPB, and TUG. A narrative synthesis of the effects of the interventions on physical performance was performed according to walking speed, SPPB, and TUG. A quantitative synthesis was performed for the skeletal muscle mass and handgrip strength outcomes. Mean differences (MD) or standardized mean differences (SMD) were calculated for each group. The calculation of the effect sizes considered the use of the raw baseline SD. A pooled estimate of the MD with 95% confidence intervals was calculated for handgrip strength (kg). A pooled estimate of the SMD with 95% confidence intervals was calculated for the appendicular muscle index and handgrip. The weighted sample size method was used to summarize effect sizes from multiple independent studies. Fixed-effects models with Mantel–Haenszel method or random-effects models with the DerSimonian-Laird method were used depending on the degree of heterogeneity. The I^2 statistic was used to assess heterogeneity. For skeletal muscle mass, the effect size was considered trivial (SMD < 0.2), small (SMD 0.2–0.5), medium (SMD 0.6–0.8), or large (SMD > 0.8) [30]. Due to the lack of data for people with Sarcopenia, the threshold of 6.5 kg was considered a minimally important clinical difference for handgrip strength [31]. Subgroups were analyzed according to intervention time. Statistical significance was considered with a p value < 0.05. In the case of missing data, the reviewers contacted the authors by email. Meta-analysis would be performed using RevMan Manager 5.4 (Copenhagen, The Nordic Cochrane Centre, The Cochrane Collaboration).

2.9. Grading of Recommendation, Assessment, Development, and Evaluation

The synthesis and quality of evidence for skeletal muscle mass and muscle strength were assessed using the Grading of Recommendation, Assessment, Development, and Evaluation (GRADE) [32]. GRADE profile allows one to categorize the evidence as high, moderate, low, or very low quality [33]. Results of the GRADE analysis are shown in Table S3.

3. Results

3.1. Study Selection

A total of 1047 articles were found through electronic searches. Before starting the screening, 567 duplicate articles were eliminated, and a total of 480 articles were reviewed by title and abstract. Subsequently, 18 articles were reviewed in full text. The causes for the exclusion of articles can be seen in Table S2 (Supplementary Materials). Finally, seven randomized clinical trials were included [34–40]. The agreement rate between reviewers reached a kappa value of 0.91. Details of the study selection process are shown in Figure 1.

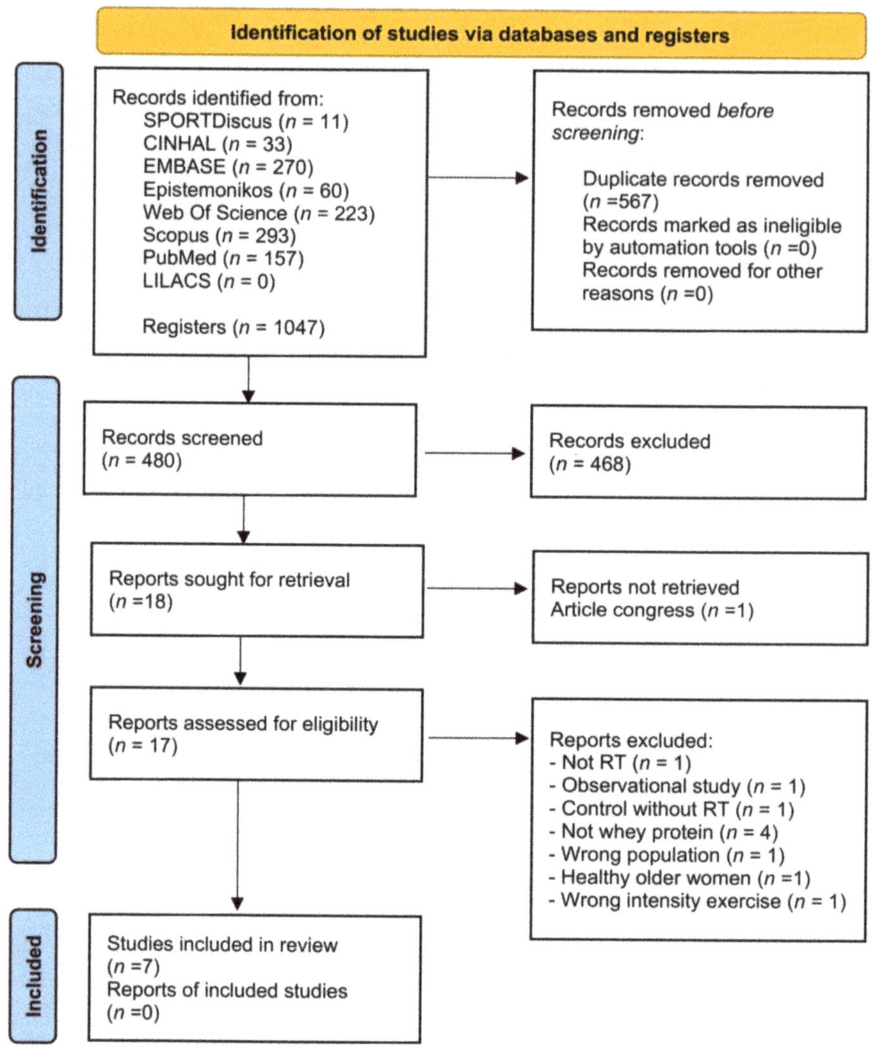

Figure 1. PRISMA flow diagram for the systematic review and meta-analysis.

3.2. Study Characteristics

Table 1 summarizes the characteristics of the studies. The general population consisted of 591 people, 399 (67.5%) women, and 192 men (32.5%). The average age of the population was 77.3 years. All studies included untrained older adults. Four studies reported the inclusion of patients with comorbidities but did not report the details [34,36,37,40]. Three studies reported the inclusion of patients with obesity, hypertension, diabetes mellitus, dyslipidemia, osteoarthritis, chronic obstructive pulmonary disease, stroke, fractures, or a history of surgery [35,38,39]. Six studies were conducted in a hospital [34–38,40], or university outpatient setting [39]. Five studies included outpatients [34,35,38–40] and two included a mixed sample of inpatients and outpatients [36,37].

Table 1. Characteristics of the studies.

Author and Year	Total Sample	Population Characteristics		Sarcopenia Diagnosis	Training Level and Co-morbidities	Time Inter-vention and Context	Intervention Characteristics	Supplement		RET	Outcome Measure	Results	
		Groups					Groups					Control	Exp
		Control	Exp					Control	Exp				
Rondanelli et al., 2016 [34]	$N = 130$ $\geq$65 years	$N = 61$ A $= 80.2 \pm$ 8.5 M = 24 W = 37	$N = 69$ A $= 80.7 \pm$ 6.2 M = 29 W = 40	DEXA < 7.26 kg/m^2 for M and <5.5 kg/m^2 for W.	NT. With comorbidities (NR). Without physical or cognitive impairment.	Twelve weeks. Outpatient in hospital, 5 days/week.	CG = RET + Placebo EG = RET + WP	A total of 32 g isocaloric maltodex-trin once a day at 12:00 pm.	A total of 22 g WP enriched with 2.5 g essential Vitamin D and AA once a day at 12:00 pm.	CG and EG = 20 min per day; Strengthen: chair exercise, toe raises, heel, knee. Knee flexion and extension exercises used weights of 0.50 and 1.50 kg. Resistance bands were used for leg extension, knee flexion, and bicep curls. Balance and Gait exercises, tandem stance, and a tandem walk, each type of exercise 8 times, BORG Intensity 12–14.	SMM: RSMM with DEXA (kg/m^2) MS: Handgrip with hand dy-namome-ter (kg)	SMM (Δ) = −0.06 (0.21, 0.90) MS (Δ) = −0.47 (−1.07, 0, 12)	SMM (Δ) = 0.21 (0.07, 0.35) [#,*] MS (Δ) = 3.20 (2.23, 4.18) [#,*]
Amasene M. et al., 2019 [37]	$N = 28$ >70 years	$N = 13$ A $= 81.7 \pm$ 6.45 M = 6 W = 7	$N = 15$ A $= 82.9 \pm$ 5.59 M = 8 W = 7	EWGSOP	NT. With comorbidities (NR), without physical or cognitive impairment.	Twelve weeks. Inpatient and outpatient in hospital. Two non-consecutive days/week.	CG = RET + Placebo EG = RET + WP	Placebo with mal-todextrin and lemon-flavored hydrox-yethylcel-lulose after each training session.	A total of 20 g WP isolate, enriched with 3 g lemon flavor leucine, after each training session.	CG and EG = 60 min per day. Adapted based on 1RM and then gradually increased the load until reaching 70% of 1RM. Strengthening limbs, 2 set x exercise, load and RM vary by participant; exercises were also practiced to improve dynamic balance.	MS: Handgrip with hand dy-namome-ter (kg/body mass) PP: SPPB total score	MS (post) = 0.3 (0.09) PP (post) = 10.3 (1.89) [#]	MS (post) = 0.4 (0.09) PP (post) = 11.3 (0.96) [#]
Nabuco H. et al., 2019 [39]	$N = 26$ $\geq$60 years	$N = 13$ A = 70.1 $\pm$ 3.9 W = 13	$N = 13$ A = 68.0 $\pm$ 4.2 W = 13	DEXA. Assessed body fat mass 35% combined with ALST less than <15.02 kg.	NT. Obesity, HT, DM or HLP. Without physical or cognitive alteration.	Twelve weeks. Outpatient at university, 3 alternate days/week.	CG = RET + Placebo EG = RET + WP	Placebo, after each training session. Maltodex-trin only on training days.	A total of 35 g hy-drolyzed WP after each training session. Only on training days.	CG and EG = Alternate conventional RET, 3 sets of 8–12 repetitions, loads adjusted individually for each exercise according to their abilities. Exercises: chest press, horizontal leg press, seated row, knee extension, preacher curl (free weights), leg curl, triceps pushdown, and seated calf raise.	MS1: Knee extension (kg). MS2: Chest press (kg) MS3: Preacher curl (kg). PP = 10-m walk test (s)	MS1 (post) = 53.4 $\pm$ 9.4 [#] MS2 (post) = 42.8 $\pm$ 7.1[#] MS3 (post) = 21.5 $\pm$ 2.9 [#] PP (post) = 6.8 $\pm$ 0.6 [#]	MS1 (post) = 51.8 [#] 10.9 [#] MS2 (post) = 44.8 $\pm$ 8.6 [#] MS3 (post) = 23.7 $\pm$ 4.3 [#] PP (post) = 6.9 $\pm$ 0.8 [#]

Table 1. *Cont.*

Author and Year	Total Sample	Population Characteristics		Sarcopenia Diagnosis	Training Level and Co-morbidities	Time Inter-vention and Context	Intervention Characteristics			RET	Outcome Measure	Results	
		Groups					Groups	Supplement					
		Control	Exp					Control	Exp			Control	Exp
Rondanelli et al., 2020 [35]	$N = 127$ ≥65 years	$N = 63$ A $= 81 \pm 5$ M = 17 W = 46	$N = 64$ A $= 80 \pm 7$ M = 26 W = 38	EWGSOP 2010.	NT. With OA, COPD, STROKE, fracture, surgery. without physical or cognitive impairment.	Eight weeks. Outpatient in hospital, 5 days/week.	CG = RET + Placebo EG = RET + WP	Isocaloric formula 40 g flavored powder with maltodextrins. Twice a day, once at breakfast and once in the afternoon.	A total of 20 g WP enriched with 2.8 g leucine, 9 g carbohy-drates, 3 g fat, 800 IU vitamin D, a mixture of vitamins, 500 mg calcium, and fibers. Twice a day, once at breakfast and once in the afternoon.	CG and EG = RET (Borg 12–14); 20 min, increased by intensity exercises up to 30 min. RET, Strengthening (5–10 min) toe raises, heel raises, knee raises, seated knee extensions, standing hip flexions, and lateral leg raises; weight-bearing ankle exercises with weights ranging from 0.5 to 1.5 kg; Resistance band leg extensions and hip flexions; double arm curls, and bicep curls. Balance, walking (5–10 min), single-leg stands, tandem stands, multi-directional weight shifts, tandem walk.	SMM1: AMM with DEXA (g). SMM2: RSMM with DEXA (kg/m^2). MS: Handgrip with dy-namome-ter (kg) PP1: SPPB total score PP2: Walk speed with 4 m test (m/s) PP3: TUG	SMM1 (Δ) $= -69.4$ (-843.7, 704.9) [#] SMM2 (Δ) $= -0.02$ (-0.35, 0.32) MS (Δ) $= -1.47$ (-2.01, -0.92) [#] PP1 (Δ) $= 0.33$ (0.19, 0.46) [#] PP2 (Δ) $= 0.06$ (0.043, 0.08) [#] PP3 (Δ) $= -0.76$ (-1.07, -0.44)	SMM1 (Δ) $= 949.8$ (783.7, 1115.8) [#] [*] SMM2 (Δ) $= 0.38$ (0.31, 0.44) [#] [*] MS (Δ) $= 3.98$ (3.20, 4.75) [#] [*] PP1 (Δ) $= 2.6$ (2.23, 2.98) [#] [*] PP2 (Δ) $= 0.06$ (0.43, 0.08) [#] PP3 (Δ) $= 2.95$ (2.41, 3.49) [#] [*]
Li Z. et al., 2020 [38]	$N = 169$ ≥60 years	CG1 = 51 A = 70 ± 3 M = 22 W = 29 CG2 = 37 A = 73 ± 5 M = 14 W = 23 CG3 = 48 A = 72 ± 6 M = 12 W = 21	$N = 59$ EG $= 33$ A $= 71.52 \pm 5.28$ M = 22 W = 37	AWGS 2014.	NT. With DM, HT or HLP.	Twelve weeks. Outpatient in two hospital centers, 3 alternate days/week.	CG1 = WP CG2 = RET CG3 = usual care EG = RET + WP	Without supple-mentation or placebo.	A total of 10 g WP 3 times/day with food. EPA (300 mg), DHA (200 mg), and vitamin D3 (250 IU) in capsules, with 2 capsules × 2 times a day, 30 min after breakfast and dinner.	CG2 and EG = 30 min + 60 min walk GE = Strengthening (20 min) and slow walking (5 min) 8RM focused on limbs using dumbbells and sandbags. Outdoor activity refers to a one-hour walk with sun exposure 3 days/week on the days opposite resistance training. The speed should be more than 800 steps in 10 min.	SMM1: AMM with M-BIA (kg). SMM2: RSMM with M-BIA (kg/m^2). MS: Handgrip with hand dynamometer (kg)	SMM1 (post) $= 15.00 \pm 3.00$ SMM2 (post) $= 6.09 \pm 0.73$ HS (post) $= 23.62 \pm 5.83$	SMM1(post) $= 16.21 \pm 3.59$ [*] SMM2 (post) $= 6.32 \pm 0.84$ [*] HS (post) $= 24.83 \pm 6.26$ [*]

Table 1. *Cont.*

Author and Year	Population Characteristics						Intervention Characteristics					Results	
	Total Sample	Groups		Sarcopenia Diagnosis	Training Level and Co-morbidities	Time Intervention and Context	Groups	Supplement		RET	Outcome Measure	Control	Exp
		Control	Exp					Control	Exp				
Amasene et al., 2021 [37]	N = 41 ≥70 years	N = 20 A = 81.2 ± 6.14 W = 20	N = 21 A = 82.9 ± 5.67 W = 21	EWGSOP 2018.	NT. With comorbidities, without physical or cognitive impairment.	Twelve weeks. Inpatient and outpatient in hospital. Two non-consecutive.	CG = RET. EG = RET + WP	Placebo with maltodextrin and lemon-flavored hydroxyethylcellulose after each training session.	A total of 20 g isolate WP enriched with 3 g leucine once daily post-exercise.	CG and EG = Supervised training, 60 min per day. Adapted based on 1RM and then gradually increased the load until reaching 70% of 1RM.	SMM1: AMM with DEXA (kg). SMM2: RSMM with DEXA (kg/m²). MS: Handgrip with hand dynamometer (kg) PP: SPPB total score	SMM1 (post) = 18.5 ± 3.6 SMM2 (post) = 7.5 ± 1.16 MS (post) = 24.5 ± 7.32 PP (post) = 10.3 ± 1.89 #	SMM1 (post) = 17.3 ± 2.78 SMM2 (post) = 6.9 ± 0.66 MS (post) = 26.6 ± 6.50 PP (post) = 11.3 ± 0.96 #
Mori et al., 2022 [40]	N = 70 ≥65 years	CG1 = 23 A = 77.6 ± 5.2 M = 4 W = 19 CG2 = 24 A = 77.8 ± 4.5 M = 16 W = 8	EG = 23 A = 77.7 ± 3.3 M = 3 W = 20	AWGS 2014.	NT. With comorbidities (NR), without physical or cognitive impairment.	Twenty-four weeks. Outpatient in hospital, 2 days/week.	CG1 = RET CG2 = WP EG = RET + WP	Without supplementation or placebo.	A total of 11 g of protein, which contains 160 kcal of energy, 2.2 g of fat, 24 g of carbohydrates, and 2300 mg of leucine per serving. It was used after 3 h of lunch.	CG1 and EG = 30–40 min. Elastic resistance band exercises, resistance exercises with body weight load 50–70% of 1RM, 2–3 sets.	SMM: RSMM with M-BIA (kg/m²). MS1: Handgrip with hand dynamometer (kg) MS2: Knee extension with hand-held dynamometer (kg) PP: Usual walking speed (m/s)	SMM (post) = 5.39 ± 0.92 # MS1 (post) = 16.8 ± 3.0 MS2 (post) = 14.6 ± 5.9 # PP (post) = 1.03 ± 0.27	SMM (post) = 5.51 ± 0.66 # MS1 (post) = 17.6 ± 3.4 # MS2 (post) = 14.6 ± 3.0 # PP (post) = 1.03 ± 0.24

Abbreviations: 1RM: one-repetition maximum; A: age; AA: amino acid; ALST: arms and legs soft tissue; AMM: appendicular muscle mass; AWGS: Asian Working Group for Sarcopenia; CG: control group; COPD: chronic obstructive pulmonary disease; DHA: docosahexaenoic acid; DM: diabetes mellitus; DEXA: Dual-Energy X-ray Absorptiometry; EG: experimental group; EPA: Eicosapentaenoic acid; EWGSOP: European Working Group on Sarcopenia in Elderly People; HLP: hyperlipidemia; HT: hypertension; M: men; M-BIA: multi-frequency bioelectrical impedance analysis; MS: muscle strength; N: number of participants; NR: not reported; NT: not trained; OA: osteoarthritis; PP: physical performance; RET: resistance exercise training; RSMM: relative skeletal muscle mass; SMM: skeletal muscle mass; SPPB: short physical performance battery; W: women; WP: whey protein. * Statistically significant differences between groups $p < 0.05$. # Statistically significant differences intragroup $p < 0.05$. Δ difference pre-post intervention.

The duration of the intervention in the studies was 8 weeks [35], 12 weeks [34,36–39], or 24 weeks [40]. The studies used isolated [36,37] or hydrolyzed WP [39], and four studies did not report the type of WP [34,35,38,40]. Supplementation doses were 10–35 g. One study reported only the total protein count but did not detail the grams or percentage of WP [40]. Daily servings of WP ranged from 1–3 times per day. One study added no other ingredients to WP supplementation, while others added leucine [34–37,40], mineral vitamins [34,35,38], and/or essential polyunsaturated fatty acids [38]. Regarding the control groups, four studies added maltodextrin formulas supplementation to RET [34–37,39], and two studies used only RET [38,40]. The WP supplementation was performed in three studies only on training days [36,37,39]. In four studies, WP supplementation was administered every day during or after meals [34,35,38,40].

Regarding RET, four studies performed progressive RET until reaching 70% of 1RM [36,37,39,40]. One study used conventional RET with an intensity of 80% of 1RM [38]. Two studies used a conventional RET with moderate intensity (Borg 12–14) [34,35]. The frequency of RET sessions varied from 2–5 days per week. The duration of the RET sessions ranged from 20 to 60 min. Studies used machine, free weight, body weight, or elastic band training.

Five studies assessed skeletal muscle mass via DEXA [33–36] or multi-frecuency bioelectrical impedance analysis [38,40]. Muscle strength was evaluated in seven studies. The studies used handgrip dynamometry [34–38,40], 1RM [39], and maximum voluntary isometric contraction [40]. Five studies evaluated physical performance through different tests: SPPB [35,36], TUG [35], 10-m walk test [39], walk speed with 4 m test [35], and usual walking speed [40]. Regarding the confounding variables, four studies [34,35,39,40] and one study [40] controlled the diet intake and the physical activity during the execution of the interventions, respectively.

3.3. Risk of Bias Assessment

Figures 2 and 3 show the results of the risk of bias assessment. As far as "overall bias" is concerned, 14.3% of the studies had a "low risk" and 85.7% had a "high risk". Regarding the randomization process, 85.7% showed low risk, and 14.3% had some concerns. In the item deviations from intended interventions, 42.9% of the studies showed a low risk, 28.6% some concerns, and 28.6% a high risk. The studies showed a low risk of bias in 57.1% and a high risk of bias in 42.9% of the items missing outcome data. All studies showed a low risk of bias in the measurement of the outcome. In relation to the selection of the reported result, 14.3% of the studies showed a low risk, 57.1% some concerns, and 28.6% a high risk.

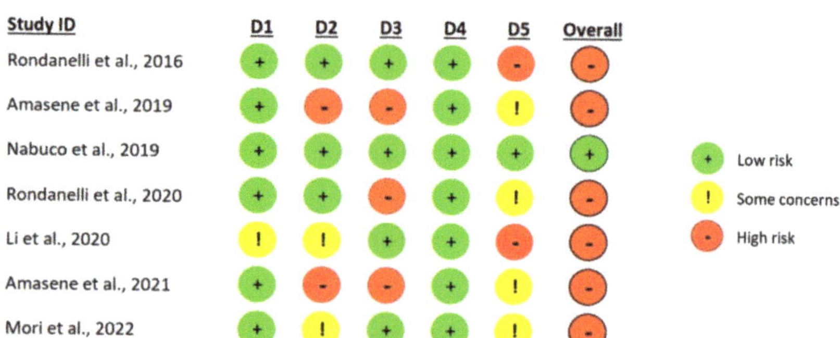

Figure 2. Risk of bias summary [34–40]. D1: Randomization process; D2: Deviations from the intended interventions; D3: Missing outcome data; D4: Measurement of the outcome; D5: Selection of the reported result.

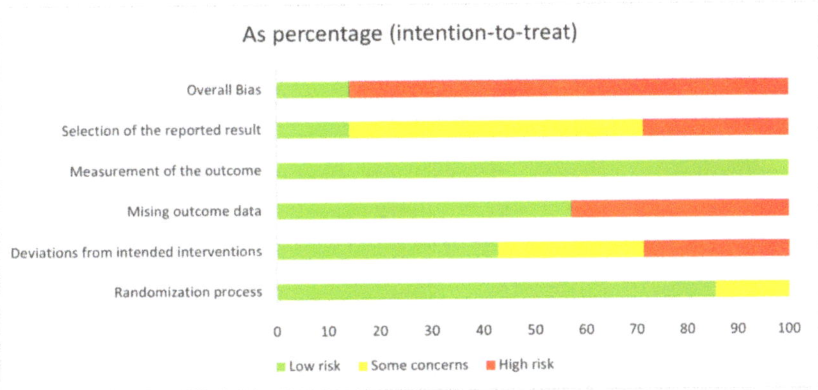

Figure 3. Risk of bias graph.

3.4. Synthesis of Results

3.4.1. Skeletal Muscle Mass: Appendicular Muscle Index

Five studies included data on appendicular muscle index to perform the meta-analysis [34–36,38,40]. The overall pooled SMD estimate showed a small effect size in favor of RET plus WP supplementation with statistically significant differences compared with RET with or without the placebo group at 4–24 weeks (SMD = 0.24; 95% CI, 0.05 to 0.42; p = 0.01). No important heterogeneity was observed (I^2 = 0%, p = 0.42) (Figure 4). There was a low quality of evidence, according to the GRADE rating. It was observed that weekly SMD varied between -0.01 and 0.13 in favor of the RET plus WP group (Table S4, Supplementary Materials). Four studies [34–36,38] with interventions lasting 4–12 showed a small effect size in favor of RET plus WP; however, there were no statistically significant differences compared with RET with or without a placebo group (SMD = 0.26; 95% CI, 0.05 to 0.47; p = 0.02), with no important heterogeneity (I^2 = 11%, p = 0.34) (Figure 5). There was a low quality of evidence, according to the GRADE rating.

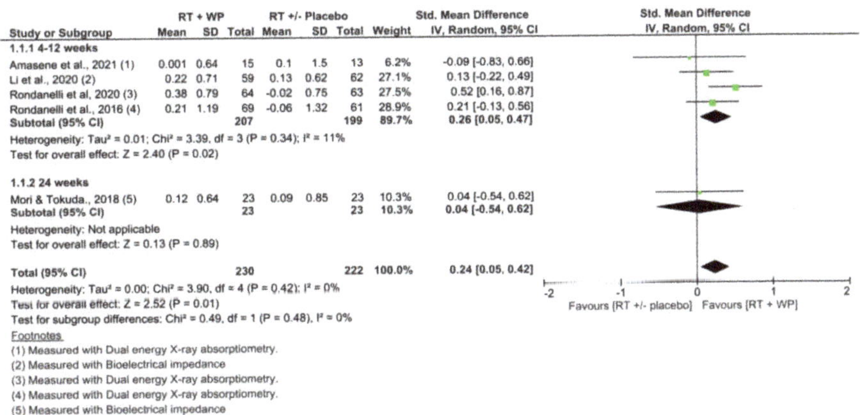

Figure 4. Comparison of RET plus WP vs. RET with or without placebo for appendicular muscle index, at 4 to 24 weeks and 4 to 12 weeks [34–36,38,40].

3.4.2. Skeletal Muscle Mass: Appendicular Muscle Mass

Three studies [34,36,38] that included data on appendicular muscle mass to perform the meta-analysis showed a small effect size in favor of RET plus WP, however, without statistically significant differences compared with RET with or without the placebo group at 4–12 weeks (SMD = 0.15; 95% CI, −0.08 to 0.39; p = 0.21). Additionally, no important heterogeneity was observed (I^2 = 0%, p = 0.63) (Figure 5). There was a low quality of evidence, according to the GRADE rating.

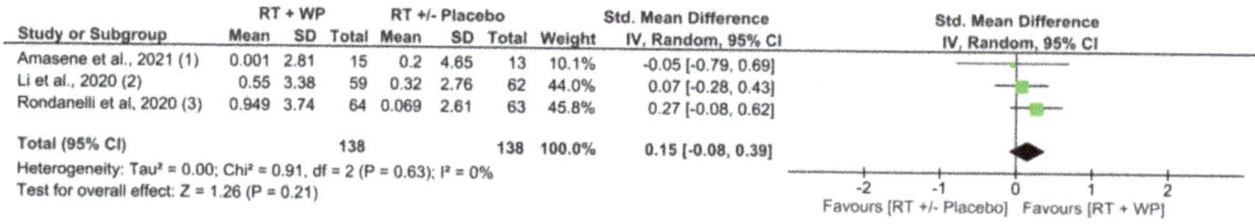

Footnotes
(1) Measured with Dual energy X-ray absorptiometry.
(2) Measured with Bioelectrical impedance.
(3) Measured with Dual energy X-ray absorptiometry.

Figure 5. Comparison of RET plus WP vs. RET with or without placebo for appendicular muscle mass at 4 to 12 weeks [34,36,38].

3.4.3. Muscle Strength

Five studies included data on handgrip strength to perform the meta-analysis [34–36,38,40]. The overall pooled MD estimate showed a difference of +2.31 kg in handgrip strength in the RET plus WP supplementation compared with the RET group with or without placebo supplementation at 4–24 weeks, with statistically significant differences (MD = 2.31 kg; 95% CI, 0.01 to 4.6; p = 0.05) and considerable heterogeneity (I^2 = 81%, p < 0.001) (Figure 6). There was a very low quality of evidence, according to the GRADE rating. Four studies [34–36,38] with interventions lasting 4–12 weeks showed a difference of +2.71 kg in handgrip strength in the RET plus WP group compared with the RET group with or without placebo supplementation at 4–24 weeks, with statistically significant differences (MD = 2.71 kg; 95% CI, 0.06 to 5.36; p = 0.05) and considerable heterogeneity (I^2 = 80%, p = 0.002) (Figure 6). There was a very low quality of evidence, according to the GRADE rating.

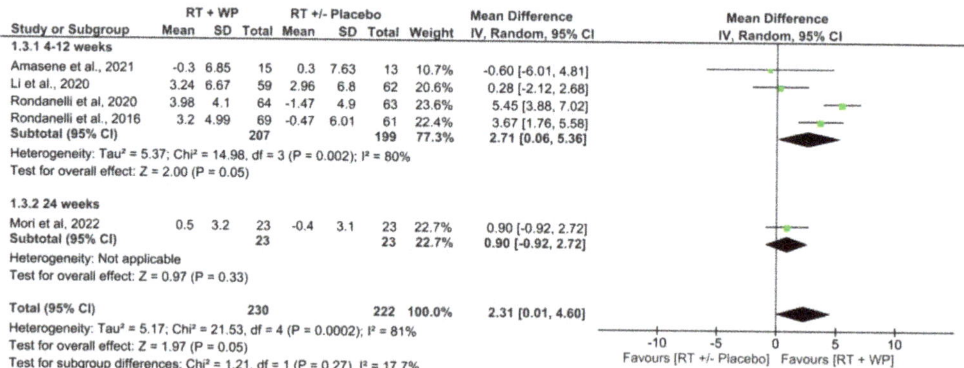

Figure 6. Comparison of RET plus WP vs. RET with or without placebo for handgrip strength at 4 to 24 weeks and 4 to 12 weeks [34–36,38,40].

3.4.4. Physical Performance

Two studies evaluated walking speed at four weeks [35] and 24 weeks [40]. Mori et al. (2022) found no difference between groups [40]. However, Rondanelli et al. (2020) showed statistically significant improvements in walking speed in favor of the RET plus WP supplementation [35]. Three studies evaluated physical performance according to SPPB at four [35] and 12 weeks [36,37]. Amasene et al. (2021) observed no difference between groups [36]. However, Rondanelli et al. (2020) showed statistically significant improvements in physical performance in favor of the RET plus WP group [35]. Only one study evaluated dynamic balance through the TUG test at four weeks [35]. The results indicated that the RET plus WP group was statistically more effective in improving dynamic balance [35].

4. Discussion

The aim of this systematic review was to determine the effectiveness of WP supplementation during RET vs. RET with or without placebo supplementation on skeletal muscle mass, muscle strength, and physical performance in older people with Sarcopenia. Our results indicate that WP supplementation associated with RET is effective in increasing skeletal muscle mass according to the appendicular muscle index, and handgrip strength. However, we did not observe differences in appendicular muscle mass between RET plus WP supplementation and RET with or without placebo supplementation. In addition, the increase in handgrip strength did not exceed the minimally important clinical difference [31]. There were insufficient data to perform a meta-analysis on physical performance. The results of the studies were discordant in relation to physical performance.

The effectiveness of adding WP supplementation during RET has been well studied in other older populations [19,20,22,24]. However, systematic reviews focused on older people with Sarcopenia are scarce. In this sense, our results differ partially from the findings of the systematic review by Chang and Choo (2023), whose aim was to evaluate the effectiveness of WP, leucine, and vitamin D supplementation in patients with Sarcopenia [41]. Based on two studies, the meta-analysis by Chang and Choo (2023) showed statistically significant differences in favor of the WP, leucine, and vitamin D group, with moderate effect sizes for skeletal muscle mass and large effect sizes for handgrip strength. The findings of Chang and Choo (2023) are likely to have overestimated effect sizes. Even though our findings are supported by a larger number of studies, the certainty of the evidence demonstrates the need to improve the methodological quality of future studies to obtain more accurate conclusions.

The small increases in skeletal muscle mass and handgrip strength in older people with Sarcopenia could be due to several factors. There are reports demonstrating that older people performing RET (with and without Sarcopenia) present a higher percentage increase in muscle strength than in skeletal muscle mass [42,43]. Additionally, these results could be explained by the characteristics of the sample, the dosage of the interventions, and the scant control of confounding variables, including physical activity and diet intake. Regarding the characteristics of the sample, 69.25% ($n = 265$) of the participants included in our meta-analysis were women. Furthermore, 2/7 of the studies included in this systematic review were conducted only on women [34,37]. This is relevant since it has been observed that older women with Sarcopenia tend to show a lower increase in skeletal muscle mass with RET [44]. On the other hand, the blunted stimulation of muscle protein synthesis rates is known as anabolic resistance, and it has been suggested as a theoretical framework to support interventions in people with Sarcopenia [45]. Aging anabolic resistance is influenced by factors such as digestion, absorption, anabolic signaling proteins, muscle perfusion, splanchnic amino acid sequestration, physical activity levels, and postprandial amino acid availability and delivery [45]. Because of these factors, it has been suggested that a greater amount of WP supplementation (~40 g) [46] together with a RET program could be more effective in increasing skeletal muscle mass. However, the dose of WP ranged from 10 to 22 g in the studies included in the meta-analysis [34–36,38,40].

We know that to build skeletal muscle mass, the amount and quality of food consumed (diet) are important. It has been reported that older people with Sarcopenia have alterations in their usual diet intake [47–50]. However, only three studies controlled diet intake during the intervention period [34,35,40]. In addition, it is well known that levels of physical activity can decrease anabolic resistance, thus increasing the effectiveness of nutritional and exercise interventions [45]. However, only the study by Mori et al. (2022) monitored the levels of physical activity and diet intake before and after the intervention. They found that levels of physical activity and diet intake did not change after the intervention in either group [40]. Their results indicated that there were no differences between RET and RET plus WP supplementation in improving handgrip strength, knee extension strength, and relative skeletal muscle mass [40].

RET programming, to obtain gains in skeletal muscle mass and muscle strength, must be individualized, progressive, and of moderate-high intensity (70–80% of 1RM). However, several studies included in this systematic review performed moderate RET or adapted strength training by performing it with elastic bands or sandbags [35,38,40]. Additionally, training frequency could also influence the effects of RET on skeletal muscle mass. For example, Rondanelli et al. (2016; 2020) conducted training sessions five times a week and found statistically significant differences in appendicular muscle mass and relative skeletal muscle mass in favor of RET plus WP supplementaion. In contrast, studies that had 2–3 weekly sessions found no differences between the groups after the intervention [35,36,38]. Finally, current evidence shows us that evaluating skeletal muscle mass through a Computed Tomography Scan or Magnetic Resonance Imaging (especially changes in the thigh of older people) could be a better way to see changes [51].

Insufficient and controversial scientific evidence exists on the effect of WP supplementation combined with RET on physical performance [35,36,40]. Rondanelli et al. (2020) showed that an intervention with RET plus WP supplementation induces positive effects on walking speed, physical performance using SPPB, and dynamic balance using TUG in older adults with Sarcopenia. This intervention also reduces hospitalization time, which is related to decreased healthcare expenditures and different adverse events associated with hospitalization time [4]. However, the other two studies did not report significant differences in favor of WP supplementation plus RET compared with placebo supplementation plus RET [36,40]. Mori et al. (2022) reported no benefit in both interventions in improving walking speed [40], and Amasene et al. (2021) [36] reported that the physical performance measured using SPPB significantly improved in both intervention groups regardless of protein-enriched supplementation. Nonetheless, in that study [36], the intervention group demonstrated a decrease in the prevalence of frailty, as evidenced by five older adults supplemented with protein who were classified as frail at baseline but were no longer so after the intervention [36]. The results reported by Amasene et al. (2021) [36] emphasize the effectiveness of RET programs alone in improving the physical performance of older adults, as also reported in a recent systematic review [52]. Therefore, the addition of WP supplementation during RET may not be necessary to achieve significant improvements in physical performance among this population, and additional studies are required in this research field to fully understand of the role of WP supplementation on physical performance during RET.

Finally, the role of WP supplementation in the treatment of Sarcopenia has been previously recommended by international consensus [1,27]. Prior research has suggested that physical exercise is beneficial for individuals with Sarcopenia; however, it may not be sufficient alone to achieve significant clinical outcomes. As a result, a combined exercise program along with supplementation of WP, essential amino acids, and vitamin D has shown significant effects when compared with exercise alone [34]. That study highlights the potential benefits of combined interventions to treat Sarcopenia and improve clinical outcomes. Although physical performance is an important clinical measurement to diagnose Sarcopenia and is recommended by international consensus, it was measured only in two studies [34,35]. Therefore, more studies to analyze the effect on all clinical measurements of Sarcopenia (skeletal muscle mass, muscle strength, and physical performance) are necessary.

Limitations

Our results should be considered with caution due to the following limitations: First, despite searching eight databases, articles in other languages than English, Spanish, or Portuguese could have been excluded. Second, the studies have shown high clinical heterogeneity in terms of WP supplementation and RET dosing. Third, due to the small number of articles included in the quantitative synthesis, the accuracy of the meta-analytic tests used may be affected. Fourth, although it is a consequence of the high clinical heterogeneity of the studies, the quantitative syntheses showed considerable statistical heterogeneity for handgrip strength. Fifth, it was not possible to perform a moderator analysis and assess publication biases due to the limited number of articles included in the meta-analyses.

5. Conclusions

RET plus WP supplementation is more effective in increasing skeletal muscle mass and handgrip strength in older people with Sarcopenia compared with RET with or without placebo supplementation. However, the effect sizes were small for skeletal muscle mass, and the handgrip strength did not exceed the minimally important clinical difference. The quality of the evidence was low to very low, according to the GRADE approach. Further studies are needed in this research field. Future research should report in detail the dosage and periodization of RET program, as well as the proportions of other ingredients that are incorporated into WP supplementation. Higher doses of WP supplementation and a higher frequency of intake could probably improve the results obtained to date. Changes in physical activity and diet intake are confounding variables that future studies should control for. Strategies to avoid participant dropout, blinding, and reporting according to clinical trial registration protocols could reduce the risk of bias.

Supplementary Materials: The following supporting information can be downloaded at: https://www.mdpi.com/article/10.3390/nu15153424/s1, Table S1: Search strategies; Table S2: Excluded studies; Table S3: Grading of Recommendation, Assessment, Development, and Evaluation (GRADE); Table S4: Weekly improvements between RET plus WP and RET with or without placebo.

Author Contributions: Conceptualization, I.C.-V.; Methodology, I.C.-V., J.P.-N., C.V.-A. and C.O.-D.; Software, I.C.-V. and J.P.-N.; Validation, I.C.-V., J.P.-N. and G.N.M.-N.; Formal Analysis, I.C.-V., J.P.-N. and W.S.-L.; Investigation, I.C.-V., J.P.-N., C.V.-A. and C.O.-D.; Resources, G.N.M.-N. and S.M.H.; Data Curation, I.C.-V., C.V.-A. and C.O.-D.; Writing—Original Draft Preparation, I.C.-V., G.N.M.-N. and W.S.-L.; Writing—Review and Editing, I.C.-V., G.N.M.-N., S.M.H. and W.S.-L.; Visualization, I.C.-V. and S.M.H.; Supervision, G.N.M.-N. and S.M.H.; Project Administration, I.C.-V. and G.N.M.-N.; Funding Acquisition, G.N.M.-N. All authors have read and agreed to the published version of the manuscript.

Funding: This work has been funded in part by the Interuniversity Center for Healthy Aging, Code RED211993 and Dirección de Investigación (DIUFRO) at Universidad de La Frontera.

Institutional Review Board Statement: Not applicable.

Informed Consent Statement: Not applicable.

Data Availability Statement: Data are contained within the article.

Conflicts of Interest: The authors declare no conflict of interest.

References

1. Cruz-Jentoft, A.J.; Bahat, G.; Bauer, J.; Boirie, Y.; Bruyère, O.; Cederholm, T.; Cooper, C.; Landi, F.; Rolland, Y.; Sayer, A.A.; et al. Sarcopenia: Revised European consensus on definition and diagnosis. *Age Ageing* **2019**, *48*, 16–31. [CrossRef] [PubMed]
2. Anker, S.D.; Morley, J.E.; von Haehling, S. Welcome to the ICD-10 code for sarcopenia. *J. Cachexia Sarcopenia Muscle* **2016**, *7*, 512–514. [CrossRef] [PubMed]
3. Beaudart, C.; Zaaria, M.; Pasleau, F.; Reginster, J.Y.; Bruyère, O. Health Outcomes of Sarcopenia: A Systematic Review and Meta-Analysis. *PLoS ONE* **2017**, *12*, e0169548. [CrossRef] [PubMed]
4. Álvarez-Bustos, A.; Rodríguez-Sánchez, B.; Carnicero-Carreño, J.A.; Sepúlveda-Loyola, W.; Garcia-Garcia, F.J.; Rodríguez-Mañas, L. Healthcare cost expenditures associated to frailty and sarcopenia. *BMC Geriatr.* **2022**, *22*, 747. [CrossRef] [PubMed]
5. Dent, E.; Morley, J.E.; Cruz-Jentoft, A.J.; Arai, H.; Kritchevsky, S.B.; Guralnik, J.; Bauer, J.M.; Pahor, M.; Clark, B.C.; Cesari, M.; et al. International Clinical Practice Guidelines for Sarcopenia (ICFSR): Screening, Diagnosis and Management. *J. Nutr. Health Aging* **2018**, *22*, 1148–1161. [CrossRef]
6. Hurst, C.; Robinson, S.M.; Witham, M.D.; Dodds, R.M.; Granic, A.; Buckland, C.; De Biase, S.; Finnegan, S.; Rochester, L.; Skelton, D.A.; et al. Resistance exercise as a treatment for sarcopenia: Prescription and delivery. *Age Ageing* **2022**, *51*, afac003. [CrossRef]

7.   Churchward-Venne, T.A.; Tieland, M.; Verdijk, L.B.; Leenders, M.; Dirks, M.L.; van Loon Luc, J.C. There Are No Nonresponders to Resistance-Type Exercise Training in Older Men and Women. *J. Am. Med. Dir. Assoc.* **2015**, *16*, 400–411. [CrossRef]

8.   Cannataro, R.; Cione, E.; Bonilla, D.A.; Cerullo, G.; Angelini, F.; D'Antona, G. Strength training in elderly: An useful tool against sarcopenia. *Front. Sports Act. Living* **2022**, *4*, 950949. [CrossRef]

9.   Prevett, C.; Moncion, K.; Phillips, S.M.; Richardson, J.; Tang, A. Role of Resistance Training in Mitigating Risk for Mobility Disability in Community-Dwelling Older Adults: A Systematic Review and Meta-analysis. *Arch. Phys. Med. Rehabil.* **2022**, *103*, 2023–2035. [CrossRef]

10. Lu, L.; Mao, L.; Feng, Y.; Ainsworth, B.E.; Liu, Y.; Chen, N. Effects of different exercise training modes on muscle strength and physical performance in older people with sarcopenia: A systematic review and meta-analysis. *BMC Geriatr.* **2021**, *21*, 708. [CrossRef]

11. Carneiro, M.A.S.; Franco, C.M.C.; Silva, A.L.; Castro-E-Souza, P.; Kunevaliki, G.; Izquierdo, M.; Cyrino, E.S.; Padilha, C.S. Resistance exercise intervention on muscular strength and power, and functional capacity in acute hospitalized older adults: A systematic review and meta-analysis of 2498 patients in 7 randomized clinical trials. *Geroscience* **2021**, *43*, 2693–2705. [CrossRef]

12. Marzuca-Nassr, G.N.; Alegría-Molina, A.; SanMartín-Calísto, Y.; Artigas-Arias, M.; Huard, N.; Sapunar, J.; Salazar, L.A.; Verdijk, L.B.; van Loon, L.J.C. Muscle Mass and Strength Gains Following Resistance Exercise Training in Older Adults 65–75 Years and above 85 Years. *Int. J. Sport Nutr. Exerc. Metab.* **2023**. *ahead of print*.

13. Cuyul-Vásquez, I.; Berríos-Contreras, L.; Soto-Fuentes, S.; Hunter-Echeverría, K.; Marzuca-Nassr, G.N. Effects of resistance exercise training on redox homeostasis in older adults. A systematic review and meta-analysis. *Exp. Gerontol.* **2020**, *138*, 111012. [CrossRef]

14. Sun, X.; Liu, W.; Gao, Y.; Qin, L.; Feng, H.; Tan, H.; Chen, Q.; Peng, L.; Wu, I.X.Y. Comparative effectiveness of non-pharmacological interventions for frailty: A systematic review and network meta-analysis. *Age Ageing* **2023**, *52*, afad004. [CrossRef]

15. Adjetey, C.; Karnon, B.; Falck, R.S.; Balasubramaniam, H.; Buschert, K.; Davis, J.C. Cost-effectiveness of exercise versus multimodal interventions that include exercise to prevent falls among community-dwelling older adults: A systematic review and meta-analysis. *Maturitas* **2023**, *169*, 16–31. [CrossRef]

16. Wang, H.; Huang, W.Y.; Zhao, Y. Efficacy of Exercise on Muscle Function and Physical Performance in Older Adults with Sarcopenia: An Updated Systematic Review and Meta-Analysis. *Int. J. Environ. Res. Public Health* **2022**, *19*, 8212. [CrossRef]

17. Zhao, H.; Cheng, R.; Song, G.; Teng, J.; Shen, S.; Fu, X.; Yan, Y.; Liu, C. The Effect of Resistance Training on the Rehabilitation of Elderly Patients with Sarcopenia: A Meta-Analysis. *Int. J. Environ. Res. Public Health* **2022**, *19*, 15491. [CrossRef]

18. Mende, E.; Moeinnia, N.; Schaller, N.; Weiß, M.; Haller, B.; Halle, M.; Siegrist, M. Progressive machine-based resistance training for prevention and treatment of *sarcopenia* in the oldest old: A systematic review and meta-analysis. *Exp. Gerontol.* **2022**, *163*, 111767. [CrossRef]

19. Cermak, N.M.; Res, P.T.; de Groot, L.C.; Saris, W.H.; van Loon, L.J. Protein supplementation augments the adaptive response of skeletal muscle to resistance-type exercise training: A meta-analysis. *Am. J. Clin. Nutr.* **2012**, *96*, 1454–1464. [CrossRef]

20. Morton, R.W.; Murphy, K.T.; McKellar, S.R.; Schoenfeld, B.J.; Henselmans, M.; Helms, E.; Aragon, A.A.; Devries, M.C.; Banfield, L.; Krieger, J.W.; et al. A systematic review, meta-analysis and meta-regression of the effect of protein supplementation on resistance training-induced gains in muscle mass and strength in healthy adults. *Br. J. Sports Med.* **2018**, *52*, 376–384. [CrossRef]

21. Cereda, E.; Pisati, R.; Rondanelli, M.; Caccialanza, R. Whey Protein, Leucine- and Vitamin-D-Enriched Oral Nutritional Supplementation for the Treatment of Sarcopenia. *Nutrients* **2022**, *14*, 1524. [CrossRef] [PubMed]

22. Kirwan, R.P.; Mazidi, M.; García, C.R.; Lane, K.E.; Jafari, A.; Butler, T.; de Heredia, F.P.; Davies, I.G. Protein interventions augment the effect of resistance exercise on appendicular lean mass and handgrip strength in older adults: A systematic review and meta-analysis of randomized controlled trials. *Am. J. Clin. Nutr.* **2022**, *115*, 897–913. [CrossRef] [PubMed]

23. Luiking, Y.C.; Deutz, N.E.P.; Memelink, R.G.; Verlaan, S.; Wolfe, R.R. Postprandial muscle protein synthesis is higher after a high whey protein, leucine-enriched supplement than after a dairy-like product in healthy older people: A randomized controlled trial. *Nutr. J.* **2014**, *13*, 9. [CrossRef] [PubMed]

24. Kuo, Y.Y.; Chang, H.Y.; Huang, Y.C.; Liu, C.W. Effect of Whey Protein Supplementation in Postmenopausal Women: A Systematic Review and Meta-Analysis. *Nutrients* **2022**, *14*, 4210. [CrossRef] [PubMed]

25. Page, M.J.; McKenzie, J.E.; Bossuyt, P.M.; Boutron, I.; Hoffmann, T.C.; Mulrow, C.D.; Shamseer, L.; Tetzlaff, J.M.; Akl, E.A.; Brennan, S.E.; et al. The PRISMA 2020 statement: An updated guideline for reporting systematic reviews. *BMJ* **2021**, *372*, n71. [CrossRef]

26. Higgins, J.P.; Thomas, J.; Chandler, J.; Cumpston, M.; Li, T.; Page, M.J.; Welch, V.A. (Eds.) *Cochrane Handbook for Systematic Reviews of Interventions*, 1st ed.; Wiley: Hoboken, NJ, USA, 2019. [CrossRef]

27. Chen, L.-K.; Woo, J.; Assantachai, P.; Auyeung, T.-W.; Chou, M.-Y.; Iijima, K.; Jang, H.C.; Kang, L.; Kim, M.; Kim, S.; et al. Asian Working Group for Sarcopenia: 2019 Consensus Update on Sarcopenia Diagnosis and Treatment. *J. Am. Med. Dir. Assoc.* **2020**, *21*, 300–307.e2. [CrossRef]

28. Ouzzani, M.; Hammady, H.; Fedorowicz, Z.; Elmagarmid, A. Rayyan—A web and mobile app for systematic reviews. *Syst. Rev.* **2016**, *5*, 210. [CrossRef]

29. Sterne, J.A.C.; Savović, J.; Page, M.J.; Elbers, R.G.; Blencowe, N.S.; Boutron, I.; Cates, C.J.; Cheng, H.Y.; Corbett, M.S.; Eldridge, S.M.; et al. RoB 2: A revised tool for assessing risk of bias in randomised trials. *BMJ* **2019**, *366*, l4898. [CrossRef]

30. Fritz, C.O.; Morris, P.E.; Richler, J.J. Effect size estimates: Current use, calculations, and interpretation. *J. Exp. Psychol. Gen.* **2012**, *141*, 2–18. [CrossRef]
31. Bohannon, R.W. Minimal clinically important difference for grip strength: A systematic review. *J Phys Ther Sci.* **2019**, *31*, 75–78. [CrossRef]
32. Guyatt, G.H.; Oxman, A.D.; Vist, G.E.; Kunz, R.; Falck-Ytter, Y.; Alonso-Coello, P.; Schünemann, H.J. GRADE: An emerging consensus on rating quality of evidence and strength of recommendations. *BMJ* **2008**, *336*, 924–926. [CrossRef]
33. Santesso, N.; Glenton, C.; Dahm, P.; Garner, P.; Akl, E.A.; Alper, B.; Brignardello-Petersen, R.; Carrasco-Labra, A.; De Beer, H.; Hultcrantz, M.; et al. GRADE guidelines 26: Informative statements to communicate the findings of systematic reviews of interventions. *J. Clin. Epidemiol.* **2020**, *119*, 126–135. [CrossRef]
34. Rondanelli, M.; Klersy, C.; Terracol, G.; Talluri, J.; Maugeri, R.; Guido, D.; Faliva, M.A.; Solerte, B.S.; Fioravanti, M.; Lukaski, H.; et al. Whey protein, amino acids, and vitamin D supplementation with physical activity increases fat-free mass and strength, functionality, and quality of life and decreases inflammation in sarcopenic elderly. *Am. J. Clin. Nutr.* **2016**, *103*, 830–840. [CrossRef]
35. Rondanelli, M.; Cereda, E.; Klersy, C.; Faliva, M.A.; Peroni, G.; Nichetti, M.; Gasparri, C.; Iannello, G.; Spadaccini, D.; Infantino, V.; et al. Improving rehabilitation in sarcopenia: A randomized-controlled trial utilizing a muscle-targeted food for special medical purposes. *J. Cachexia Sarcopenia Muscle* **2020**, *11*, 1535–1547. [CrossRef]
36. Amasene, M.; Cadenas-Sanchez, C.; Echeverria, I.; Sanz, B.; Alonso, C.; Tobalina, I.; Irazusta, J.; Labayen, I.; Besga, A. Effects of Resistance Training Intervention along with Leucine-Enriched Whey Protein Supplementation on Sarcopenia and Frailty in Post-Hospitalized Older Adults: Preliminary Findings of a Randomized Controlled Trial. *JCM* **2021**, *11*, 97. [CrossRef]
37. Amasene, M.; Besga, A.; Echeverria, I.; Urquiza, M.; Ruiz, J.R.; Rodriguez-Larrad, A.; Aldamiz, M.; Anaut, P.; Irazusta, J.; Labayen, I. Effects of Leucine-Enriched Whey Protein Supplementation on Physical Function in Post-Hospitalized Older Adults Participating in 12-Weeks of Resistance Training Program: A Randomized Controlled Trial. *Nutrients* **2019**, *11*, 2337. [CrossRef]
38. Li, Z.; Cui, M.; Yu, K.; Zhang, X.-W.; Li, C.-W.; Nie, X.-D.; Wang, F. Effects of nutrition supplementation and physical exercise on muscle mass, muscle strength and fat mass among sarcopenic elderly: A randomized controlled trial. *Appl. Physiol. Nutr. Metab.* **2021**, *46*, 494–500. [CrossRef]
39. Nabuco, H.C.G.; Tomeleri, C.M.; Fernandes, R.R.; Sugihara, P., Jr.; Cavalcante, E.F.; Cunha, P.M.; Antunes, M.; Nunes, J.P.; Venturini, D.; Barbosa, D.S.; et al. Effect of whey protein supplementation combined with resistance training on body composition, muscular strength, functional capacity, and plasma-metabolism biomarkers in older women with sarcopenic obesity: A randomized, double-blind, placebo-controlled trial. *Clin. Nutr. ESPEN* **2019**, *32*, 88–95. [CrossRef]
40. Mori, H.; Tokuda, Y. De-Training Effects Following Leucine-Enriched Whey Protein Supplementation and Resistance Training in Older Adults with Sarcopenia: A Randomized Controlled Trial with 24 Weeks of Follow-Up. *J. Nutr. Health Aging* **2022**, *26*, 994–1002. [CrossRef]
41. Chang, M.C.; Choo, Y.J. Effects of Whey Protein, Leucine, and Vitamin D Supplementation in Patients with Sarcopenia: A Systematic Review and Meta-Analysis. *Nutrients* **2023**, *15*, 521. [CrossRef]
42. Vikberg, S.; Sörlén, N.; Brandén, L.; Johansson, J.; Nordström, A.; Hult, A.; Nordström, P. Effects of Resistance Training on Functional Strength and Muscle Mass in 70-Year-Old Individuals with Pre-sarcopenia: A Randomized Controlled Trial. *J. Am. Med. Dir. Assoc.* **2019**, *20*, 28–34. [CrossRef] [PubMed]
43. Nabuco, H.C.G.; Tomeleri, C.M.; Sugihara Junior, P.; Fernandes, R.R.; Cavalcante, E.F.; Antunes, M.; Ribeiro, A.S.; Teixeira, D.C.; Silva, A.M.; Sardinha, L.B.; et al. Effects of Whey Protein Supplementation Pre- or Post-Resistance Training on Muscle Mass, Muscular Strength, and Functional Capacity in Pre-Conditioned Older Women: A Randomized Clinical Trial. *Nutrients* **2018**, *10*, 563. [CrossRef] [PubMed]
44. Zhang, Y.; Zou, L.; Chen, S.-T.; Bae, J.H.; Kim, D.Y.; Liu, X.; Song, W. Effects and Moderators of Exercise on Sarcopenic Components in Sarcopenic Elderly: A Systematic Review and Meta-Analysis. *Front. Med.* **2021**, *8*, 649748. [CrossRef] [PubMed]
45. Burd, N.A.; Gorissen, S.H.; van Loon, L.J.C. Anabolic Resistance of Muscle Protein Synthesis with Aging. *Exerc. Sport Sci. Rev.* **2013**, *41*, 169. [CrossRef]
46. Churchward-Venne, T.A.; Holwerda, A.M.; Phillips, S.M.; van Loon, L.J.C. What is the Optimal Amount of Protein to Support Post-Exercise Skeletal Muscle Reconditioning in the Older Adult? *Sports Med.* **2016**, *46*, 1205–1212. [CrossRef]
47. Verlaan, S.; Maier, A.B.; Bauer, J.M.; Bautmans, I.; Brandt, K.; Donini, L.M.; Maggio, M.; McMurdo, M.E.; Mets, T.; Seal, C.; et al. Sufficient levels of 25-hydroxyvitamin D and protein intake required to increase muscle mass in sarcopenic older adults—The PROVIDE study. *Clin. Nutr.* **2018**, *37*, 551–557. [CrossRef]
48. Bo, Y.; Liu, C.; Ji, Z.; Yang, R.; An, Q.; Zhang, X.; You, J.; Duan, D.; Sun, Y.; Zhu, Y.; et al. A high whey protein, vitamin D and E supplement preserves muscle mass, strength, and quality of life in sarcopenic older adults: A double-blind randomized controlled trial. *Clin. Nutr.* **2019**, *38*, 159–164. [CrossRef]
49. Lin, C.C.; Shih, M.H.; Chen, C.D.; Yeh, S.L. Effects of adequate dietary protein with whey protein, leucine, and vitamin D supplementation on sarcopenia in older adults: An open-label, parallel-group study. *Clin. Nutr.* **2021**, *40*, 1323–1329. [CrossRef]
50. Kemmler, W.; von Stengel, S.; Kohl, M.; Rohleder, N.; Bertsch, T.; Sieber, C.C.; Freiberger, E.; Kob, R. Safety of a Combined WB-EMS and High-Protein Diet Intervention in Sarcopenic Obese Elderly Men. *Clin. Interv. Aging* **2020**, *15*, 953–967. [CrossRef]

51. Fuchs, C.J.; Kuipers, R.; Rombouts, J.A.; Brouwers, K.; Schrauwen-Hinderling, V.B.; Wildberger, J.E.; Verdijk, L.B.; van Loon, L.J. Thigh muscles are more susceptible to age-related muscle loss when compared to lower leg and pelvic muscles. *Exp. Gerontol.* **2023**, *175*, 112159. [CrossRef]
52. Talar, K.; Hernández-Belmonte, A.; Vetrovsky, T.; Steffl, M.; Kałamacka, E.; Courel-Ibáñez, J. Benefits of Resistance Training in Early and Late Stages of Frailty and Sarcopenia: A Systematic Review and Meta-Analysis of Randomized Controlled Studies. *J. Clin. Med.* **2021**, *10*, 1630. [CrossRef]

MDPI
St. Alban-Anlage 66
4052 Basel
Switzerland
www.mdpi.com

Nutrients Editorial Office
E-mail: nutrients@mdpi.com
www.mdpi.com/journal/nutrients